An Introduction to Practical Formal Methods using Temporal Logic

An Introduction to Practical Formal Methods using Temporal Logic

Michael Fisher

*Department of Computer Science,
University of Liverpool, UK*

A John Wiley & Sons, Ltd., Publication

Library of Congress Cataloging-in-Publication Data

Fisher, Michael, 1962 –
  An introduction to practical formal methods using temporal logic / Michael Fisher.
      p. cm.
  Includes bibliographical references and index.
  ISBN 978-0-470-02788-2 (cloth)
  1. Temporal automata. 2. Logic, Symbolic and mathematical. I. Title.
  QA267.5.T45F57 2011
  511.3 – dc22

                                                            2010046372

A catalogue record for this book is available from the British Library.

Print ISBN: 978-0-470-02788-2
e-Pdf ISBN: 978-0-470-98011-8
o-Book ISBN: 978-0-470-98010-1
e-Pub ISBN: 978-1-119-99146-5

Typeset in 10/12pt Times by Laserwords Private Limited, Chennai, India
Printed and bound in Singapore by Markono Print Media Pte Ltd

Contents

Preface

I look to the future because that's where I'm going to spend the rest of my life.
  – George Burns

In this book, I have tried to introduce temporal logics and then show why they might be useful in system specification, verification and development. That I can even attempt this is due to the work of very many outstanding researchers who have tackled this area over the last 30 years. I am enormously indebted to these people for their research on defining, refining and expanding this subject and so making my task here at least possible. (For an indication of who at least some of the key researchers are, just look at the References section at the end of this book.) Never has the motto 'standing on the shoulders of giants' seemed more appropriate.

As well as reading, and hearing, about the work of such experts, I have been very fortunate to have collaborated directly with quite a few of them. Though I cannot list everyone I have worked with (the references will again identify some of them), I would like to thank all of them for helping me to understand the area a little better. I have always learnt a great deal from all of them, and value their continued collaboration.

Finally, I would like to thank those who have provided detailed comments on earlier drafts of this book. Their advice and suggestions have been invaluable: Rafael Bordini; Stéphane Demri; Jürgen Dix; Clare Dixon; Valentin Goranko; Anthony Hepple; Koen Hindriks; Gerard Holzmann; Alexei Lisitsa; Alessio Lomuscio; Michel Ludwig; Stephan Merz; Mark Ryan; Sven Schewe; and Mike Wooldridge. Thanks to all of you for recognizing my misunderstandings, highlighting my omissions and correcting my mistakes.

This book is dedicated to Joan, for making it all possible, and to Sue, Christopher and James, for making it all worthwhile.

For more examples and resources used in this book, please visit my website at http://www.csc.liv.ac.uk/~michael/TLBook.

Michael Fisher
Liverpool, 2010.

1

Introduction

Time is an illusion, lunchtime doubly so.
  – Douglas Adams

Time plays a central role in our lives. In describing the world, or our activities within it, we naturally invoke temporal descriptions. Some of these are explicit, such as 'next week' or 'in 5 minutes', while others implicitly acknowledge the passing of time, for example 'during', 'did' or 'will do'. Not surprisingly, it is also important to be able to describe temporal aspects within the world of Computer Science: computations naturally proceed through time, and so have a *history* of activity; computational processes take time to act; some processes must finish before others can start; and so on. Consequently, being able to understand, and reason about, temporal concepts is central to Computer Science.

In this book, we will explain how some of these temporal notions can be described and manipulated. This, in turn, will allow us to carry out a temporal analysis of certain aspects of computation. To be precise in our temporal descriptions, we will use *formal logic*. These not only provide a concise and unambiguous basis for our descriptions, but are supported by many well-developed tools, techniques and results that we can take advantage of.

This book will provide an introduction to work concerned with formal logic for capturing temporal notions, called *temporal logic*, together with some of its applications in the formal development and analysis of computational systems. The name 'temporal logic' may sound complex and daunting. Indeed, the subject can sometimes be difficult because it essentially aims to capture the notion of time in a logical framework. However, while describing potentially complex scenarios, temporal logic is often based on a few simple, and fundamental, concepts. We aim to highlight these in this book.

An Introduction to Practical Formal Methods Using Temporal Logic, First Edition. Michael Fisher.
© 2011 John Wiley & Sons, Ltd. Published 2011 by John Wiley & Sons, Ltd.

As we might expect, this combination of expressive power and conceptual simplicity has led to the use of temporal logic in a range of subjects concerned with computation: Computer Science, Electronic Engineering, Information Systems and Artificial Intelligence. This representation of dynamic activity via temporal formalisms is used in a wide variety of areas within these broad fields, for example Robotics [176, 452], Control Systems [317, 466], Dynamic Databases [62, 110, 467], Program Specification [339, 363], System Verification [34, 122, 285], and Agent-Based Systems [207, 429]. Yet that is not all. Temporal logic also has an important role to play in Philosophy, Linguistics and Mathematics [222, 470], and is beginning to be used in areas as diverse as the Social Sciences and Systems Biology.

But why is temporal logic so useful? And is it really so simple? And how can we use practical tools based on temporal logic? This book aims to (at least begin to) answer these questions.

1.1   Aims of the book

Our aims here are to

- provide the reader with some of the background to the development and use of temporal logic;

- introduce the foundations (both informal and formal) of a simple temporal logic; and

- describe techniques and tools based on temporal logic and apply them to sample applications.

This book is *not* deeply technical. It simply aims to provide sufficient introduction to a number of areas surrounding temporal logic to enable either further, in-depth, study or the use of some of the tools described. Consequently, we would expect the readership to consist of those studying Computer Science, Information Systems or Artificial Intelligence at either undergraduate or postgraduate level, or software professionals who wish to expand their knowledge in this area. Since this is an *introductory* text, we aim to provide references to additional papers, books and online resources that can be used for further, and deeper, study. There are also several excellent, more advanced, textbooks and monographs that provide much greater technical detail concerning some of the aspects we cover, notably [34, 50, 122, 224, 299, 327, 339, 363, 364].

While there are very few proofs in this book, some of the elements are quite complex. In order to support the reader in understanding these aspects, we have often provided both exercises and pointers to further study in each chapter. We have interspersed exercises throughout the text, and sometimes provide a further selection of exercises at the end of each chapter, with answers in Appendix B. In addition, further resources can be found on the Web pages associated with this book:

http://www.csc.liv.ac.uk/~michael/TLBook

This URL provides links not only to additional material related to the book, but also contains pointers to a range of systems that are, at least in part, based on temporal logic.

1.2   Why temporal logic?

As computational systems become more complex, it is often important to be able to describe, clearly and unambiguously, their behaviour. Formal languages with well-defined semantics are increasingly used for this purpose, with *formal logic* being particularly prominent. This logic not only presents a precise language in which computational properties can be described, but also provides well-developed logical machinery for manipulating and analysing such descriptions.

For example, it is increasingly important to *verify* that a computational system behaves as required. These requirements can be captured as a formal specification in an appropriately chosen formal logic, with this specification then providing the basis for *formal verification*. While a system can be *tested* on many different inputs, formal verification provides a comprehensive approach to potentially establishing the correctness of the system in *all* possible situations. Verification within formal logic is aided by a logic's machinery, such as proof rules, normal form and decision procedures. Alternatively, we may wish to use the logical specification of a system in other ways, such as treating it as a *program* and directly executing it. Again, the well-developed logical machinery helps us with this.

Though logical specifications are clearly an important area to develop, the increased complexity of contemporary computational systems has meant that specifications in terms of traditional logic can become inappropriate and cumbersome. Consequently, much of the recent work concerning the use of formal logic in Computer Science has concentrated on developing logic that provides an appropriate level of abstraction for representing complex dynamic properties. It is precisely for this reason that *temporal logic* has been developed. Temporal logic has been used in Linguistics since the 1960s. In particular, temporal logic was originally used to represent tense in natural language [420]. However, in the late 1970s, temporal logic began to achieve a significant role in the formal specification and verification of concurrent and distributed systems [411, 412]. This logic is now at the heart of many specification, analysis and implementation approaches.

1.2.1   Motivation: evolution of computational systems

The way computational systems are designed and programmed has evolved considerably over the last 40 years. Correspondingly, the abstractions used to characterize such systems have changed during that time. When formal approaches to program development were initially envisaged, the key abstraction was that of a *transformational system* [260]. Transformational systems are essentially those whose behaviour can be described in terms of each component's input/output behaviour:

In other words, each component in a system receives some input, carries out some operation (typically on data structures), and terminates having produced some output. The

Formal Methods that have been developed for such systems describe the data structures and the behaviour of operations (via pre- and post-conditions) on these structures. Specification notations particularly relevant to this type of system were developed in the late 1960s and came to prominence in the 1970s. Typical examples include Floyd-Hoare Logics [214, 274, 418], weakest precondition semantics [146], VDM [304], Z [135], and (more recently) B [7, 340, 446], as well as the functional programming metaphor.

While the use of Formal Methods for transformational systems has been very effective in many areas, it became clear in the 1970s that an increasing number of systems could not easily be categorized as 'transformational'. Typically, this was because the components were either non-terminating, continuously reading input (not just at the beginning of computation), continuously producing output (not just at the end), or regularly interacting with other concurrent or distributed components. These have been termed *reactive systems* [260] and can be visualized in a more complex way, for example:

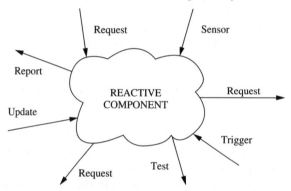

This diagram highlights the fact that multiple inputs can be received, and multiple outputs can be generated, by reactive systems. Such systems are typically interacting, evolving, non-terminating systems.

Formal Methods for *reactive systems* often require more sophisticated techniques than the pre- and post-conditions provided in notations such as VDM or Z. In particular, in the late 1970s, *temporal logic* was applied to the specification of reactive systems, with this approach coming to prominence in the 1980s [363, 414]. It is widely recognized that *reactive* systems [260], as described above, represent one of the most important classes of systems in Computer Science and, although the analysis of such systems is difficult, it has been successfully tackled using temporal representations [168, 411, 460], where a number of useful concepts, such as *safety*, *liveness* and *fairness* properties can be formally, and concisely, specified [363]. Such a logical representation of a system then permits the analysis of the system's properties via logical methods, such as *logical proof*. A specific proof method for deciding whether a temporal formula is true or false is one of the aspects that we will examine later in this book.

1.3   What is temporal logic?

Temporal logic is an extension of classical logic[1], specifically adding operators relating to time [168, 196, 210, 224, 225, 279, 433, 460, 478]. Modal logic [67, 68, 109, 226, 291]

---

[1] A very brief review of classical logic is provided in Appendix A.

provides some of the formal foundations of temporal logic, and many of the techniques used in temporal logic are derived from their modal counterparts. In addition to the operators of classical logic, temporal logic often contains operators such as '$\bigcirc$', meaning *in the next moment in time*, '$\square$', meaning *at every future moment*, and '$\lozenge$', meaning *at some future moment*. These additional operators allow us to construct formulae such as

$$\square(try\_to\_print \;\Rightarrow\; \lozenge\neg try\_to\_print)$$

to characterize the statement that

> "*whenever* we try to print a document then at *some* future moment we will *not* try to print it".

The flexibility of temporal logic allows us also to provide formulae such as

$$\square(try\_to\_print \;\Rightarrow\; \bigcirc(printed \lor try\_to\_print))$$

meant to characterize

> "*whenever* we try to print a document *then*, at the *next* moment in time, *either* the document will be printed *or* we again try to print it"

and

$$\square(printed \;\Rightarrow\; \bigcirc\square\neg try\_to\_print)$$

meaning

> "*whenever* the document has been printed, the system will *never* try to print it (ever again)".

Given the above formulae then, if we try to print a document, i.e.

$$try\_to\_print$$

we *should* be able to show that, eventually, it will stop trying to print the document. Specifically, the statement

$$\lozenge\square\neg try\_to\_print$$

can be inferred from the above formulae. We will see later how to establish automatically that this is, indeed, the case.

Although there are many different temporal logics [168, 196, 279], we will mainly concentrate on one very popular variety that is:

- *propositional*, with no explicit first-order quantification;

- *discrete*, with the underlying model of time being isomorphic to the Natural Numbers (i.e. an infinite, discrete sequence with distinguished initial point); and

- *linear*, with each moment in time having at most one successor.

Note that the infinite and linear constraints ensure that each moment in time has *exactly* one successor, hence the use of just one form of '$\bigcirc$' operator. If we allow several immediate successors, then we typically require other operators. (More details concerning such logics will be provided in Chapter 2.)

1.4   Structure of the book

The book comprises four parts, of very different sizes.

- In the first part, we introduce temporal logic (Chapter 2), and show how it can be used to specify a variety of computational systems (Chapter 3).

- In the second part, we then describe techniques that use these temporal specifications, namely deductive verification (proof; Chapter 4), algorithmic verification (model checking; Chapter 5), and model building (direct execution; Chapter 6).

- In the third part (Chapter 7), we provide an overview of some of the areas where temporal-based formal techniques are being used, not only in Computer Science, but also in Artificial Intelligence and Engineering.

- Finally, in Appendices A and B, we provide a review of classical logic and sample answers to exercises, respectively.

In the first part, we essentially introduce the basic concepts of temporal logic and temporal specification. Throughout the chapters in this first part we provide a range of examples and exercises to reinforce the topics. In the second part, comprising chapters describing verification and execution approaches, we split each chapter into:

- an *introductory* section conveying the motivation for, and principles of, the approach being tackled;

- details of a particular *technique* epitomizing the approach;

- an overview of a particular *system* implementing the technique described; and

- a selection of *advanced topics* concerning this area and an overview of *alternative* systems tackling this problem.

Again, within this structure, examples and (some) exercises will be provided.

To give an idea of the substructure of Chapters 4, 5 and 6, we can give the following broad distribution of topics.

- Chapter 4 (Deduction)

INTRODUCTORY:	the idea behind deductive verification and temporal proof.
TECHNIQUE:	clausal temporal resolution.
SYSTEM:	TSPASS.
ADVANCED:	including alternative approaches, such as tableaux and extensions.

- Chapter 5 (Model Checking)

  INTRODUCTORY: the idea behind algorithmic temporal verification and
                    model checking.
  TECHNIQUE: the automata-theoretic view of model checking.
  SYSTEM: Spin.
  ADVANCED: including alternative approaches, such as extended,
                    bounded, or abstracted model checking.

- Chapter 6 (Execution)

  INTRODUCTORY: model construction through direct execution.
  TECHNIQUE: METATEM execution algorithm.
  SYSTEM: Concurrent MetateM.
  ADVANCED: including alternative approaches, such as temporal
                    logic programming and interval-based execution.

Finally, in Chapter 7, we provide an overview of selected applications where temporal-based formal methods have been, or are being, used. Some use the techniques described in the book, others use alternative temporal approaches, but all help to highlight the wide applicability of temporal methods.

2

Temporal logic

Time flies like an arrow. Fruit flies like a banana.
  – Groucho Marx

In this Chapter, we will:

- provide the basic intuition behind temporal logics (Section 2.1);

- define the syntax of a simple propositional temporal logic (Section 2.2);

- examine the formal semantics of this temporal logic (Section 2.3);

- explore some classes of temporal logic formulae that are useful in reactive system specification (Section 2.4);

- revisit the question 'what is temporal logic' (Section 2.5);

- introduce a *normal form* for our temporal logic that will be useful in Chapters 4 and 6 (Section 2.6);

- discuss the link between propositional temporal logic and finite state automata, something that will be essential in Chapter 5 (Section 2.7);

- provide pointers to some more advanced work in this area (Section 2.8); and

- give a final selection of exercises (Section 2.9), the solutions to which can be found in Appendix B.

To begin with we will only describe a simple propositional, discrete, temporal logic. However, much of the temporal framework we develop below is applicable to other variants of temporal logic we will see later.

An Introduction to Practical Formal Methods Using Temporal Logic, First Edition. Michael Fisher.
© 2011 John Wiley & Sons, Ltd. Published 2011 by John Wiley & Sons, Ltd.

2.1 Intuition

In classical propositional logic[1], formulae are evaluated within a single fixed world (or state). For example, a proposition such as

"it is Monday"

is either *true* or *false*. Such propositions are then combined using constructs (or connectives) such as '$\wedge$', '$\neg$', and '$\Rightarrow$' [132]. Classical propositional logic is thus used to describe *static* situations. The meaning of statements in such a logic can then be modelled by mapping basic (also called *atomic*) propositions to Boolean values (i.e., **T** or **F**). For example,

$$[a \mapsto \mathbf{T},\ b \mapsto \mathbf{F},\ c \mapsto \mathbf{T}]$$

is a particular structure (called a 'model') satisfying the propositional formula

$$a \wedge (b \vee c).$$

Here, the model states that b is **False**, yet c is **True**. This means that $b \vee c$ is **True** and, combined with the fact that a is **True**, means that, overall, $a \wedge (b \vee c)$ is **True** for this allocation of truth values to atomic propositions. There are, of course, other models that satisfy this formula, for example

$$[a \mapsto \mathbf{T},\ b \mapsto \mathbf{T},\ c \mapsto \mathbf{F}].$$

For any such model, M, we can write that

$$M \models a \wedge (b \vee c)$$

meaning that 'the model M satisfies the formula $a \wedge (b \vee c)$'.

When we use temporal logic, however, evaluation of formulae takes place within a *set* of worlds, rather than a fixed world. Thus, 'it is Monday' may be true in some worlds, but not in others. An *accessibility relation* is then used to navigate between worlds, and this relation is then interpreted as a *temporal* relation[2], thus characterizing a particular model of *time*. So, as we move from one world to another, the intuition is that we are moving *through time*. Commonly, propositional, discrete, linear *temporal* logic extends the descriptive power of propositional logic in order to be able to describe sequences (hence: *linear*) of distinct (hence: *discrete*) worlds, with each world being similar to a classical (propositional) model. So, we can equivalently describe the basis of our model of time in terms of a sequence of worlds, a sequence of states or a sequence of propositional models. Each state in the sequence is taken as modelling a different moment in time; hence the name *temporal logic*.

In order to allow us to describe navigation through our model of time via this accessibility relation, the set of classical operators ('$\wedge$', '$\neg$', etc.) is extended with various

---

[1] A very brief review of classical propositional logic is provided in Appendix A.

[2] The general study of *modal logic* concerns different interpretations of accessibility between worlds, see [68, 97, 226, 292].

temporal operators. In particular, in the variety of temporal logic we will investigate, the *next-time* operator ('○') allows us to move to the 'next' moment in time (modelled by the *next* accessible world).

So, though we can still have propositional formulae such as

$$a \wedge (b \vee c),$$

in such a propositional temporal logic, we can also express properties of the *rest* of the worlds being modelled. For example, the temporal formula

$$(a \wedge b) \wedge \bigcirc(a \vee b)$$

describes a situation where '$a \wedge b$' is satisfied in the current world (or state) and '$a \vee b$' is satisfied in the next world/state. Thus, the following might be a model for the above formula.

$$\text{state 1}: \quad [\, a \mapsto \mathbf{T}, \;\; b \mapsto \mathbf{T} \,]$$
$$\text{state 2}: \quad [\, a \mapsto \mathbf{F}, \;\; b \mapsto \mathbf{T} \,]$$
$$\cdots \qquad\qquad \cdots$$

As can be seen from this, the view of time as a *sequence* of moments/states/worlds is central to our approach. The basic temporal logic that we will consider first is based on exactly this discrete, linear view of time. In this logic, called *LTL*, *PLTL*, or as we will term it, *PTL* [223], the underlying model of time is a discrete, linear order isomorphic to the Natural Numbers, $\mathbb{N}$. Pictorially, this model of time is:

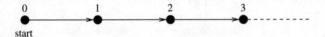

Here, each of the black circles represents a classical propositional state, and the arrows represent the accessibility relation, in our case the 'step' to the next moment in time. Note that we also have one state identified as the 'start of time'.

Example 2.1 *If we consider days within the week, beginning with Monday, we might get the discrete sequence*

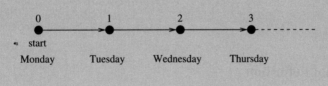

This example also shows the potentially cyclic nature of temporal patterns because we would expect Monday to again occur after every Sunday, Tuesday to occur after every Monday, and so on.

2.2 Syntactic aspects

We now move on to a more detailed explanation of the formal syntax available within PTL. In PTL, as well as classical propositional operators, we use temporal operators referring to moments in the *future*:

Formula	Intuitive Meaning
$\bigcirc \varphi$	φ is true in the *next* moment in time
$\square \varphi$	φ is true in *all* future moments
$\diamondsuit \varphi$	φ is true in *some* future (or present) moment
$\varphi U \psi$	φ continues being true *up until* some future moment when ψ is true
$\varphi W \psi$	φ continues being true *unless* ψ becomes true

If we ensure that there is always an initial moment, as in the model above, then we can also introduce the '**start**' operator, which is only true at this *beginning of time*. The formal semantics of these temporal operators, together with the standard propositional calculus operators, will later be defined with respect to an appropriate model structure. For the moment, however, let us consider how we might encode several simple examples using this syntax.

Example 2.2

- *"If a message is sent to a receiver, then the message will eventually be received"*:

$$send\_msg \Rightarrow \diamondsuit receive\_msg$$

- *"It is always the case that, if either 'have_passport' or 'have_ticket' is false, then, in the next moment in time 'board_flight' will also be false"*:

$$\square \left((\neg have\_passport \vee \neg have\_ticket) \Rightarrow \bigcirc \neg board\_flight \right)$$

- *"If something is born, then it is living up until the point in time that it becomes dead"* (note that we will explain the detailed semantics of the 'until' operator, U, later):

$$born \Rightarrow living U dead$$

2.2.1 Formal definition

We now describe the formal syntax of PTL; see [223] for further details. Formulae in PTL are constructed from the following elements.

- A finite set of propositional symbols, PROP, typically being represented by lowercase alphanumeric strings, such as p, q, r, *trigger*, *terminate_condition2*, *lunch*, ...

- Propositional connectives: **true**, **false**, $\neg$, $\vee$, $\wedge$, $\Leftrightarrow$, and $\Rightarrow$.

- Temporal connectives: $\bigcirc$, $\Diamond$, $\Box$, **start**, U, and W.

- Parentheses, '(' and ')', generally used to avoid ambiguity.

The set of well-formed formulae of PTL, denoted by WFF, is now inductively defined as the smallest set satisfying the following rules.

- Any element of PROP is in WFF.

- **true**, **false** and **start** are in WFF.

- If φ and ψ are in WFF, then so are

$$\neg\varphi \quad \varphi \lor \psi \quad \varphi \land \psi \quad \varphi \Rightarrow \psi \quad \varphi \Leftrightarrow \psi \quad (\varphi)$$
$$\Diamond\varphi \quad \Box\varphi \quad \varphi U \psi \quad \varphi W \psi \quad \bigcirc\varphi.$$

Example 2.3 *The following are all legal* WFF *of PTL*

$$pU(q \land \Diamond r) \qquad a \Rightarrow \Box\bigcirc(bWc) \qquad (f \land \bigcirc g)U\Diamond\Box\neg h$$

whereas the following are not

$$p\Diamond q \qquad\qquad (Ur) \qquad\qquad a \Rightarrow \Box b\bigcirc c$$

Terminology. A *literal* is defined as either a propositional symbol or the negation of a propositional symbol. An *eventuality* is defined as a formula of the form $\Diamond\varphi$, where φ is any PTL formula, and a *state formula* is a WFF containing *no* temporal operators.

Exercise 2.1 *Which of the following are not legal* WFF *of PTL, and why?*

(a) april $\lor$ $\bigcirc U$ *may*

(b) may $\lor$ $\bigcirc((april W may))$

(c) $\bigcirc july$ $\land$ *august* $(\Diamond september)$

We will see later that it is often both convenient and intuitive to describe a system's behaviour using a set of PTL formulae in the form of implications. To begin to get a feel for this, consider the following:

Example 2.4 *All the following implications are legal* WFF *of PTL (where each name is a proposition).*

$$
\begin{array}{rcl}
monday &\Rightarrow& \bigcirc tuesday \\
begin &\Rightarrow& \Diamond finish \\
july &\Rightarrow& \Diamond(december \land winter) \\
sunset &\Rightarrow& \bigcirc(night W dawn) \\
born &\Rightarrow& \Diamond\Box old \\
monday &\Rightarrow& sad U saturday
\end{array}
$$

We will examine this *implicational form* further in Section 2.6, but for the moment we will continue to look at the formal basis of PTL.

Exercise 2.2 *How might we represent the following statements in PTL?*

(a) *"In the next moment in time, 'running' will be true and, at some time after that, 'terminated' will be true."*

(b) *"There is a moment in the future where either 'pink' is always true, or 'brown' is true in the next moment in time."*

(c) *"In the* second *moment in time, 'hot' will be true."*

2.3   Semantics

As we have seen, temporal logic is an extension of classical logic, whereby time becomes an extra parameter modifying the truth of logical statements. Models for temporal (and modal) logics are typically 'Kripke Structures' [325] of the form

$$\mathcal{M} \ = \ \langle S, \ R, \ \pi \rangle$$

where

- S is the set of *moments* in time (our accessible worlds or states),

- R is a temporal accessibility relation (in the case of PTL, this characterizes a sequence that is linear, discrete and has finite past), and

- $\pi : S \mapsto \mathbf{P}(\text{PROP})$ maps each moment/world/state to a set of propositions[3] (i.e. those that are true in that moment/world/state).

However, in the case of PTL, which has a linear, discrete basis that is isomorphic to $\mathbb{N}$, this model structure is often simplified from the above to

$$\mathcal{M} \ = \ \langle \mathbb{N}, \ \pi \rangle$$

where

- $\pi : \mathbb{N} \mapsto \mathbf{P}(\text{PROP})$ maps each Natural Number (representing a moment in time) to the set of propositions true at that moment.

Note.   Sometimes we simplify this structure even further to

$$\mathcal{M} \ = \ \langle s_0, \ s_1, \ s_2, \ s_3, \ \ldots \rangle$$

where each s_i is the set of propositions satisfied at the ith moment in time.

---

[3] '**P**' is the *powerset* operator, which takes a set, say X, and generates a new set containing all possible subsets of X. For example, $\mathbf{P}(\{a, b, c\}) = \{\emptyset, \{a\}, \{b\}, \{c\}, \{a, b\}, \{a, c\}, \{b, c\}, \{a, b, c\}\}$.

Generally, however, we will use the $\mathcal{M} = \langle \mathbb{N}, \pi \rangle$ variety, for example

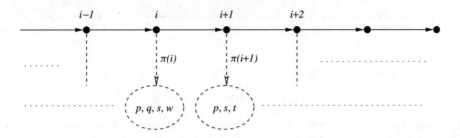

Each such structure represents a particular scenario. So, to assess the truth of a WFF in a temporal structure, we first need to define a *semantics* for WFF over such structures. In our case, a semantics is provided by an interpretation relation '$\models$' of type

$$(\mathcal{M} \times \mathbb{N} \times \text{WFF}) \rightarrow \mathbb{B}$$

where '$\mathbb{B}$' is, as usual, the set of Boolean values $\{\mathbf{T}, \mathbf{F}\}$.

Thus, the '$\models$' function maps a temporal structure, a particular moment in that structure and a well-formed temporal formula on to either $\mathbf{T}$ or $\mathbf{F}$. So, for a structure, $\mathcal{M}$, temporal index, i, and formula, φ, then

$$\langle \mathcal{M}, i \rangle \models \varphi$$

is true (i.e. evaluates to $\mathbf{T}$) if, and only if, φ is satisfied at the temporal index i within the model $\mathcal{M}$. If this is the case, then $\mathcal{M}$ can be legitimately called a *model* for φ, when interpreted at moment i and φ is *satisfied* on this model. Recall that, a formula is said to be *valid* if it is satisfied in *all* possible interpretations; while it is *unsatisfiable* if it is satisfied in *no* interpretation.

So, the interpretation relation for PTL essentially provides the semantics for the logic. Below, we explain the semantics for the key operators in PTL.

Semantics of propositions.   We begin with the semantics of basic (atomic) propositions[4]:

$$\langle \mathcal{M}, i \rangle \models p \quad \text{iff} \quad p \in \pi(i) \quad \text{(for any } p \in \text{PROP)}$$

Recalling that $\mathcal{M} = \langle \mathbb{N}, \pi \rangle$ we can simply 'look up' the propositional symbol in the given temporal structure (via '$\pi(i)$') in order to see whether it is satisfied or not at that particular moment in time, i.

Classical operators.   The semantics of the standard classical operators is as expected:

$$\langle \mathcal{M}, i \rangle \models \quad \neg\varphi \qquad \text{iff} \quad \text{it is } not \text{ the case that } \langle \mathcal{M}, i \rangle \models \varphi$$
$$\langle \mathcal{M}, i \rangle \models \quad \varphi \wedge \psi \quad \text{iff} \quad \text{both } \langle \mathcal{M}, i \rangle \models \varphi \text{ and } \langle \mathcal{M}, i \rangle \models \psi$$
$$\langle \mathcal{M}, i \rangle \models \quad \varphi \vee \psi \quad \text{iff} \quad \langle \mathcal{M}, i \rangle \models \varphi \text{ or } \langle \mathcal{M}, i \rangle \models \psi$$
$$\langle \mathcal{M}, i \rangle \models \quad \varphi \Rightarrow \psi \quad \text{iff} \quad \text{if } \langle \mathcal{M}, i \rangle \models \varphi \text{ then } \langle \mathcal{M}, i \rangle \models \psi$$

---

[4] From now on, 'iff' is an abbreviation for 'if, and only if'.

Temporal operators.   Recall that the '**start**' operator can only be satisfied at the 'begin-ning of time':

Thus, its semantics reflects this:

$$\langle \mathcal{M}, i \rangle \models \textbf{start} \quad \text{iff} \quad (i = 0)$$

The '**next**' operator, however, provides a constraint on the next moment in time, and uses the temporal accessibility relation to look one step forward:

Since we are using the Natural Numbers, $\mathbb{N}$, as a basis, the semantics simply involves incrementing the temporal index we are examining:

$$\langle \mathcal{M}, i \rangle \models \bigcirc \varphi \quad \text{iff} \quad \langle \mathcal{M}, i + 1 \rangle \models \varphi$$

Example 2.5 *Once we have the next-time operator, '$\bigcirc$', then we can specify how the current situation changes when we move to the next moment in time. For example:*

(a) $(sad \wedge \neg rich) \Rightarrow \bigcirc sad$

(b) $(x\_equals\_1 \wedge added\_3) \Rightarrow \bigcirc (x\_equals\_4)$

Aside.   A note about the 'end of time'. We are using $\mathbb{N}$ as our underlying model of time, but another popular choice is to use a linear, discrete sequence, with distinguished beginning, but with a *finite* length.

In this case, the semantics of the temporal operators must be modified to take the future nature of the model into account. For example, the 'next' operator examines the next moment in the sequence if there is one, but typically defaults to **true** if there is no such 'next' moment. Thus the, seemingly contradictory, formula '$\bigcirc$ **false**' is actually satisfied at the last state in a finite sequence! We will not say too much more about this aspect here, but note that these finite length models are often useful as approximations to $\mathbb{N}$ when the full infinite sequence is too difficult to deal with. Thus, techniques such as *bounded model checking* [65] assess properties of temporal sequences by taking increasingly long approximations of the infinite sequence. Checking properties of such finite sequences is generally *much* simpler than checking the properties of the full infinite sequence [123] and so, if an error can be found at an early stage by considering a finite approximation, then this can be of great benefit (see Section 5.4.7).

Returning to the semantics of the temporal operators, the 'sometime in the future' operator, '$\Diamond$', describes a constraint on the future. It is indeterminate, meaning that while we can be sure that φ *will* be true either now or in the future, we can not be sure exactly *when* it will become true.

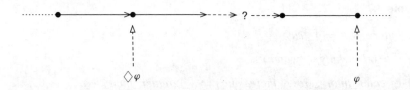

Consequently, the semantics of '$\Diamond\varphi$' involves some indeterminacy about which temporal index 'φ' will be true at. We only know for certain that there *must be* such a moment.

$$\langle \mathcal{M}, i \rangle \models \Diamond\varphi \quad \text{iff} \quad \text{there exists } j \text{ such that } (j \geq i) \text{ and } \langle \mathcal{M}, j \rangle \models \varphi$$

Note that there is a choice in the semantics above about whether to take $j \geq i$ or $j > i$. We generally take the first of these; using the latter ensures that, for example, '$\Diamond\varphi$' refers to φ as occurring strictly in the future, rather than in either the present *or* future. Thus, there is an alternative operator, $\Diamond^+$:

$$\langle \mathcal{M}, i \rangle \models \Diamond^+\varphi \quad \text{iff} \quad \text{there exists } j \text{ such that } (j > i) \text{ and } \langle \mathcal{M}, j \rangle \models \varphi$$

where $\Diamond\varphi \Leftrightarrow (\varphi \vee \Diamond^+\varphi)$. Note, however, that $\Diamond^+\varphi$ is equivalent to $\bigcirc\Diamond\varphi$ and so can still be defined within PTL.

Example 2.6

(a) $(\neg resigned \wedge sad) \Rightarrow \Diamond famous$

(b) $sad \Rightarrow \Diamond happy$

(c) $is\_monday \Rightarrow \Diamond^+ is\_friday$

Now we turn to properties that we require to hold *throughout* the future. The 'always in the future' operator, '$\square$', provides us with the ability to represent invariant properties, that is properties that are true at all moments in time from now on.

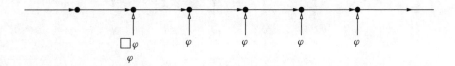

Thus, the semantics of '□' refers to *all* future moments in time, that is all $j \geq i$, where i is the current temporal index:

$$\langle \mathcal{M}, i \rangle \models \Box\varphi \quad \text{iff} \quad \text{for all } j, \text{ if } (j \geq i), \text{ then } \langle \mathcal{M}, j \rangle \models \varphi$$

Again we can, if necessary, define '□⁺':

$$\langle \mathcal{M}, i \rangle \models \Box^+\varphi \quad \text{iff} \quad \text{for all } j, \text{ if } (j > i), \text{ then } \langle \mathcal{M}, j \rangle \models \varphi$$

Example 2.7

(a) lottery_win ⇒ □*rich*

(b) □¬*(electricity* ∧ *water)*

(c) (greater_than_zero ∧ *increasing)* ⇒ □*greater_than_zero*

Exercise 2.3 *By examining the formal semantics of* □*, show that* □p ⇒ □□p.

Importantly, '□' is the dual of '◇' and so a theorem of PTL (and often an axiom) is that

$$\Box\varphi \Leftrightarrow \neg\Diamond\neg\varphi .$$

Intuitively, this makes sense as, when something is always true, then it must never be false, and *vice versa*.

Example 2.8 *If 'wet' is equivalent to '¬dry', then*

$$\Box wet \Leftrightarrow \neg\Diamond dry$$

In contrast to the ○, □ and ◇ operators that apply to just one subformula, we now turn to *binary* temporal operators that describe the ongoing relationship between *two* subformulae. Not surprisingly, these increase the expressive power of the logical language although, as we will see in Section 4.2.1, we can remove them by introducing additional atomic propositions.

The 'until' operator, 'U', allows us to describe behaviour over finite intervals within the overall sequence of states. Thus, $\varphi U \psi$ characterizes the situation where some property, 'φ' persists until a moment in time (which is *guaranteed* to occur) where another property, 'ψ' becomes true. On our linear models, we might represent this as follows.

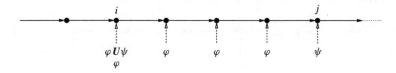

Thus, the semantic definition we must use here is more complex than those we have seen so far, characterizing a finite interval of time over which 'φ' is true.

$$\langle \mathcal{M}, i \rangle \models \varphi U \psi \quad \text{iff} \quad \text{there exists } j \text{ such that } (j \geq i) \text{ and } \langle \mathcal{M}, j \rangle \models \psi \text{ and}$$
$$\text{for all } k, \text{ if } (i \leq k < j), \text{ then } \langle \mathcal{M}, k \rangle \models \varphi$$

Here, we find the (relevant) moment in time when 'ψ' occurs (i.e., moment 'j') and then, at all moments (represented by 'k') between now and j, φ must be true.

So, in the above diagram, the second argument to the 'U' is guaranteed to occur. Thus, $\varphi U \psi$ implies $\Diamond \psi$. Consequently, though we are not sure *when* ψ will occur in the future, we are certain that it will. In addition, we know that from now until whenever ψ occurs, φ must continually be true.

Example 2.9

(a) *start_lecture $\Rightarrow$ talk U end_lecture*

(b) *request $\Rightarrow$ continue_replying U acknowledgement*

(c) *less_than_7 $\Rightarrow$ increasing U more_than_7*

The 'unless' operator, 'W', is very similar to the 'until' operator, except that, in $\varphi W \psi$, ψ is *not* guaranteed to occur and so the persistent property can, in fact, persist forever. Thus, each of the following scenarios are legitimate:

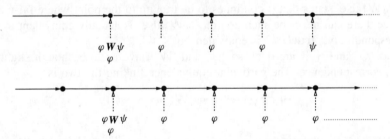

In Figure 2.11, ψ never occurs and so φ persists forever. Consequently, 'W' is often termed the '*weak* until' operator. Correspondingly, the semantics of 'W' can be defined by the two possibilities:

$$\langle \mathcal{M}, i \rangle \models \varphi W \psi \quad \text{iff} \quad \text{either} \quad \langle \mathcal{M}, i \rangle \models \varphi U \psi \quad \text{or} \quad \langle \mathcal{M}, i \rangle \models \Box \varphi$$

Example 2.10

(a) *commence $\Rightarrow$ executing W error*

(b) *stay_in_room W fire_alarm*

Note that, in these examples, neither 'error' nor 'fire_alarm' is required to occur.

Exercise 2.4

(a) As we have seen,
$$\varphi W \psi \Leftrightarrow (\varphi U \psi \lor \Box \varphi).$$

Fill in the missing formulae below using $\varphi W \psi$ (and no explicit 'U') as part of the right-hand side:
$$\varphi U \psi \Leftrightarrow \ldots\ldots$$

(b) Does the temporal formula
$$(l \land a) U (l \land ((l \land b) U c))$$

imply $l U c$? Justify your answer, either informally or by appealing to the semantics given.

Hint: if we know that $(l \land a) U (l \land ((l \land b) U c))$ is satisfied in a particular model, does it follow that $l U c$ must also be satisfied in that model?

In summary, the temporal operators U ('until') and W ('unless') characterize intervals within the temporal sequence during which certain properties hold. While the 'U' guarantees a *finite* interval (since $\varphi U \psi$ implies $\Diamond \psi$), the 'W' operator can represent either finite or infinite intervals of behaviour. For example, 'φU **false**' is unsatisfiable, since we can *never* find a moment at which '**false**' can be made true, while 'φW **false**' is satisfiable. In fact, 'φW **false**' states that φ continues to be true up to the point where **false** becomes true. Since there can never be such a point, 'φW **false**' is actually equivalent to '$\Box \varphi$'.

Correspondingly, '**true**$U \psi$' is equivalent to '$\Diamond \psi$'.

Just as '$\Diamond$' and '$\Box$' are dual, so 'U' and 'W' have a close, though slightly more complex, correspondence. The particular equivalence linking the two is
$$\neg(\varphi U \psi) \Leftrightarrow (\neg\psi) W (\neg\varphi \land \neg\psi).$$

Example 2.11 If '*living* U *dead*' is not *true*, that is $\neg(living U dead)$, then we can derive
$$(\neg dead) W (\neg living \land \neg dead).$$

If *living* and *dead* are opposites, that is $(\neg living \land \neg dead)$ reduces to **false**, then the above reduces to
$$(\neg dead) W \textbf{false}.$$

This, as we have seen, is equivalent to
$$\Box \neg dead.$$

living and *dead* are defined as being opposites, so the above is also
$$\Box living.$$

Thus, if '*living* U *dead*' is not *true*, then $\Box living$ must be *true*.

Finally, a note that in PTL the two core operators usually chosen are '$\bigcirc$' and 'U'. With these we can define all the other temporal operators (and so they can be considered as a functionally complete set of operators). However, as we will see later, with suitable introduction of new propositions we can reduce the operators we need to consider down to '$\bigcirc$' and '$\Diamond$'.

2.3.1   Axiom systems

Theorems describe the universal truths of a formal language, while an axiomatic system (or axiomatization) provides rules for generating new universal truths (i.e. theorems) from given ones. A theorem is typically defined as a formula prefixed by the '$\vdash$' symbol, for example '$\vdash$ *true_thing*'. There are many axiom systems for PTL, each providing a set of basic theorems together with *inference rules* for generating new theorems. Readers interested in such axiomatizations, or soundness and completeness results, should consult, for example [168, 223, 346, 363, 460]. Rather than going into details here, we will just present some sample axioms (with simple instances) for PTL:

$\vdash \Box(\varphi \Rightarrow \psi) \Rightarrow (\Box\varphi \Rightarrow \Box\psi)$

for example, "if it is always the case that when it rains it pours then, if it always rains, it always pours."

Note that '$\Box$' takes precedence over '$\Rightarrow$'.

$\vdash (\text{start} \wedge \Box\varphi) \Rightarrow \Box\bigcirc\varphi$

for example, "if, from the beginning, the sky is always pink, then the sky will be pink whenever it is examined in the future."

$\vdash \Box\neg\varphi \Leftrightarrow \neg\Diamond\varphi$

for example, "if it is always dark then it is not the case that it is sometimes light."

$\vdash \bigcirc(\varphi \wedge \psi) \Leftrightarrow \bigcirc\varphi \wedge \bigcirc\psi$

for example, "tomorrow I will be both happy and hungry if, and only if, both tomorrow I will be happy and tomorrow I will be hungry."

$\vdash (\varphi U (\psi \wedge \varphi U \varrho)) \Rightarrow (\varphi U \varrho)$

for example, "if I am sleeping from now up until the time the alarm goes off and continue sleeping up until my doorbell rings, then I am sleeping from now up until the time my doorbell rings."

Many of the above are obvious, but we must still state them as axioms since the axiomatization must allow us to generate *all* true statements within the logic.

Note, however, that the *key* axiom describing how this (discrete, linear) form of temporal logic works is the *induction axiom*:

$$\vdash \Box(\varphi \Rightarrow \bigcirc\varphi) \Rightarrow (\varphi \Rightarrow \Box\varphi)$$

We will see later in what fashion this corresponds to induction, but some simple intuition is given by this example, which is an instance of the induction axiom:

if, whenever I am happy, I will certainly be happy on the next day, then, if I am happy now, then I will always be happy in the future.

2.4   Reactive system properties

As mentioned in Chapter 1, a key motivation for the development of temporal logic has been the specification of dynamic properties of reactive systems. To give the reader an initial indication of such specifications (there will be more on this in Chapter 3), we here provide some interesting classes of properties, such as *safety* and *liveness* [363, 334], useful in system specification.

2.4.1   Safety properties

These are properties of a system corresponding to the requirement that 'something bad *will not* happen'. Such properties are described using temporal formulae of the form '$\Box \neg \ldots$', where the '$\ldots$' formula is usually non-temporal or at least just contrains local states. Typical examples are[5]:

$\Box$ $\neg(reactor\_temperature > 1000)$

$\Box$ $\neg(one\_way \wedge \bigcirc other\_way)$

$\Box$ $\neg(\bigcirc\bigcirc(x = 0) \wedge \bigcirc\bigcirc(y = z \div x))$

All of these formulae describe some situation that must *not* occur at any moment in time.

2.4.2   Liveness properties

Rather than describing situations that must *never* occur, as safety properties do, *liveness* properties capture the fact that some particular situation *must* eventually occur. Just as safety properties are characterized by the phrase 'something bad *will not* happen', so we can characterize liveness properties as 'something good *will* happen' [363]. These properties are usually described by temporal formulae of the form '$\Diamond \ldots$', again where '$\ldots$' is usually a non-temporal formula. Typical examples are:

$\Diamond$ *rich*

$\Diamond$ *terminate*

$\Diamond$ $(x > 5)$

Again, these constrain any temporal model to have at least one future moment in time when the required proposition (e.g. *rich*, *terminate* or $x > 5$) will occur.

As you can imagine, most specifications combine safety and liveness properties. In addition, there is a specific style of combination of '$\Box$' and '$\Diamond$' that relates to another class of properties, namely *fairness* properties [218]. Before we consider such properties, it is useful to look at the meaning of one particular combination of '$\Box$' and '$\Diamond$'.

2.4.3   Characterising 'infinitely often'

The combination of '$\Box$' and '$\Diamond$' operators as '$\Box\Diamond$' is often referred to as representing 'infinitely often'. But why? Let us look at the semantics of one particular instance of

---

[5] In some of the examples from now on we will take some liberties with our PTL syntax and allow both numbers and arithmetical statements (with obvious semantics). For example, we consider *reactor_temperature* > 1000 to be a proposition.

this, namely $\Box\Diamond p$. Specifically, let us assume that $\Box\Diamond p$ is to be evaluated at moment i within a model $\mathcal{M}$.

If we use the semantics given earlier, we can evaluate the truth of $\Box\Diamond p$, as follows:

$$\langle\mathcal{M}, i\rangle \models \Box\Diamond p \quad \text{iff for all } j, \text{ if } (j \geq i), \text{ then } \langle\mathcal{M}, j\rangle \models \Diamond p$$

Now, expanding $\langle\mathcal{M}, j\rangle \models \Diamond p$ further we get

$$\langle\mathcal{M}, i\rangle \models \Box\Diamond p \quad \text{iff for all } j, \text{ if } (j \geq i), \text{ then}$$
$$\text{there exists } k \text{ such that } (k \geq j) \text{ and } \langle\mathcal{M}, k\rangle \models p$$

Now, assume that we choose both a $j \geq i$, and a $k \geq j$ where $\langle\mathcal{M}, k\rangle \models p$. Since we quantify over all j's, then we can now choose another moment, say l, such that $l > k$, which requires us to satisfy p again in the future, and so on... So, whichever point in time we are at, there will always be another point in the future where p *must* occur. Since there are an infinite number of points on our timeline, then p must occur *infinitely often*.

Aside.   Although this is just one example of a combination of temporal operators with interesting properties and uses, there are many others, not least the dual of 'infinitely often', that is '$\Diamond\Box \ldots$'. This is sometimes termed *'eventually stable'* and has an important role in the temporal characterization of stability [212, 361, 362].

Exercise 2.5 *Using our PTL semantics show why $\Box\Diamond q$ does* not *imply $\Box q$. In particular, provide a structure (a temporal sequence) describing a counter-example to the property $\Box\Diamond q \Rightarrow \Box q$.*

2.4.4   Fairness properties

Once we have the 'infinitely often' concept then we can characterize properties that do not just happen 'always' or 'never' or 'at least once' (such as safety and liveness properties). We can now capture the idea that a property happens *some* of the time. This is particularly important if we wish to attempt something continuously; in this case we would like to know whether it will succeed never, once, every time or just infinitely often. Such properties are often termed *fairness* properties [218, 363].

Fairness properties are particularly useful when describing the scheduling of processes, the response to messages etc. For example, we can characterize one form of fairness (called *strong fairness*) as

> "if something is attempted/requested infinitely often, then it will be successful/allocated infinitely often".

Such properties typically involve the above '$\Box\Diamond \ldots$' construct, and yet there are many forms of fairness [218], for example:

$\Box\Diamond attempt \Rightarrow \Box\Diamond succeed$

... "if we attempt something infinitely often, then we will succeed infinitely often".

$\Box\Diamond attempt \Rightarrow \Diamond succeed$

... "if we attempt something infinitely often, then we will succeed at least once".

$\Box attempt \Rightarrow \Box\Diamond succeed$

... "if we attempt something continuously, then we will succeed infinitely often".

$\Box attempt \Rightarrow \Diamond succeed$

... "if we attempt something continuously, then we will succeed at least once".

And so on. There is a large range of different views of fairness, many of which can be captured using some form of temporal logic. It is important to note that the above represent constructs of a range of different powers. For example, one of the weaker forms is

$$\Box attempt \Rightarrow \Diamond succeed.$$

Here, even though attempts are continuously made, we are only *guaranteed* to succeed once. A much stronger form is that of

$$\Box\Diamond attempt \Rightarrow \Box\Diamond succeed$$

whereby there will be an infinite number of successes (but not necessarily continuously successful) even though attempts are not made continuously. Again, there has been much work comparing the different combinations of temporal operators [361].

Example 2.12 *Once we have our 'toolkit' of classes of formulae, as above, we can specify a variety of different types of properties of computational systems. Consider the properties we might require of a simple resource allocation system, specifically the access to a shared printer on a network. Here, processes make requests of a central print server by sending relevant requests, namely a 'print_request' message with its single argument being the name of the requesting process. The print server carries out allocation by sending 'printing' messages, again with one argument identifying the process for which printing is being undertaken.*

Now, using temporal logic[6], we can describe a range of properties of the system. Three of these are outlined below:

$\Diamond(\exists x.printing(x))$ *Liveness*

     "eventually, printing will be allowed for some *process"*

$\Box\neg(printing(a) \wedge printing(b))$ *Safety*

     "printing for processes 'a' and 'b' can never occur simultaneously"

$\forall y.\Box\Diamond print\_request(y) \Rightarrow \Box\Diamond printing(y)$ *Fairness*

     "if a process makes a print request infinitely often, then printing for that process will occur infinitely often"

---

[6] For brevity we use first-order temporal notation but, because the quantification occurs over a fixed domain, this is effectively still PTL. Again, consult Appendix A to clarify this first-order notation.

Exercise 2.6 *Imagine that we had the constraint* $\Diamond attempt \Rightarrow \Diamond\Box succeed$. *What does this mean in terms of how successful our attempts are?*

2.5 What is temporal logic?

Now that we have seen both syntax and semantics for PTL, together with a variety of examples expressed in its language, we can reconsider the philosophical question 'what is temporal logic'. While we have given a formal description of the PTL logic, there are a number of alternative formalizations providing different, and interesting, ways to view PTL. In this section we will briefly examine a few of these because this is useful in shedding light on the precise nature of (propositional) temporal logic.

Answer 1: Temporal logic is a fragment of first-order logic

> "*PTL corresponds to a specific, decidable (PSPACE-complete) fragment of classical first-order logic (with arithmetic operations).*"

To give a flavour of this, we now provide an alternative view of the semantics for PTL, this time in classical logic. Primarily this is achieved by representing temporal propositions as classical predicates parameterized by the moment in time being considered [309]. Below we select a few temporal formulae and, assuming they are to be evaluated at the moment i, show how these formulae can be represented in classical logic. Note that our previous model structure, '$\mathcal{M}$', is now encoded within the arithmetical and first-order aspects.

$i \models$	**start**	is represented by	$i = 0$
$i \models$	$\bigcirc p$	is represented by	$p(i+1)$
$i \models$	$\Diamond p$	is represented by	$\exists j.(j \geq i) \wedge p(j)$
$i \models$	$\Box p$	is represented by	$\forall j.(j \geq i) \Rightarrow p(j)$

This gives an obvious way to deal with temporal logic formulae, namely to translate them into first-order logic over $\mathbb{N}$ and deal with their translated versions. Such approaches have indeed been used to handle PTL formulae, though they are not without difficulties [270].

Example 2.13 *Using the above translation,*

$$\Box q \Rightarrow \Diamond q$$

becomes

$$\forall i. \; \left((\forall j.(j \geq i) \Rightarrow q(j)) \;\; \Rightarrow \;\; (\exists k.(k \geq i) \wedge q(k)) \right)$$

> *Through some (first-order) logical manipulation this can be shown to be true as long as quantification occurs over a non-empty domain. Since the domain here corresponds to the set of moments in time, then the domain should actually be infinite.*

Answer 2: Temporal logic characterizes simple induction

> *"PTL captures a simple form of arithmetical induction."*

Recall that the key axiom describing how PTL works is the induction axiom:

$$\vdash \Box(\varphi \Rightarrow \bigcirc\varphi) \Rightarrow (\varphi \Rightarrow \Box\varphi)$$

Now, viewing PTL as a fragment of classical logic (as in Answer 1 above), rooted at $i = 0$, this induction axiom can be reformulated as

$$[\forall i.\varphi(i) \Rightarrow \varphi(i+1)] \Rightarrow [\varphi(0) \Rightarrow \forall j.\varphi(j)]$$

which easily transforms into

$$[\varphi(0) \wedge \forall i.\varphi(i) \Rightarrow \varphi(i+1)] \Rightarrow \forall j.\varphi(j)$$

This should now be familiar as the *arithmetical induction principle*, that is, if we can

1. show $\varphi(0)$, and

2. show that, for any i, if we already know $\varphi(i)$, then we can establish $\varphi(i+1)$

then $\varphi(i)$ is true for *all* elements of the domain.

Answer 3: Temporal logic is a multi-modal logic

> *"PTL can be seen as a multi-modal logic, comprising two modalities, [1] and [*], which interact closely."*

Aside. Modal logic is a vast area of research describing the general forms of logic based on the notion of propositions having distinct truth values in different states/worlds/points and using *modal operators* to navigate between these worlds. Typically, modal logics have just two operators: '[ ]', meaning 'in all accessible worlds'; and '⟨ ⟩', meaning 'in some accessible world'. This can be generalized to multiple accessibility relations and so multiple modal operators, essentially allowing each modal operator to be parameterized, for example $[\alpha]$, $\langle\alpha\rangle$, $[\beta]$, ..., etc. There is a vast array of literature on modal logics; for example see [67, 68, 109, 226, 291].

Now, PTL can be seen as a multi-modal logic, comprising two modalities, [1] and [*], which interact closely. (Effectively, '[1]' is a modal operator that corresponds to the 'next' relation, while '[*]' is a modal operator that corresponds to the 'always' relation. Thus, in our syntax, '[1]' is usually represented as '$\bigcirc$', while '[*]' is usually represented as '$\Box$'.) So, now, the induction axiom in PTL

$$\vdash \Box(\varphi \Rightarrow \bigcirc\varphi) \Rightarrow (\varphi \Rightarrow \Box\varphi)$$

can now be viewed as the *interaction* axiom

$$\vdash [*](\varphi \Rightarrow [1]\varphi) \Rightarrow (\varphi \Rightarrow [*]\varphi)$$

in a modal logic with two modalities. There are here two distinct accessibility relations (characterized by [1] and [∗]), but with quite a strong relationship between them. Specifically, [∗] represents the reflexive transitive closure of [1]. Multi-modal logics with such interactions are typically much more complex than the simple combination of two modal logics without such interactions [226]. This explains, in part, why PTL is more complex than a straightforward combination (fusion) of two simple modal logics [330].

Answer 4: Temporal logic is a logic describing sequences

"PTL can be thought of as a logic over sequences."

As mentioned earlier, the models for PTL are infinite sequences. So, a sequence-based semantics can be given for PTL:

$$s_i, s_{i+1}, \ldots \models \bigcirc p \quad \text{if, and only if, } s_{i+1}, \ldots \models p$$
$$s_i, s_{i+1}, \ldots \models \Diamond p \quad \text{if, and only if, there exists a } j \geq i \text{ such that } s_j, \ldots \models p$$
$$s_i, s_{i+1}, \ldots \models \Box p \quad \text{if, and only if, for all } j \geq i \text{ then } s_j, \ldots \models p$$

In this way, PTL can be seen as a logic for describing such infinite sequences. Later in this chapter we mention different models of time, such as trees and partial-orders, that can form the basis for other varieties of temporal logics. Such logics describe properties of trees or partial orders.

Answer 5: Temporal logic characterizes a class of ω-automata

"PTL can be seen as a syntactic characterization of certain finite-state automata over infinite words (ω-automata)."

As we will see in Section 2.7, models of PTL can be seen as strings accepted by a class of finite automata – Büchi Automata [95, 167, 454]. Consequently, temporal formulae might be used to describe certain automata. Intuitively:

- formulae such as $\Box(p \Rightarrow \bigcirc q)$ give constraints on possible transitions between automaton states;

- formulae such as $\Box\Diamond r$ give constraints on *accepting states* within an automaton, that is states that must be visited infinitely often; and

- formulae such as $s \Rightarrow \Box t$ describe global invariants within an automaton.

We will see in Chapter 5 that the link between PTL and Büchi Automata is a key part of several temporal verification techniques.

All of the above interpretations are viable and provide differing views of PTL. Indeed there are undoubtedly other views as well. As we will see during the rest of this book, the way we view temporal logic depends greatly on what we wish to do with our temporal logic formulae.

2.6   Normal form

While we have only presented simple examples, it should be clear that PTL formulae can become quite complex and difficult to understand. Both from a user point of view, and from a 'mechanisation' point of view, it is often useful to replace one complex formula by several simpler formulae. Of course the new, simpler, formulae should have a behaviour equivalent to the original. Though there are many forms this replacement could take, we consider a particularly useful *normal form* where formulae are represented as

$$\Box \bigwedge_{i=1}^{n} R_i$$

where each of the R_i, termed a *rule*, is an implication in the style

formula about current behaviour $\Rightarrow$ formula about current and future behaviour .

In later chapters, this normal form will be used as the basis for both deduction and execution methods. However, it is also useful in simplifying PTL formulae for human readership, especially as we remove many of the more complex temporal operators (we will see how to do this in Chapter 4).

For the moment, however, we will just define this normal form. It is called *Separated Normal Form (SNF)* [195, 208]. As well as being in the rule form described above, formulae in SNF are also restricted in terms of their temporal operators. Thus, in SNF, each R_i must be of one of the following forms:

$$\mathbf{start} \quad \Rightarrow \quad \bigvee_{b=1}^{r} l_b \qquad \text{(an } \textit{initial} \text{ rule)}$$

$$\bigwedge_{a=1}^{g} k_a \quad \Rightarrow \quad \bigcirc \bigvee_{b=1}^{r} l_b \qquad \text{(a } \textit{step} \text{ rule)}$$

$$\bigwedge_{a=1}^{g} k_a \quad \Rightarrow \quad \Diamond l \qquad \text{(a } \textit{sometime} \text{ rule)}$$

Note, here, that each k_a, l_b, or l is simply a literal.

It is also important to realize that the rules in SNF, R_i, do not contain the temporal operators '$\Box$', 'U' or 'W'. As we will see later, these are removed in the translation process being replaced by SNF rules that simulate their behaviour.

This normal form gives a simple and intuitive description of what is true at the beginning of time (via *initial* rules), what must be true when moving from one moment to the next (via *step* rules), and what constraints exist on future moments in time (via *sometime* rules). Intuitively, this makes sense – if we are to describe what happens in time, all we need is to describe what happens at the start, what happens between now and a successive moment, and what should happen at other (unspecified) moments in the future.

We also note that, later, SNF rules will be used as the basis for a temporal *proof* technique called *clausal temporal resolution*. In this case, we will also term SNF rules as *SNF clauses*.

Example 2.14 *The example formulae below correspond to the three different types of SNF rules.*

$$
\begin{array}{rrcl}
\text{INITIAL:} & \mathbf{start} & \Rightarrow & losing \vee hopeful \\
\text{STEP:} & (losing \wedge \neg hopeful) & \Rightarrow & \bigcirc losing \\
\text{SOMETIME:} & hopeful & \Rightarrow & \Diamond \neg losing
\end{array}
$$

Later we will consider how SNF can be used, and how an arbitrary PTL formula can be transformed into SNF. For the moment, we just note the important property that *any* PTL formula can be transformed into a set of SNF rules that have (essentially) equivalent behaviours, specifically the preservation of satisfiability. This is achieved at the expense of only a polynomial increase in both the size of the representation and the number of atomic propositions used within the representation [195, 208]. Thus, certain PTL operators, such as 'U', are removed at the expense of introducing new atomic propositions. See Section 4.2.1 for details of this translation process.

2.7   Büchi automata and temporal logic

As we have already seen, models for PTL are essentially infinite, discrete, linear sequences, with an identified start state. Thus, each temporal formula corresponds to a set of models on which that formula is satisfied.

Now, look again at these models. If we think of each particular state within a model, then we can see that each such state is of finite size. Since there are only a finite number of propositions and a finite set of values for those propositions, then we can actually define a finite set of all possible states and use a (new) distinct symbol to represent each one of these states. Now, with each state represented by a symbol, our models become sequences of such symbols, that is, strings.

While it is no easier to handle an infinite number of strings than to handle an infinite number of temporal models, viewing models as strings allows us to utilize the large amount of previous work on *finite automata*. In particular, we can define a finite automaton that accepts *exactly* the strings we are interested in and so we can use finite automata to represent temporal models. As pioneers in this area saw, the automata needed are a specific form of automata over infinite strings, often termed 'ω-automata' [442, 454, 486]. Before describing such automata, let us first recap the concept of (finite) automata over *finite* words (strings), which are more commonly used in Computer Science, particularly as language recognisers [13].

Recap: Finite automata. A *finite-state automaton* that can accept *finite* words/strings can be defined as [289]:

$$FSA = \langle A, S, \delta, I, F \rangle$$

Here

- A is a *finite* alphabet of symbols,

- S is a *finite* set of states,

- $\delta \subseteq S \times A \times S$ is a transition relation,

- $I \subseteq S$ is a set of initial states, and

- $F \subseteq S$ is a set of final states.

Such an automaton accepts finite strings constructed from the alphabet A. Thus, a sequence of symbols is accepted if we can begin in an initial state (i.e. a state in I), read symbols successively from the sequence, each time moving to states prescribed by δ and, after reading the last symbol in the sequence, end in one of the final states (i.e. a state in F).

Aside: Product of finite state automata. Given two finite-state automata, $FSA_1 = \langle A_1, S_1, \delta_1, I_1, F_1 \rangle$ and $FSA_2 = \langle A_2, S_2, \delta_2, I_2, F_2 \rangle$, then the *product* $FSA_1 \times FSA_2$ is the new automaton, $FSA_{1\times 2}$ defined as:

$$\langle A_1 \times A_2, S_1 \times S_2, \delta, I_1 \times I_2, F_1 \times F_2 \rangle$$

where $((\langle s1, s2 \rangle, a, \langle t1, t2 \rangle) \in \delta$ if, and only if, $(s1, a, t1) \in \delta_1$ and $(s2, a, t2) \in \delta_2$ for $a \in A_1 \cap A_2$.

Now, in order to model the infinite execution sequences seen within PTL, we require a form of finite automata that is able to recognize *infinite* words/strings. There are a number of types of automata that fit this requirement, termed ω-automata, but the variety we will use are *Büchi Automata* [95]. A Büchi Automaton is essentially a finite state automaton, as above,

$$BA = \langle A, S, \delta, I, F \rangle$$

where the concept of *acceptance* of a string now has to take into account *infinite* behaviour and is defined as follows. For any (infinite) run, say ρ, then this run is accepted by BA if, and only if,

at least one element of F appears infinitely often along ρ.

Büchi Automata accept exactly the type of sequences we have been talking about earlier. In particular, a Büchi Automaton, B_φ, can be used to represent all the possible models of the formula φ. There are many different mechanisms used for generating such an automaton[7] from a PTL formula φ, for example:

---

[7] Apparently an alternative approach is described (in French) in [374].

- we can recursively take φ apart, constructing an automaton for each of the simpler components and combining these (typically using union and product operations; see later) to generate B_φ;

- since PTL has the *finite model property* we can finitely describe all possible propositional configurations in models for φ and therefore all possible states in B_φ – so, we might generate all these configurations/states to begin with and use φ to help us define δ, I and F within B_φ; or

- transform φ into SNF and then use the fact that SNF is quite close to the automaton form [76] to generate B_φ – *initial* clauses essentially describe initial states, *step* clauses capture a form of transition function, and (unconstrained) *sometime* clauses describe the set of accepting states.

While these often reduce to the same techniques, the second is a more common approach and, indeed, much work has been carried out on the efficient generation of Büchi Automata from PTL formulae [451, 455]; see, for example, the LTL2AUT tool [134]. However, we will briefly look at the first approach, since it gives some intuitive understanding of the relationship between temporal formulae and automata. We will begin with simple examples of automata generated from PTL formulae.

2.7.1   Simple Büchi automata corresponding to temporal formulae

Temporal formulae can be described using Büchi Automata, as follows. (A very useful resource here is the Büchi Store [96] which provides a repository of useful Büchi Automata.) We start by characterizing the alphabet that our automaton will recognize as all possible combinations of proposition symbols that we might see. Thus, each 'letter' in our alphabet is a subset of $\mathbf{P}(\text{PROP})$. For example, that our automaton can move from one state to another by recognizing '$\{a, b, c\}$' means that the propositions a, b and c must be true at that moment.

Thus, the formula '$\bigcirc a$' can be represented by the automaton (and, thus, any sequence of states that satisfies $\bigcirc a$ must be a sequence accepted by this automaton)

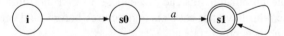

Here, from the initial state, **i**, we move forward one step with *any* combination of symbols from the alphabet to reach state **s0**. Now, we can move forward to **s1** while accepting any set of propositions containing 'a'. Such an automaton accepts *infinite* sequences, so we must add a transition (with no label) from **s1** back to itself, thus capturing an infinite 'tail' with no constraints (i.e. any trailing sequence of symbols is accepted).

There are several points to note here.

- This is a simple, deterministic (i.e. there is no choice of which state to move to) automaton.

- The 'accepting' states, that is those in F, are graphically represented by a double circle (and, typically, not all states are accepting).

- While we previously represented models of temporal formulae with 'information' (such as which propositions are true) located at states/nodes, as we see from above this information is now captured on the label of the transition between states. If a transition does not have a specific label, *any* symbol can be accepted on that transition.

- We will often, as above, write the elements within the set recognized (e.g. '*a*') rather than providing the whole set as a label (e.g. '{*a*}').

Now, if we consider the formula '$\bigcirc (a \vee \bigcirc b)$', we generate a correspondingly branching automaton:

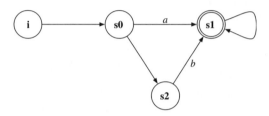

Again, we have the cycle on state **s1** accepting any infinite tail.

The automaton translation of the formula '$\Box a$' has more interesting infinite behaviour:

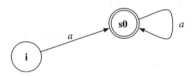

Thus, now, the 'infinite tail' has to always accept an '*a*'.

Aside: Automata descriptions.   Readers familiar with Büchi Automata will notice that, in the above diagram, we could simply have combined the 'i' and 's0' states to give just one state within the automaton. Instead we choose to give non-minimal automata that are clearer to those who have not seen such structures before, and often achieve this by separating initial and accepting states.

Example 2.15 *While a pictorial representation of the formula '$\Box a$' is given above, the automaton representation of this is $\langle A, S, \delta, I, F \rangle$ where*

$A = \mathbf{P}(\{a, b, c\})$
$S = \{\mathbf{i}, \mathbf{s0}\}$
$\delta = \{(\mathbf{i}, a, \mathbf{s0}), (\mathbf{s0}, a, \mathbf{s0})\}$
$I = \{\mathbf{i}\}$
$F = \{\mathbf{s0}\}$

*Here, we consider just three propositional symbols: a; b; and c. Again, we omit explicit set notation and use transitions such as (**i**, a, **s0**) to represent a transition that recognizes any element from the alphabet that contains a.*

The formula '◇a' exhibits an automaton in which the accepting state becomes particularly important:

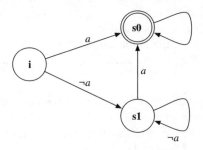

Notice here that there is an infinite path through this structure which continually makes *a* false (i.e. **i**, **s1**, **s1**, **s1**, ...). However, an accepting state does *not* occur infinitely often on this path and so the path is not accepted by the automaton. All acceptable paths must eventually visit **s0** and, as soon as they do, they are condemned to cycle through **s0** forever.

Example 2.16 *The automaton representation of '◇a' above, for example over the alphabet {a, b, c}, is*

$A = \mathbf{P}(\{a, b, c\})$
$S = \{\mathbf{i}, \mathbf{s0}, \mathbf{s1}\}$
$\delta = \{(\mathbf{i}, a, \mathbf{s0}), (\mathbf{i}, \neg a, \mathbf{s1}), (\mathbf{s0}, \textit{true}, \mathbf{s0}), (\mathbf{s1}, \neg a, \mathbf{s1}), (\mathbf{s1}, a, \mathbf{s0})\}$
$I = \{\mathbf{i}\}$
$F = \{\mathbf{s0}\}$

In the above (and sometimes later), we will abuse the notation slightly by using '¬ a' as a label. There is no such letter in the alphabet – we just use this as shorthand for 'any subset of {a, b, c} that does not contain a'. Similarly, we will use 'true' as a shorthand for 'any letter from the alphabet'.

Finally, the formula '□a ∨ □b' exhibits a different form of complex behaviour:

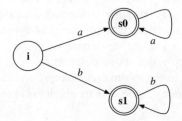

This shows one of the ways of composing automata. If we have one automaton accepting models of □a and another accepting models of □b, then the above structure is essentially the union of the two automata (and, as we have seen, captures models of '□a ∨ □b'). We will see later that conjunction (i.e. '∧') can be much more complex.

In all the above automata, the key aspect is that each must accept at least one infinite sequence (visiting F infinitely often). This leads us to a mechanism for deciding the truth of temporal formulae. Given φ which we wish to show to be valid, then

φ is valid   iff $\neg\varphi$ is unsatisfiable
            iff $B_{\neg\varphi}$, the automaton accepting all models of $\neg\varphi$, is empty
            iff $B_{\neg\varphi}$ has no infinite accepting sequences

Thus, in order to check the validity of φ, we might construct $B_{\neg\varphi}$, the Büchi Automaton capturing all possible models of $\neg\varphi$. If $B_{\neg\varphi}$ has *no* infinite accepting sequences, then we know that φ is valid. We can add this technique to those that we will see later for deciding PTL formulae (see Chapter 4).

Exercise 2.7 *Describe a Büchi Automaton corresponding to* '**start** $\Rightarrow a\,U\,b$'.

2.7.2   Checking Büchi emptiness

Essentially, checking the emptiness of a Büchi Automaton is a cyclic process carrying out each of the following deletions on the automaton structure:

- remove edges/transitions with *inconsistent* labelling – we have not seen many such examples yet, but imagine we had a labelling for an edge/transition which gave both 'a' and '$\neg\,a$' as a label;

- remove nodes/states which have *no* edges/transitions from them – deletion is allowed here since no infinite path (and all our accepting paths are infinite) can come through this node/state; and

- remove terminal, non-accepting sets of nodes/states.

This last element needs a little explanation. A terminal *strongly-connected component* in a directed graph is a set of nodes such that all transitions from any of these nodes lead to other nodes in the set. So, we can think of such sets as subgraphs where, once entered, it is impossible to escape from. Now, we can search for strongly connected components in our automaton structure using Tarjan's efficient (linear complexity) algorithm [468]. If we find one of these terminal strongly-connected components in which *none* of the nodes are accepting, then this means that any acceptable path cannot get 'stuck' in one of these subgraphs. So, we can delete this whole subgraph.

Thus, we continually apply the above deletion steps. If they eventually lead to an empty set of nodes/states, then the automaton is empty and cannot accept any sequences. Otherwise, we reach a non-empty automaton to which the above cannot be applied further.

2.7.3   Automata products

Since we later have need for the product operation on Büchi Automata, and since this leads towards a way to construct larger Büchi Automata, then we will outline this aspect here.

Given $BA_1 = \langle A_1, S_1, \delta_1, I_1, F_1 \rangle$ and $BA_2 = \langle A_2, S_2, \delta_2, I_2, F_2 \rangle$ then

$$BA_1 \times BA_2 = \langle A_1 \times A_2, S_1 \times S_2, \delta', I_1 \times I_2, F_1 \times F_2 \rangle$$

where, for $a \in A_1 \cap A_2$,

$$(\langle s_1, s_2 \rangle, a, \langle s_1', s_2' \rangle) \in \delta' \text{ if, and only if, } (s_1, a, s_1') \in \delta_1 \text{ and } (s_2, a, s_2') \in \delta_2$$

Example 2.17 *Consider two automata over the alphabet* $\mathbf{P}(\{a, b\})$, *one describing* '$\square a$' *the other describing* '$\square b$':

$$BA_{\square a} = \langle \mathbf{P}(\{a, b\}), \{\mathbf{si}, \mathbf{s0}\}, \{(\mathbf{si}, a, \mathbf{s0}), (\mathbf{s0}, a, \mathbf{s0})\}, \{\mathbf{si}\}, \{\mathbf{s0}\} \rangle$$

$$BA_{\square b} = \langle \mathbf{P}(\{a, b\}), \{\mathbf{ti}, \mathbf{t0}\}, \{(\mathbf{ti}, b, \mathbf{t0}), (\mathbf{t0}, b, \mathbf{t0})\}, \{\mathbf{ti}\}, \{\mathbf{t0}\} \rangle$$

Now, the product of these, that is $BA_{\square a} \times BA_{\square b}$ *gives us the automaton for* $BA_{\square(a \wedge b)}$:

$$BA_{\square(a \wedge b)} = \left\langle \begin{array}{l} A = \mathbf{P}(\{a, b\}), \quad S = \{(\mathbf{si}, \mathbf{ti}), (\mathbf{si}, \mathbf{t0}), (\mathbf{s0}, \mathbf{ti}), (\mathbf{s0}, \mathbf{t0})\}, \\ \delta = \{((\mathbf{si}, \mathbf{ti}), ab, (\mathbf{s0}, \mathbf{t0})), ((\mathbf{s0}, \mathbf{t0}), ab, (\mathbf{s0}, \mathbf{t0}))\}, \\ I = \{(\mathbf{si}, \mathbf{ti})\}, \quad F = \{(\mathbf{s0}, \mathbf{t0})\} \end{array} \right\rangle$$

Here, 'ab' is a shorthand for any subset containing both *a and b. Finally, note that, since* $\square(a \wedge b)$ *is equivalent to* $\square a \wedge \square b$, *then the automaton generated also represents* $BA_{(\square a \wedge \square b)}$.

So, from the above, it looks like the Büchi Automaton for any temporal formula $\varphi \wedge \psi$ can simply be generated by taking the product of the Büchi Automaton for φ with the Büchi Automaton for ψ. Yet this is *not* the case! The intuitive reason is that the product of two Büchi Automata A and B must have accepting states exactly where *both* A and B *simultaneously* have accepting states. In our example above this is, indeed, the case. However, it is perfectly legitimate for A and B to have accepting states at different places – in this case, the product automaton will have *no* accepting states. The example below, from [484], clarifies this.

> **Example 2.18** *Consider the following automata, each accepting an infinite sequence of a's:*
>
>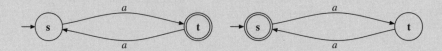
>
> *So, we would expect that the product also accepts an infinite sequence of a's. However, the product is empty as there are* no *accepting states that are reached simultaneously in the two automata separately!*

Yet, this is not as bad as it seems. Even though the set of sequences accepted by $A \times B$ is a strict subset of those accepted by both A and B, it is known that we can construct another Büchi Automaton, C, such that the sequences accepted by C are *exactly* those accepted by both A and B [111]. This takes a little manipulation of a product-like combination of A and B, ensuring that, in C, the accepting states from both A and B are visited infinitely often (though not necessarily simultaneously).

Aside: Products are useful. While the above might imply that taking the product of two Büchi Automata is not so useful, it is in fact the basis for very many tools and methods. First of all, for simple examples such as the $\square a \wedge \square b$ one above, then the product does what we expect. In addition, there is an important subcase where the the product automaton *does* accept the intersection of sequences accepted by each automaton separately. This is the case where, when we take $A \times B$, then *all* the states within A are accepting! In this case we do not have the problems above and can take the pure product. We will see in Chapter 5 that this is the basis of *model checking*. Here, A is an automaton representing all legal runs of a system, and each state in A is an accepting state. Then, as B will represent all possible *bad* executions, the product, $A \times B$ will accept only 'bad' executions of our system.

> **Example 2.19** *To see how choices combine, consider the following two automata, representing '$\bigcirc a \vee \bigcirc b$' and '$(x \wedge \bigcirc y) \vee (y \wedge \bigcirc x)$', respectively (again, we will abuse the notation slightly to provide a simple explanation):*
>
>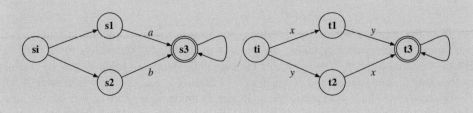

The product of these two automata is

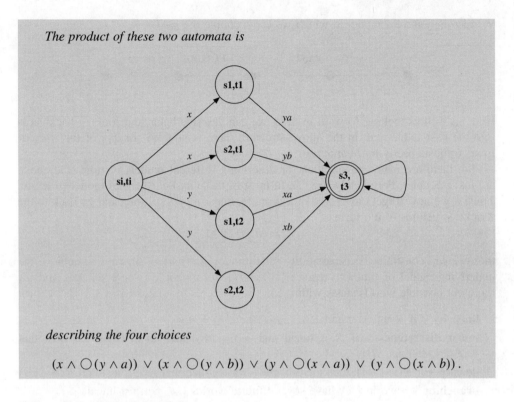

describing the four choices

$$(x \wedge \bigcirc(y \wedge a)) \vee (x \wedge \bigcirc(y \wedge b)) \vee (y \wedge \bigcirc(x \wedge a)) \vee (y \wedge \bigcirc(x \wedge b)).$$

2.8   Advanced topics

In this chapter we have covered a number of aspects relating to the foundations of (one form of) temporal logic. However, there are a *great many* other aspects that are being, or have been, actively researched [196, 279]. In this section, we give a brief overview of, and references to, a selection of these. The interested reader can then explore these areas separately.

2.8.1   Varieties of temporal model

There are *many* different varieties of temporal accessibility relation, generally corresponding to some intuitive model of time. Two of the most common models are linear and branching structures. As described above, linear accessibility means that at most one other world is accessible from the current one, while branching means that there may be more than one:

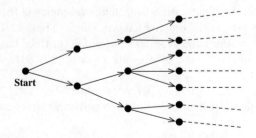

We also may have a model in which we can examine the *past*.

Here, as well as looking forward in time, we can describe behaviour looking backwards into the *past*. Note that, in the above structure there is now no concept of the *start* of time; we just specify an *infinite past*.

So, there are many possibilities for describing different models of time (the above are just a couple). Fortunately, we can fit most of these into our general model structure. Thus, if we now forget our choice of a discrete, linear model of time, and go back to the Kripke Structures of the form

$$\mathcal{M} = \langle S, R, \pi \rangle$$

then we can constrain the accessibility relation (R) in a variety of ways to achieve the underlying model of time we require [168, 479], for example (where w_1, w_2, and w_3 represent possible worlds/states within S):

linearity – if $R(w_1, w_2)$ and $R(w_1, w_3)$, then $w_2 = w_3$

linear discreteness – if R is linear and $R(w_1, w_2)$, then there is *no* w_3 such that $R(w_1, w_3)$ and $R(w_3, w_2)$

density – if $R(w_1, w_2)$, then there is *always* a w_3 such that $R(w_1, w_3)$ and $R(w_3, w_2)$

branching – a world may have several future worlds (i.e. not just linear)

finite past – there is a w_1 such that there is no w_2 such that $R(w_2, w_1)$

And so on.

Example 2.20 *If we consider the* linear discreteness *condition above, we develop the following constraint on the worlds/states within S and relation R within* $\mathcal{M}$:

$$\forall w_1, w_2 \in S.R(w_1, w_2) \Rightarrow$$
$$\neg(\exists w_3 \in S. (w_1 \neq w_3) \wedge (w_2 \neq w_3) \wedge R(w_1, w_3) \wedge R(w_3, w_2))$$

Thus, if w_2 is accessible directly from w_1 then there is no distinct w_3 that is accessible from w_1 and from which w_2 is accessible. Intuitively, this means that w_2 is the next world to w_1 and that no other world, say w_3, can appear between them.

A wide variety of temporal logics has been developed based on a number of these underlying structures, for example *branching* temporal logics [116, 170, 172, 435, 441], *dense* temporal logics [46, 313], *bounded* temporal logics [104, 123], temporal sequences with *gaps* [31, 224], *circular* time [434], *partial-order* (or trace-based) logics [265, 310, 404, 473], temporal intervals [242], real-time models [23, 341, 494] and *alternating* temporal logics [26, 105].

Just as there is a wide variety of temporal structures, a large range of temporal languages have been developed, usually linked to particular structural properties. We will mention a few of these in the sections below.

2.8.2   Branching temporal logic

As we saw above, an alternative to the linear model of PTL is to use a branching interpretation. Basically, the same operators as found in PTL are used in such *branching-time temporal logic*, but certain *path operators* are also added:

A – "on all future paths starting here"
E – "on some future path starting here"

These quantify over paths through the relevant tree structure.

Branching-time temporal logic is very popular, particularly in the model-checking area [114, 122].

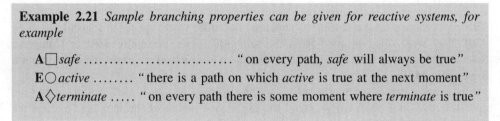

Example 2.21 *Sample branching properties can be given for reactive systems, for example*

A□*safe* .... *"on every path, safe will always be true"*
E○*active* .... *"there is a path on which active is true at the next moment"*
A◇*terminate* ..... *"on every path there is some moment where terminate is true"*

Note that there are two main varieties of branching-time temporal logic, which are syntactically distinguished as follows. The logic CTL [116] enforces the restriction that each temporal operator must be prefixed by a path operator. A typical CTL formula is

$$\mathbf{A}\Diamond(\mathbf{E}\Diamond p \wedge \mathbf{E}\Box q).$$

Notice how the path/temporal operators appear in pairs.

While CTL has many useful properties, such as low model-checking complexity, it lacks expressiveness. Consequently, a more powerful logic, called CTL* [172], was developed. This allows *any* combination of path or temporal operator, with a typical CTL* formula being

$$\mathbf{A}\Box\Diamond\mathbf{E}\mathbf{A}p.$$

However, there is a price to pay for this power, with CTL* being generally quite complex to deal with, in terms of both decision procedures [172] and axiomatization [435].

There has been much debate about the relative merits of linear and branching temporal logics [170, 335]. An advantage of the branching-time approach (especially CTL) is that the model-checking problem (see later) is of polynomial complexity whereas model checking for linear versions is often exponential. However, for an argument that the situation is not so clear cut, see [485].

An advantage of linear model structures is that the satisfiability problem for most versions of PTL is PSPACE, while in the branching (CTL*) case, including the next-time operator, satisfiability is commonly much harder (2-EXPTIME).

There is clearly a choice to be made here, but an important (non-logical) aspect is that many people find the linear, PTL style, to be much more intuitive and easier to understand. Indeed, Nain and Vardi have shown how a linear view is indeed the most natural, and the most useful, form from a technical as well as an aesthetic point of view [387, 388].

2.8.3   Past-time temporal logic

Another variation on the underlying temporal structure is to include the notion of *past* time. In this case, we can utilize a set of operators which are essentially the past-time counterparts of the PTL operators introduced earlier:

$\bullet\varphi$	φ was true in the *last* moment in time.
$\blacksquare\varphi$	φ was true in *all* past moments.
$\blacklozenge\varphi$	φ was true in *some* past moment.
$\varphi \mathcal{S} \psi$	φ has been true *since* some past moment when ψ was true.

Though such past-time connectives appeared in earlier works on *tense* logics [100, 309, 420], they were often omitted from the temporal logics originally used in Computer Science. However, researchers found it convenient to re-introduce past-time into temporal logics [45, 347].

In many cases, particularly when a finite history is considered, adding past-time operators does not necessarily extend the expressive power of the language and, as we will see later, such operators can all be removed from PTL by translation into normal forms such as SNF. However, the direct use of past-time operators can allow much greater succinctness [366].

2.8.4   Fixpoints

Even in the short description given so far, we can see that many temporal operators have been defined and developed. The simplest syntax for discrete linear temporal logic involves just the '$\square$', '$\diamondsuit$', and '$\bigcirc$' operators. This gives a logic that is usable, but that is not easily able to represent more complex properties such as '$p\,\mathcal{U}\,q$'. Kamp showed that the addition of the '$\mathcal{U}$' operator (and '$\mathcal{S}$', its past-time counterpart) gives us a 'reasonable' level of expressive power (corresponding to first-order descriptions over linear orders) [309]. However, Wolper later showed that there are useful computational patterns (such as something being true at every *even* moment in time) that require even stronger operators [500]. This then led to a whole range of research publications developing and analysing such operators, including *quantifiers over Boolean-valued variables* (see Section 2.8.7), additional *grammar* operators [500] and *fixpoint operators* [69]. These classes are all strictly more expressive than PTL.

In the next section we will consider quantification but, to give an idea here of what a more powerful temporal language looks like, we will briefly outline the use of *fixpoints* (aka *fixed points*) in temporal logics [39, 40, 221, 483].

Aside: What are fixpoints?   In general, fixpoints are solutions to recursive equations, such as $x = f(x)$. In basic algebra, we know that $a = 0$ is a solution of $a = (a * 5)$. Similarly, some such equations have no finite solutions, for example $a = \infty$ is a solution of $a = (1 + a)$. Conversely, there can often be several solutions to fixpoint equations. For example, $a = 0$ and $a = 1$ are both solutions of $a = (a * a)$. We often wish to distinguish these solutions, so we might identify them by taking either

ν, representing the maximal (greatest) fixpoint, or

μ, representing the minimal (least) fixpoint.

For example, we write down $\nu a.(a * a)$ to denote the maximal fixpoint of $a = (a * a)$; similarly with 'μ'. So, now, the solution[8] of $\nu a.\ (a * a)$ is $a = 1$, and the solution of $\mu a.\ (a * a)$ is $a = 0$.

It is important to note that, in some cases *no* fixpoint exists. Or even only one exists and so the maximal and minimal versions coincide.

These two operators, μ (least fixpoint) and ν (greatest fixpoint), have been transferred to temporal logics allowing the development of various fixpoint temporal logics [39, 40, 221, 483]. Utilizing fixpoint operators allows us to develop sophisticated temporal descriptions based on quite simple syntax.

Aside: What are temporal fixpoints?   Just as fixpoints are solutions of algebraic equations (above), in the temporal world they represent solutions of formulae. So, a fixpoint solution to the formula '$\psi \Leftrightarrow (\varphi \wedge \bigcirc \psi)$' is some formula '$\psi$' that makes the statement true. Indeed, taking $\psi \equiv \square \varphi$ does make this formula true, since

$$\square \varphi \Leftrightarrow (\varphi \wedge \bigcirc \square \varphi)$$

while $\psi \equiv \bigcirc \varphi$ does not, since

$$\bigcirc \varphi \not\Leftrightarrow (\varphi \wedge \bigcirc \bigcirc \varphi).$$

These operators form the basis of a temporal logic called νTL in which the basic temporal operators are the 'next' and fixpoint operators and with which all the standard temporal operators can be defined, for example:

$$\begin{aligned} \square \varphi &\equiv \nu \xi.(\varphi \wedge \bigcirc \xi) \\ \lozenge \varphi &\equiv \mu \xi.(\varphi \vee \bigcirc \xi) \end{aligned}$$

Here, $\square \varphi$ is defined as the maximal fixpoint (ξ) of the formula $\xi \Leftrightarrow (\varphi \wedge \bigcirc \xi)$. Thus, the maximal fixpoint above effectively defines $\square \varphi$ as the 'infinite' formula

$$\varphi \wedge \bigcirc \varphi \wedge \bigcirc \bigcirc \varphi \wedge \bigcirc \bigcirc \bigcirc \varphi \wedge \dots$$

Note that the *minimal* fixpoint of $\xi \Leftrightarrow (\varphi \wedge \bigcirc \xi)$ is '**false**', since putting **false** in place of 'ξ' is legitimate and **false** is the minimal solution.

So, because there may be many different fixpoint solutions, it is vital to be able to distinguish between them. Again, this is essentially what the 'least fixpoint' and 'greatest fixpoint' operators allow us to do. Thus,

$$\xi \Leftrightarrow (\psi \vee (\varphi \wedge \bigcirc \xi))$$

---

[8] One might argue that actually the maximal solution of $a = (a * a)$ is $a = \infty$.

has several solutions. The minimal (smallest) one involves satisfying ψ as quickly as possible, while the maximal (greatest) one allows for the possibility that $\varphi \wedge \bigcirc \ldots$ repeats forever. The formula $\xi \Leftrightarrow (\psi \vee (\varphi \wedge \bigcirc \xi))$ describes a sequence of 'φ' terminated by a 'ψ', and the two solutions give different views of how quickly the solutions must be found. We have actually already seen these two solutions before:

$$\varphi U \psi \equiv \mu \xi.(\psi \vee (\varphi \wedge \bigcirc \xi))$$
$$\varphi W \psi \equiv \nu \xi.(\psi \vee (\varphi \wedge \bigcirc \xi))$$

The semantic definition of the fixpoint operators and particular restrictions on the temporal functions used can be found in [39, 40, 221, 483]. These operators are typically used in the temporal semantics of complex systems, particularly for the description of loops or procedure calls in programming language semantics (see Section 3.4.5), while their proof methods and model-checking techniques have also been developed [88, 136, 461].

2.8.5   A little complexity

Unsurprisingly, there has been a great deal of work on analysing the different temporal logics one might use. One important aspect of this is the computational complexity of deciding whether formulae are valid or not and deciding whether formulae are satisfied over some temporal structure (i.e. *model checking*). Since both of these aspects will be tackled from a practical viewpoint later, we just briefly list the relevant complexity results below. The interested reader is directed towards articles containing *much* more detail [88, 136, 166, 171, 224, 453, 461, 483].

	Model Checking, i.e.$M \models \varphi$	Validity, i.e. $\vdash \varphi$
PTL	PSPACE	PSPACE
CTL	Polynomial Time	EXPTIME
CTL*	PSPACE	2EXPTIME
νTL	PSPACE	PSPACE

Note that these complexity results are given in terms of the size of the descriptions. For a little intuition about the complexity classes:

$$\text{Polynomial} \subseteq \text{NP} \subseteq \text{PSPACE} \subseteq \text{EXPTIME} \subseteq \text{EXPSPACE} \subseteq \text{2EXPTIME} \subseteq \ldots$$

with (we believe!) each complexity class being exponentially more complex than the previous one.

2.8.6   Quantification: QPTL

An alternative way of extending PTL is to add *quantification*. Here, the usual first-order quantifier symbols, '$\forall$' and '$\exists$', can be added. A first (small) step on this route is to allow quantification, but only over Boolean-valued variables (i.e., propositions of the language). Thus, using such a logic, called *quantified propositional temporal logic* (QPTL) [499], it is possible to write formulae such as

$$\exists p.p \wedge \bigcirc \bigcirc p \wedge \Diamond \square \neg p$$

where p is a propositional variable. In certain cases [39, 454, 487, 500], QPTL turns out to be equivalent to the fixpoint extension considered above [39] as well as ETL, which extends PTL with grammar operators [500].

Aside: The philosophy of differing styles of quantification. The particular form of quantification used here, termed the *substitutional interpretation* [252], can be defined as:

$$\langle M, s \rangle \models \exists p.\varphi \quad \textit{if, and only if,} \quad \textit{there exists a model } M' \textit{ such that}$$
$$\langle M', s \rangle \models \varphi \textit{ and } M' \textit{ differs from } M$$
$$\textit{in at most the valuation given to } p$$

Haack engages in a thorough discussion of the philosophical arguments between the proponents of the above and the *objectual interpretation* of quantification, more standard in classical logic [252]:

$$\langle M, s \rangle \models \exists p.\varphi \quad \textit{if, and only if,} \quad \textit{there exists a proposition } q \in \text{PROP}$$
$$\textit{such that } \langle M, s \rangle \models \varphi(p/q)$$

where $\varphi(p/q)$ is the formula φ with p replaced by q throughout.

2.8.7   Quantification: FOTL

Rather than quantifying over Boolean-valued (propositional) variables, what if we use standard (first-order logic inspired) quantification over arbitrary sets? Adding such first-order (and, in the sense above, objectual) quantification to temporal logic, for example PTL, is very appealing. However, it is fraught with danger! A full first-order temporal logic is very convenient for describing many scenarios, but is so powerful that we can write down formulae that capture a form of arithmetical induction from which it is but a short step to being able to represent full arithmetic [1, 373, 463, 464]. Consequently, full first-order temporal logic is incomplete; in other words, the set of valid formulae is *not* recursively enumerable when considered over models such as the Natural Numbers.

While some work was carried out on methods for handling, where possible, such specifications [363], first-order temporal logic was generally avoided. Even 'small' fragments of first-order temporal logic, such as the *two-variable monadic* fragment, are not recursively enumerable [373, 280]. Consequently, alternative directions have been explored. One that has led to significant research activity is to simplify the basis for the temporal logic and then to extend with first-order aspects. This route has been followed successfully in research on the Temporal Logic of Actions (TLA) [338, 3, 339].

In recent years, however, work on first-order extensions of PTL has been revived. In particular, work by Hodkinson *et al.* showed that *monodic* fragments of first-order temporal logics could have complete axiomatizations and could even be decidable [280]. A monodic temporal formula is one whose temporal subformulae have, at most, one free variable. Thus,

$$\forall x.p(x) \Rightarrow \bigcirc(\exists y.q(x, y))$$

is monodic, while

$$\forall x.\exists y.p(x) \Rightarrow \bigcirc q(x, y)$$

is not. Wolter and Zakharyaschev showed that any set of valid *monodic* formulae is finitely axiomatizable over a temporal model based on the Natural Numbers [502]. Intuitively, the monodic fragment restricts the amount of information transferred between temporal states so that, effectively, only individual elements of information are passed between temporal states. This avoids the possibility of describing the evolution through time of more complex items, such as relations, and so retains desirable properties of the logic. These results have led on to considerable recent work analysing and mechanizing monodic fragments of FOTL [143, 280, 318] which, in turn, has led to new applications [181, 211, 227].

Aside: equality.   Within FOTL we would often like to use *equality* over the elements of the domain. However, we must be very careful. First of all, a typical use of equality might be in formulae such as

$$\forall w.\forall z. \; (a(w) \wedge b(z)) \Rightarrow \bigcirc(w = z) \, ,$$

but this *is not* a monodic formula because the '$\bigcirc(w = z)$' subformula has two free variables, w and z. A monodic use of equality in FOTL might be

$$\forall w. \; a(w) \Rightarrow \bigcirc(\exists v.b(v) \wedge (w = v)) \, .$$

However, we must again be careful. Even when monodic formulae are used the addition of equality (or function symbols) can easily lead to the loss of recursive enumerability from these monodic fragments [142, 502]. It is necessary to restrict consideration to certain fragements *within* monodic FOTL to recover these properties [278].

2.8.8   Temporal Horn Clauses

We can restrict SNF, or even a general temporal clausal form, further using the usual Horn Clause restriction, namely that each clause should involve at most one positive literal. This particular fragment of PTL (and of first-order temporal logic) has been primarily studied in the context of the TEMPLOG executable temporal logic (see [1, 5, 54] and Section 6.4.3). To give some idea of the type of rules that are allowed in Temporal Horn Clauses [5], we will carry out the usual categorization of *unit*, *definite* and *negative* Horn Clauses. (See Appendix A for further details.)

- Recall that a *definite* clause has exactly one positive literal. We can subcategorize definite clauses with respect to the form of the positive literal as follows (in each case, b_i is one of **start**, q, $\bigcirc q$, $\square q$ or $\diamondsuit q$, where q is a proposition, and p is a proposition). Note that we revert to *implication form* for simplicity.

  - *definite clause:* $\bigwedge_i b_i \Rightarrow p$

  - *definite next clause:* $\bigwedge_i b_i \Rightarrow \bigcirc p$

  - *definite permanent clause:* $\bigwedge_i b_i \Rightarrow \square p$

  - *definite eventuality clause:* $\bigwedge_i b_i \Rightarrow \diamondsuit p$

- A positive temporal Horn Clause has no negative literals and, as in classical logic, represents a *fact*.

- A negative Horn Clause, such as $\neg a \lor \neg b$ can be represented as the implication $(a \land b) \Rightarrow$ **false** and, again, as the *goal*

$$\leftarrow a, b \, .$$

Thus, the Horn fragment of temporal logic is constructed in much the same way as the Horn fragment of classical logic (see Appendix A), the only difference being that clauses are also categorized as either *initial* or *global*, as in SNF. Not surprisingly, the Logic Programming paradigm that had been successfully applied to classical Horn Clauses (providing PROLOG), was in turn applied to a variety of Temporal Horn Clauses. This is explored further in Section 6.4.3.

2.8.9   Interval logics and granularities

There are a range of *interval* temporal logics that have been developed for use in Computer Science and Artificial Intelligence; although we cover one variety in the next section and another in Chapter 7, the interested reader might begin by consulting [14, 241, 382, 449]. Essentially, these logics use intervals of time as their basis, rather than time points (as is the case in PTL). Thus propositions are evaluated over some interval, rather than at a specific time point. An interval is a sequence of time with duration, for example a week, and so we can assess the truth values of propositions within this interval. To do this we use operators derived from natural language counterparts, such as 'throughout', 'during', and 'by the end'. Thus, we might assess the following of our interval representing a week: 'it is warm throughout that week'; 'the package arrived during that week'; and 'my bank account is in credit by the end of that week'.

There are a number of excellent articles covering much more than we can here: introductory articles, such as [358, 491]; surveys of interval problems in Artificial Intelligence, such as [163, 232]; and the comprehensive survey of interval and duration calculi by Goranko *et al.* [242].

Significant work has been carried out on temporal hierarchies of differing granularities, for example in [90, 217, 378, 422], with comprehensive descriptions being given in [173, 62]. The idea here is to explore different *granularities* of time. Thus, while in PTL we have a fixed length of step to the next state, we might wish to consider different durations or, more likely, small steps that make up a larger next step. Thus, if we say *march* $\Rightarrow \bigcirc april$, then we might also want to describe particular days within a month, or even hours within a day. This requires distinct granularities of time to deal with all these aspects. Finally, the work on interval temporal logics has also led to alternative views of granularity and projection [87, 248, 249, 384].

2.8.10   Interval Temporal Logic (ITL)

The interval logic developed by Moszkowski *et al.* in the early 1980s was close, in spirit, to the PTL being developed at that time [223]. Moszkowski's logic is called ITL and was originally devised in order to model digital circuits [257, 381]. Although the basic temporal model is similar to that of PTL given earlier, ITL formulae are interpreted in a subsequence (defined by $\sigma_b, \ldots, \sigma_e$) of, rather than at a point within, the model σ. Thus, basic propositions (such as p) are evaluated at the *start* of an interval:

$$\langle \sigma_b, \ldots, \sigma_e \rangle \models p \quad \text{if, and only if,} \quad p \in \sigma_b$$

Now, the semantics of two common PTL operators can be given as follows:

$$\langle \sigma_b, \ldots, \sigma_e \rangle \models \Box\varphi \quad \text{if, and only if,} \quad \text{for all } i, \text{ if } b \leq i \leq e \text{ then } \langle \sigma_i, \ldots, \sigma_e \rangle \models \varphi$$

$$\langle \sigma_b, \ldots, \sigma_e \rangle \models \bigcirc\varphi \quad \text{if, and only if,} \quad e > b \text{ and } \langle \sigma_{b+1}, \ldots, \sigma_e \rangle \models \varphi$$

A key aspect of ITL is that it contains the basic temporal operators of PTL, together with the *chop* operator, ';', which is used to fuse intervals together (see also [438, 488]). Thus:

$$\langle \sigma_b, \ldots, \sigma_e \rangle \models \varphi;\psi \quad \text{if, and only if,} \quad \text{there exists } i \text{ such that } b \leq i \leq e \text{ and both}$$
$$\langle \sigma_b, \ldots, \sigma_i \rangle \models \varphi \text{ and } \langle \sigma_i, \ldots, \sigma_e \rangle \models \psi$$

This powerful operator is both useful and problematic (in that the operator ensures a high complexity logic). It is useful in that it allows intervals to be split based on their properties; for example '$\Diamond$' can be derived in terms of ';', that is

$$\Diamond\varphi \equiv \textbf{true}; \varphi$$

meaning that there is some (finite) subinterval during which **true** is satisfied that is followed (immediately) by a subinterval in which φ is satisfied. If we ensure that the first subinterval is non-empty, then this defines '$\Diamond^+$'.

Further simple examples of formulae in ITL are given below, together with explanations.

- p persists through the current interval: $\Box p$

- We can define steps within an interval: $up \wedge \bigcirc down \wedge \bigcirc\bigcirc up \wedge \bigcirc\bigcirc\bigcirc down$

- We can construct sequences of intervals: $\Box january; \bigcirc \Box february; \bigcirc \Box march; \ldots$

- p is **false** for a while, then **true** for a while: $\Box\neg p; \bigcirc \Box p$.

As mentioned in the last section, there has been significant work on granularity within ITL, particularly via the *temporal projection* operation [87, 248, 249, 384].

2.8.11   Real-time logic and duration calculi

In describing real-time aspects, a number of languages can be developed [21]. For instance, our basic temporal logic can be extended with annotations allowing us to express real-time constraints [322]. Thus, 'I will finish writing this section within 5 time units' might be represented by:

$$\Diamond_{\leq 5} finish .$$

In addition, there is the possibility of explicitly relating to clocks (and clock variables) within a temporal logic [399]. Consequently, there are a great many different real-time temporal logics (and axiomatizations [447]. There are several excellent surveys of work in this area, including those by Alur and Henzinger [21, 22], Ostroff [400] and Henzinger [266].

In a slightly different direction, the *duration calculus* [107, 165] was introduced in [508], and can be seen as a combination of an interval temporal representation with real-time aspects. It has been used in many applications involving real-time systems, with behaviours mapping on to the dense underlying temporal model.

2.9 Final exercises

Exercise 2.8

(a) *We wish to say that 'in three moments of time, the variable x will have value 0 and, from that moment on, the variable y will always be greater than 1.'*

  How might we represent this in our discrete, linear temporal logic (PTL), assuming we can use basic arithmetical operations and that statements such as 'x = 0' can be used as propositions?

(b) *Give a semantic definition for a 'sometime in the past' operator within a discrete, linear model in the style of semantics seen in Section 2.3.*

(c) *In PTL, we can conjoin next-formulae such as*

$$p \wedge \bigcirc p \wedge \bigcirc\bigcirc p \wedge \bigcirc\bigcirc\bigcirc p \wedge \ldots$$

  How many such formulae do we have to conjoin together to give the same behaviour as $\square p$?

Exercise 2.9

(a) *The PTL formula $\square(\varphi \Rightarrow \lozenge\psi)$ says that the formula $\varphi \Rightarrow \lozenge\psi$ is always true. If we also know that $\lozenge\varphi$ is true now, then how often will ψ be forced to be true in the future? Now, if we instead know that $\square\varphi$ is true, how often will ψ be forced to occur?*

(b) *Does the temporal formula '$\square(p \Rightarrow \lozenge q) \wedge \square p$' imply $\square q$ in PTL? If so, explain why. If not, provide a counter-example.*

(c) *Is $\bigcirc\square\bigcirc\varphi \Rightarrow \lozenge\bigcirc\lozenge\varphi$ valid in PTL, that is, is it true in all possible models?*

(d) *Is $\square\lozenge\varphi \Rightarrow \lozenge\bigcirc\varphi$ valid in PTL, that is, is it true in all possible models?*

Exercise 2.10 *In PTL we describe the models by* $\langle \mathbb{N}, \pi \rangle$. *What would be different if we instead used*

(a) $\langle \mathbb{Z}, \pi \rangle$, *where* $\mathbb{Z}$ *is the Integers, or*

(b) $\langle \mathbb{R}, \pi \rangle$, *where* $\mathbb{R}$ *is the Real Numbers?*

Exercise 2.11 *Look at tools (other than* `Spin`, *which we examine later) for generating automata from temporal logic formula, notably Gastin's LTL2BA:*

   `http://www.lsv.ens-cachan.fr/~gastin/ltl2ba`

and Schneider's tool for generating state-machine structures from temporal formulae

`http://es.informatik.uni-kl.de/teaching/tools/SymbolicFSM.html`

Finally, also look at the Büchi Store

   `http://buchi.im.ntu.edu.tw`

3

Specification

A man with a watch knows what time it is. A man with two watches is never sure.
  – Segal's Law

In this Chapter, we will:

- introduce the idea of using temporal logic to represent simple system behaviours (Section 3.1);

- extend this to provide a temporal semantics for a basic imperative programming language (Section 3.2);

- show how to link temporal descriptions of separate elements, particularly to describe distributed, concurrent, communicating systems (Section 3.3);

- highlight a selection of more advanced work including more complex programming languages, alternative models of concurrency, and general properties of temporal semantics (Section 3.4);

- present a final selection of exercises (Section 3.5), the solutions to which can again be found in Appendix B; and

- provide a brief roadmap of the forthcoming chapters (Section 3.6).

3.1 Describing simple behaviours

This chapter concerns temporal *specification* of behaviours. More specifically, we will mainly address ways of describing the behaviours of *programs*. But what is the link

An Introduction to Practical Formal Methods Using Temporal Logic, First Edition. Michael Fisher.
© 2011 John Wiley & Sons, Ltd. Published 2011 by John Wiley & Sons, Ltd.

between a program and a specification given as a temporal formula? The following diagram attempts to capture this relationship. Here, a program can be *executed*, generating an execution sequence. Execution sequences are again just structures isomorphic to the Natural Numbers, together with a mapping from each point to the set of conditions that are true at that point and, hence, execution sequences can correspond to appropriate temporal structures. There can be multiple possible execution sequences for each program, since the program might contain certain non-determinism. A temporal specification is a temporal formula and, as we have seen already, such a formula itself also corresponds to a set of model sequences again isomorphic to the Natural Numbers. Thus, both programs and specifications describe sets of sequences. Much of this chapter concerns providing appropriate semantics so that these two mechanisms correspond.

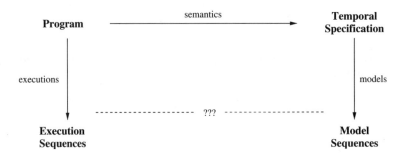

Let us begin by considering again the role of sequences in corresponding to both programs and specifications in a little more detail.

3.1.1   Sequences

It is important that we are able to interpret temporal formulae and visualize the possible execution sequences associated with each specification. Thus, when we look at a temporal specification, the possible executions of the program being specified correspond to the possible models for the specification.

For example, if we were able to generate a program such as

    do(a); do(b); do(c)

then we might expect the temporal specification for such a program to be

$$\textbf{start} \;\Rightarrow\; \bigcirc do(a) \wedge \bigcirc \bigcirc do(b) \wedge \bigcirc \bigcirc \bigcirc do(c)$$

and the models (temporal sequences) for this formula to be something like

These models correspond, in turn, to potential executions of the program. But, can they capture all *reasonable* patterns of behaviour?

We will look at a specific programming language later, but for the moment let us consider a more complex (abstract) specification such as the temporal formula

$$\Box \left[\begin{array}{ccc} (\textbf{start} & \Rightarrow & c) \\ \wedge & (c & \Rightarrow & \bigcirc b) \\ \wedge & (b & \Rightarrow & \bigcirc a) \end{array} \right]$$

This is meant to describe some sequence of execution. So, c will be true at the beginning of the (execution) sequence and, whenever c becomes true, then b will be true at the next moment. Similarly, whenever b is true, then a will be true in the next moment. Typical models for this formula look like

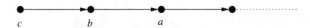

Notation   We will be using specifications of the following form quite a lot.

$$\Box \left[\begin{array}{ccc} (\varphi_1 & \Rightarrow & \psi_1) \\ \wedge & (\varphi_2 & \Rightarrow & \psi_2) \\ \wedge & (\ldots & \Rightarrow & \ldots) \end{array} \right]$$

Consequently, we will usually omit the conjunctions and use the following notation.

$$\Box \left[\begin{array}{ccc} (\varphi_1 & \Rightarrow & \psi_1) \\ (\varphi_2 & \Rightarrow & \psi_2) \\ (\ldots & \Rightarrow & \ldots) \end{array} \right]$$

from now on. Note that all these specifications are generally in SNF (see Section 2.6). Returning to our running example, if we now extend our earlier specification to be

$$\Box \left[\begin{array}{ccc} \textbf{start} & \Rightarrow & c \\ c & \Rightarrow & \bigcirc b \\ b & \Rightarrow & \bigcirc a \\ a & \Rightarrow & \bigcirc b \end{array} \right]$$

we get *infinite* cyclic behaviour:

with the 'a' *then* 'b' alternating infinitely after the initial 'c'. This happens because, as soon as b becomes true in a particular moment, then $b \Rightarrow \bigcirc a$ requires that a be true in the next moment. Then, at that next moment, $a \Rightarrow \bigcirc b$ forces b to be true again at the subsequent moment, and so on, indefinitely. Similarly,

$$\Box \left[\begin{array}{ccc} \textbf{start} & \Rightarrow & c \\ c & \Rightarrow & \bigcirc b \\ b & \Rightarrow & \bigcirc a \\ a & \Rightarrow & \bigcirc c \end{array} \right]$$

reproduces a 'cba' cycle infinitely.

From the above, we can see how such specifications can characterize executions of non-terminating, cyclic programs. But, what about other patterns of behaviour?

3.1.2   Choices

Consider a slightly more complex example in detail:

$$\Box \left[\begin{array}{rcl} \textbf{start} & \Rightarrow & a \\ \textbf{start} & \Rightarrow & b \\ (a \wedge b) & \Rightarrow & \bigcirc(c \vee d) \\ c & \Rightarrow & \bigcirc c \\ d & \Rightarrow & \bigcirc e \end{array} \right]$$

Here, the key formula we are interested in is

$$(a \wedge b) \Rightarrow \bigcirc(c \vee d)$$

which says that when 'a' and 'b' are both true then, in the next moment, there is a *choice* of making at least one of 'c' or 'd' true.

Note that this is a *non-deterministic* choice. Thus, there are several possible models of the whole formula, one being that we choose to make c true, and so '$c \Rightarrow \bigcirc c$' ensures that 'c' continues forever:

Another model is that we choose to make 'd' true, thus forcing 'e' to be true in the next moment:

A third model is that we actually make *both* 'c' and 'd' true, producing both the above behaviours:

Not surprisingly, choices in program execution (for example, branches due to conditional statements) are typically specified using disjunctive formulae, and so different executions (taking different branches) correspond to distinct models.

Exercise 3.1 *What pattern of behaviour does the following specify?*

$$\Box \left[\begin{array}{rcl} \textbf{start} & \Rightarrow & x \\ x & \Rightarrow & \bigcirc(w \wedge y \wedge z) \\ z & \Rightarrow & \bigcirc a \\ (z \wedge y \wedge w) & \Rightarrow & \bigcirc b \\ (a \wedge b) & \Rightarrow & \bigcirc x \end{array} \right]$$

One of the most useful aspects of temporal logic is that its models can be viewed as representations of the behaviour through time of 'real-life' systems. For example, the first use of temporal logic in Computer Science [411] was based on the observation that temporal models can be seen as execution sequences of programs. As we will see in the next section, this observation allows us to provide the semantics of programming languages using temporal logic.

3.2   A semantics of imperative programs

Returning to the diagram at the beginning of the chapter, we see that a *semantics* plays a key role in relating programs to specifications. In general, the purpose of a semantics is to assign some 'meaning' to each syntactic object of a particular language. In the study of programming languages, a semantics is required to assign a mathematical meaning to each syntactically correct program. A formal semantics is desirable as it makes reasoning about the properties of a program far easier.

Formal semantics of programming languages have been presented in many different forms, for example operational semantics [410, 498] and denotational semantics [444, 462]. Now the aim of a *temporal semantics* [412, 413] is to model the meaning of a program by providing a suitable temporal formula (or a set of models that satisfy such a formula). So a temporal semantics can be seen as a function that takes a program and returns a temporal formula (i.e. a specification) representing the meaning of that program.

3.2.1   A first temporal semantics

To see how we might go about producing a temporal semantics, we will examine a few program constructs and show what varieties of temporal formulae might be generated to represent these. Let us define a temporal semantics through a semantic function '[[ ]]' that maps program fragments on to their (temporal) meaning (i.e. a temporal formula).

Let us begin with simple assignment (to Boolean variables):

$$[\![\texttt{x:=true}]\!] \quad \equiv \quad \bigcirc x.$$

Thus, we model the assignment of 'true' to the Boolean variable 'x' by making the corresponding proposition 'x' true in the next state (i.e. after the assignment operation has taken place).

Rather than considering such isolated program statements, we are more likely to have a sequence of program statements, constructed through sequential composition, ';'. So:

$$[\![\texttt{x:=true; S}]\!] \quad \equiv \quad \bigcirc (x \wedge [\![\texttt{S}]\!])$$

Thus, we tackle the semantics of the assignment and then move on to consider the next program statement (in the case above, some arbitrary statement 'S') at the next moment in time. So, using the above two possibilities, we can provide a temporal semantics for sequences of simple (Boolean) assignments.

We will see later that we can also tackle more complex program constructs. However, we will now just mention one last feature of temporal semantics, namely their representation of parallel activities. So, parallel statements (i.e. two things happening at once; in this case, 'S' and 'T') might simply be modelled by[1]

$$[\![S \parallel T]\!] \quad \equiv \quad [\![S]\!] \wedge [\![T]\!],$$

This simplicity is one of the appealing features of using a temporal logic and, indeed, the flexibility to represent many different forms of parallel activity was one of the advantages that initially stimulated work on temporal logic in Computer Science [412, 413].

Example 3.1 *We can use the above approach to give a (temporal) description of the meaning of the following program, where the parallel composition operation is synchronous:*

```
(x:=true; x:=false) || (y:=true; y:=false)
```

The temporal description of this program is simply:

$$\bigcirc(x \wedge \bigcirc\neg x) \wedge \bigcirc(y \wedge \bigcirc\neg y).$$

This can be rewritten to:

$$\bigcirc(x \wedge y) \wedge \bigcirc\bigcirc(\neg x \wedge \neg y).$$

Now, recall that the models satisfying this formula are of the form

$$
\begin{array}{ll}
\text{state } 0: & \ldots \\
\text{state } 1: & [\, x \mapsto \mathbf{T}, \; y \mapsto \mathbf{T} \,] \\
\text{state } 2: & [\, x \mapsto \mathbf{F}, \; y \mapsto \mathbf{F} \,] \\
\ldots & \ldots
\end{array}
$$

and so such models describe possible execution sequences for the above program:

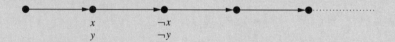

However, the following model does not *satisfy the above formula and so could not be an execution sequence for the program described above*

$$
\begin{array}{ll}
\text{state } 0: & \ldots \\
\text{state } 1: & [\, x \mapsto \mathbf{T}, \; y \mapsto \mathbf{T} \,] \\
\text{state } 2: & [\, x \mapsto \mathbf{T}, \; y \mapsto \mathbf{F} \,] \\
\text{state } 3: & [\, x \mapsto \mathbf{F}, \; y \mapsto \mathbf{T} \,] \\
\ldots & \ldots
\end{array}
$$

---

[1] As we will discuss later, this only works so neatly when we have synchronous concurrency, that is the separate elements are working on exactly the same clocks and so have the same view of a 'next' step.

Let us again note that the above semantics of parallel composition is very rigid; it assumes that the assignments in each parallel statement occur *simultaneously*. We will discuss the problem of asynchrony a little more later. Before that, however, we will return to individual imperative programs, ignoring parallelism until later in this chapter.

3.2.2   A simple imperative programming language

So let us now consider a *very simple* programming language based on assignments and simplified control structures, and then show how a temporal semantics might be provided for this. (See work by Manna and Pnueli [360, 363] for more details concerning semantics of this kind.) We will assume that the set of program variables is given by VARS, the set of program constants (including numbers) is given by CONS, and that the standard arithmetic functions are available. The elements of our simplified programming language are as follows:

Assignment: x : = v
     ⤳ after executing this statement, variable x has the value v.

Note that the value, v, may be composed of elements of CONS, VARS or arithmetic expressions.

Example 3.2 *The following are legal statements in our programming language.*

```
    y := 2

    zz := y-1

    wow := zz+y
```

Sequential composition: S1; S2
     ⤳ after successfully completing statement S1, statement S2 will be executed.

Termination: end
     ⤳ finish the program.

Example 3.3 *The following is a legal program in our programming language.*

```
a := 2;  b := a-1;  c := b+a;  end
```

Branching: if Test then S1 else S2
     ⤳ if the Boolean-valued Test evaluates to true, then execute S1, otherwise execute S2.

Looping: while Test do S
     ⤳ while Test is true, continue executing S

This is quite a simple imperative programming language and the varieties of programs it can be used for should be fairly familiar.

Example 3.4

```
x := 0;
while (x<4) do  x := x+1;
end
```

Here, the variable x is initially assigned the value zero and, during the while loop, the variable's value is gradually incremented until it reaches 4. Then the program ends.

Since the purpose of a semantics is to assign some 'meaning' to each statement within a particular language, we will now represent the meaning of the statements in our simple programming language in terms of temporal formulae, thus providing a *temporal semantics* for the language. As before, our aim is to define a function

$$[\![\ \_\ ]\!] : Prog \mapsto PTL$$

that provides a temporal formula for each program generated using the constructs identified above. However, we can only give the reader a 'flavour' of these temporal semantics because we will omit some complexities.

Thus, $[\![S]\!]$ is used to denote the temporal formula representing the semantics of statement S. In the description below, we also note that most of the semantic definitions are recursive. Specifically, the semantic definition for the 'while' construct will be defined as a recursive formula. As we discussed earlier, the full temporal semantics is likely to describe this as a (minimal) temporal fixpoint. However, for our purposes here we will ignore this aspect and just define its semantics as a recursive equation.

We also use operators not defined within core PTL, such as equality between terms (using '=') and existential quantification (using '∃'). Similarly, we assume variables within the logic can have non-Boolean values and match the variables used within the program, such as x.

Now we will look at some temporal semantics for each of the constructs within our language, beginning with 'end'. Note that we do not provide a separate semantics for sequential composition, ';', as we did above. Instead we incorporate ';' into each of the possible language constructs giving the semantics of assignment; S, if...then...else...; S, while...do...od; S etc., for all the possible language constructs.

Semantics (end)

$$[\![end]\!] \equiv \textbf{true}$$

Simply finish the program and allow *unconstrained* behaviour afterwards.

Perhaps surprisingly, the semantics of assignment is not at all trivial and so we will leave that until later.

Semantics (if...then...else)

$$[\![(\text{if Test then S1 else S2}); \text{ S3}]\!] \equiv$$
$$(\text{Test} \Rightarrow [\![\text{S1};\text{S3}]\!]) \wedge ((\neg\text{Test}) \Rightarrow [\![\text{S2};\text{S3}]\!])$$

For simplicity, we here assume that Test is a simple Boolean test (with no side-effects) that is easily evaluated.

Example 3.5 *Consider the simple statement*

```
(if (x>2) then y:=3 else y:= 5); S
```

where S is some arbitrary program statement. The semantics of this is

$$[\![(\text{if } (x>2) \text{ then } y:=3 \text{ else } y:=5); \text{ S}]\!] \equiv$$
$$((x > 2) \Rightarrow [\![y:=3;\text{S}]\!]) \wedge (\neg(x > 2) \Rightarrow [\![y:=5;\text{S}]\!])$$

Semantics (while)

$$[\![(\text{while Test do S1}); \text{ S2}]\!] \equiv$$
$$[\![(\text{if Test then S1}; (\text{while}\ldots) \text{ else S2})]\!]$$

So, the semantics of a while statement is reduced to that of an if statement with the original while statement embedded within it.

Example 3.6 *Consider*

```
(while (x<4) then x:=x+1); S
```

For this example, the semantics is

$$[\![(\text{while } (x<4) \text{ then } x:=x+1); \text{ S}]\!] \equiv$$
$$((x < 4) \Rightarrow [\![x:=x+1; ((\text{while } (x<4) \text{ then } x:=x+1); \text{ S}))]\!])$$
$$\wedge ((x > 4) \Rightarrow [\![\text{S}]\!])$$

Note that we will say more about infinite loops and infinite recursion in Section 3.4.

Aside. As mentioned already, if we truly examine the semantics of 'while' above, we see that we will generate a temporal fixpoint formula as described in Section 2.8.4. Indeed, the temporal semantics of both loops and recursive procedures are intimately linked to fixpoints.

Semantics (assignment)

Finally, we come to the semantics of assignment. Surprisingly, this is quite subtle and requires a complex semantic description. So, first we will give this semantics, then we will endeavour to explain it.

$$[\![x:=v;\ s]\!] \equiv \exists w.\ (x = w) \wedge \bigcirc (x = v(x/w) \wedge [\![s]\!])$$

Note here that $v(x/w)$ is the expression v in which occurrences of x have been replaced by w. Thus, if v is '$x + 3 + (2x)$', then $v(x/w)$ is '$w + 3 + (2w)$'.

The reason for the slightly complex formulation of the semantics of assignment, and the need for some first-order notation, is that in assigning to a variable in the current state, we need a record of the value that variable had previously. This is what the existential quantifier is used for, namely to 'remember' the previous value of the variable being assigned to. The following examples will hopefully make this clearer.

Example 3.7 *Let us look at the statement* x: = x+1. *In our programming language we need to record the previous value of* '*x*', *add one to it, and then assign this to the variable* '*x*'. *The temporal semantics works in a similar way and so, once we apply the semantics to our assignment, we get:*

$$[\![x:=x+1;\ s]\!] \equiv \exists w.\ (x = w) \wedge \bigcirc ((x = w + 1) \wedge [\![s]\!])$$

Notice how the expression '$x + 1$' *has had the current value of x (namely w) substituted into it, giving* $w + 1$. *Thus, while x has the value w in the current moment, it will have the value* $w + 1$ *in the next moment.*

Example 3.8 *Now we will consider a simpler assignment, namely* x: = 2. *Here we find that the semantic description is quickly simplified. We begin with:*

$$[\![x:=2;\ s]\!] \equiv \exists w.\ (x = w) \wedge \bigcirc (x = 2) \wedge [\![s]\!]$$

We know that x is a variable, so it must always have a value, that is $\square (\exists w.(x = w))$. *Thus, the above reduces simply to*

$$[\![x:=2;\ s]\!] \equiv \bigcirc (x = 2) \wedge [\![s]\!]$$

that is, in the next moment in time, x will have the value 2. Notice that, in this simple case, the existential part of the semantics was redundant.

Finally, note that although the above constrains how the value of 'x' changes, it has said nothing about how the values of *other* variables change at the same time. We will say a little more about this in Section 3.4.3

Tackling example 3.4

We will now examine how several semantic rules can be applied to a simple program. The program is non-trivial, so there are many issues concerning temporal semantics that we will not consider here. We simply aim to provide some intuition on how such temporal semantics might be used in practice.

First, recall the code of Example 3.4:

```
x:=0;
while (x<4) do x:=x+1;
end
```

Intuitively, we expect that the variable x successively steps through 0, 1, 2, 3, and 4, and then the program ends.

So, we now begin by applying $[\![\ ]\!]$ to our program, that is

$$[\![\texttt{x:=0; while (x<4) do x:=x+1; end}]\!].$$

Next, we expand the semantics of x: = 0, giving

$$\bigcirc((x = 0) \wedge [\![\texttt{while (x<4) do x:=x+1; end}]\!]).$$

Then expanding the while construct into an if...then branch:

$$\bigcirc((x = 0) \wedge ((x < 4) \Rightarrow [\![\texttt{x:=x+1; while...}]\!]) \wedge ((x \geq 4) \Rightarrow [\![\texttt{end}]\!]))$$

However, the test can easily be evaluated because we know the value of x at this point (namely x = 0). Consequently, we expand the '$x < 4$' branch:

$$\bigcirc((x = 0) \wedge [\![\texttt{x:=x+1; while...}]\!])$$

Through the assignment x:=x+1 this leads to x having the value 1 at the next state:

$$\bigcirc((x = 0) \wedge \bigcirc((x = 1) \wedge [\![\texttt{while...}]\!]))$$

And then we expand the while loop again.

This process continues as the value of x is increased, until we reach the point where x = 4 and we have expanded the while structure into an if...then choice again:

$$\bigcirc((x = 0) \wedge \bigcirc((x = 1) \wedge \bigcirc((x = 2) \wedge \bigcirc((x = 3) \wedge \bigcirc((x = 4) \wedge$$
$$((x < 4) \Rightarrow [\![\texttt{x:=x+1; while...}]\!]) \wedge ((x \geq 4) \Rightarrow [\![\texttt{end}]\!]))))))$$

Now, since x = 4, this time we take the '$x \geq 4$' branch, discarding the other option:

$$\bigcirc((x = 0) \wedge \bigcirc((x = 1) \wedge \bigcirc((x = 2) \wedge \bigcirc((x = 3) \wedge \bigcirc((x = 4) \wedge [\![\texttt{end}]\!]))))$$

The last part of the program we evaluate is the 'end' statement:

$$\bigcirc((x = 0) \wedge \bigcirc((x = 1) \wedge \bigcirc((x = 2) \wedge \bigcirc((x = 3) \wedge \bigcirc((x = 4) \wedge \mathbf{true}))))$$

Finally, a little simplification gives:

$$\bigcirc(x = 0) \wedge$$
$$\bigcirc\bigcirc(x = 1) \wedge$$
$$\bigcirc\bigcirc\bigcirc(x = 2) \wedge$$
$$\bigcirc\bigcirc\bigcirc\bigcirc(x = 3) \wedge$$
$$\bigcirc\bigcirc\bigcirc\bigcirc\bigcirc(x = 4)$$

> **Example 3.9** *Typical properties we might want to show for the above program include:*
>
> $$\text{SAFETY:} \qquad \Box \neg (x > 4)$$
> $$\text{LIVENESS:} \qquad \Diamond (x = 4)$$

There is, of course, much more we can say here because there has been a great deal of work on the temporal semantics of programs. While we will give some further pointers to such work later in this chapter, we have shown how a temporal formula can be used to describe (exactly) the behaviour of a program. Though the program was simple, more complex programs can be described in a similar way.

Exercise 3.2 *Concerning the temporal semantics of basic imperative programs.*

  (a) *Give a temporal semantics (i.e. a temporal formula characterizing the execution sequences) of this program:*

```
z:=3; z:=z+5; z:=z/4; end
```

  (b) *Give a temporal formula capturing a semantics of the following statement in a standard imperative programming language:*

```
(if (y>2) then y:=1 else y:=y+2); end
```

  (c) *What temporal formula would we write to describe the property that, at some point in the execution of the program in part (b) the variable y is guaranteed to have a value less than 5? Is this a liveness property, a safety property or neither?*

So, now we have seen a little of how temporal formulae can describe dynamic behaviours (by capturing the relevant execution sequences) and programs (through a temporal semantics). There is, of course, much more that we do not consider here. However, in the following sections we will now abstract away from detailed specifications of individual behaviours. Instead we will assume that we just have a temporal logic formula that describes the behaviour of one element (where 'element' could mean object, agent, thread, process, subsystem etc.). Then we can move on to examine how we might link these element specifications together.

3.3   Linking specifications

Consider a specification, $Spec_1$, for an element E_1, together with another specification, $Spec_2$, for element E_2. In programming languages, the elements can be executed together in a variety of ways (e.g. interleaved, concurrent). In addition, the ways in which information can be passed between the elements should be identified (e.g. message-passing, shared-variables). So, in order to formalize more interesting programs, we must decide on and describe the forms of concurrency and communication used in our systems. (Note that, if no such communication occurs between elements, then the elements cannot have any significant effect on each other though we may still be interested in the evolution of the combined system.)

3.3.1   Concurrency and communication

Throughout the following, we will assume we have several independent elements, E_1, $E_2, \ldots, E_n$. Again, these elements may be processes, objects, threads, agents etc. We will primarily tackle the simplest mechanism for combining elements, namely *synchronous concurrency*. Before we do this, we will examine what this means.

Interleaving versus true concurrency.   This choice concerns whether our elements to be combined can actually run at the *same* time or not.

- *True Concurrency* allows the elements to be executing at the same time, but independently:

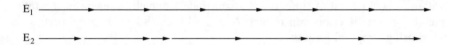

  Notice that E_1 and E_2 can be active at the same time.

- *Interleaving* ensures that only *one* of the elements can execute at any one time, and so periods of activity for one element are interspersed (or *interleaved*) with periods of activity for the other.

  Note that a typical way to characterize interleaving in temporal logic would be

  $$\Box(executing(E_1) \lor executing(E_2)) \land \Box \neg(executing(E_1) \land executing(E_2)).$$

  The first part of this ensures that it is always the case that either E_1 or E_2 are executing; the second part ensures that they cannot be executing at the same time.

In what follows, we will primarily choose the *true concurrency* option.

Synchronous versus asynchronous execution.   Assuming the elements can execute at the same time then we must also decide whether they work on the same internal clock or not. Typically, this involves a choice between *synchronous* and *asynchronous* execution.

- *Synchronous* execution ensures that all elements have exactly the same clock and work on the same notion of *next moment*.

- *Asynchronous* execution allows the elements to have different clocks with one element potentially working *faster* than another.

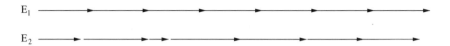

Again, we will mainly use *synchronous*, *true concurrency*, though we will explore the temporal representation of some of the other possibilities later in this chapter.

Before we move on to the temporal representation of concurrent elements we must also explore what, if any, forms of communication we are interested in. Again there are many possibilities, but we will just mention a very popular variety here.

Communication through message passing.   As the name suggests, communication here is achieved by one element *sending a message* to another element. Depending on the properties of the communication medium, this message might arrive (a) very quickly, (b) within a fixed period of time, (c) at some (indeterminate) moment in the future, or (d) possibly never (if errors can occur). As we will see, all of these properties can be specified using temporal formulae.

Usually, with message-passing, the message arrives at its destination and is recorded immediately. Typically, *send* and *receive* predicates are used to represent the sending and receiving of messages. As we shall see below, it is possible for an element to agree only to receive a message in certain situations. Thus, various forms of message-passing communication can be specified; the three most common being as follows:

- *Point-to-point* message-passing occurs between two elements with one (the sender) dispatching the message and another (the receiver) being the recipient.

- Another popular variety is *broadcast* message-passing whereby the sender dispatches a message *without* a specific destination in mind. All elements that are able to detect this message can receive it. This has the advantage that information can be dispersed to a large number of elements without a sender having to know *all* the names/addresses of the recipients.

- An increasingly useful variety is *multicast* message-passing where restrictions are made on broadcast so that messages go to a *subset* of all possible receivers. Importantly, the sender still does not need to know the names/addresses of all the recipients.

We will see how these might be specified in the rest of this chapter.

3.3.2   Specifying element linkage

Before we consider the temporal specification of message-passing communication, let us return to the basic problem of describing the concurrent activity of two elements. Recall that we have

$Spec_1$, a temporal specification of element E_1, and
$Spec_2$, a temporal specification of element E_2.

Now, we want to specify a system comprising elements E_1 and E_2 executing *concurrently* and *synchronously*, but for the moment *without any* communication.

It turns out that, in this simple case, the specification of the combined system is also simple. Specifically, we just conjoin the specifications, that is

$$Spec_1 \wedge Spec_2$$

Example 3.10 *Consider two elements, specified as follows:*

$$Spec_1 \;=\; a \wedge \bigcirc b \wedge \bigcirc\bigcirc c \wedge \bigcirc\bigcirc\bigcirc(a \vee b)$$
$$Spec_2 \;=\; x \wedge \bigcirc\bigcirc y \wedge \bigcirc\bigcirc\bigcirc z$$

Then the synchronous, concurrent combination of the elements can be specified by

$$Spec_1 \wedge Spec_2 = (a \wedge x) \wedge \bigcirc b \wedge \bigcirc\bigcirc(c \wedge y) \wedge \bigcirc\bigcirc\bigcirc(z \wedge (a \vee b))$$

Notice how, in the above example, there is *no* interaction between the propositions in $Spec_1$, namely $\{a, b, c\}$, and the propositions in $Spec_2$, namely $\{x, y, z\}$. When we consider models/executions over the combined alphabet $\{a, b, c, x, y, z\}$, we note that a model for $Spec_1 \wedge Spec_2$ must be a model for both $Spec_1$ and $Spec_2$ separately. To see this, let us consider a slightly simpler example.

Example 3.11 *Given two formulae, '$p \wedge \bigcirc q$' and '$r \wedge \bigcirc \neg r$', we can examine the models of each formula separately. Thus, sample models of $p \wedge \bigcirc q$ (over the alphabet $\{p, q, r\}$) are*

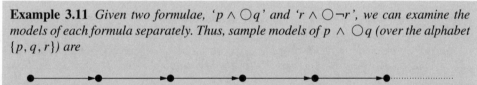

Sample models of $r \wedge \bigcirc \neg r$ (again over the alphabet $\{p, q, r\}$) are

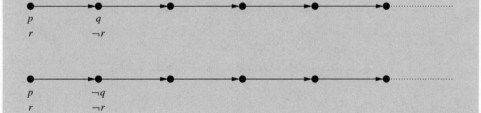

Now, notice that the only model for $(p \wedge \bigcirc q) \wedge (r \wedge \bigcirc \neg r)$ *from the above sample models above is:*

Importantly, if both formulae have models separately and if there is no explicit interaction between the alphabets of the two formulae, then the conjoined formula will also have a model.

Semantics example

Now, before we move on to considering communication between elements we will provide a more concrete version of the above example, but this time using specific programs. This will reinforce the above, but perhaps in a more recognizable way.

Example 3.12 *We consider two elements E_1 and E_2. This time, however, we will give simple program code (for example in our earlier programming language) for each:*

- E_1 = `x:=0; x:=x+1; x:=5; end`

- E_2 = `y:=7; y:=y-1; y:=0; end`

Note that, when these elements execute concurrently (and synchronously) the variables within each one do not interact.

So, let us begin by considering the possible models (or execution sequences) of each element separately. Sample models (over all variables, x and y) for E_1 include:

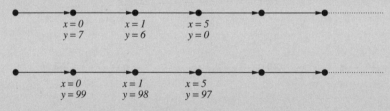

Notice how the values of x are constrained by the code of E_1, while y is unconstrained. Meanwhile, sample models (again over all variables, x and y) for E_2 include:

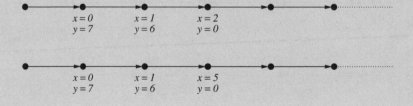

Now, among all these, the only common model/execution for the synchronous, concurrent combination of E_1 and E_2 is:

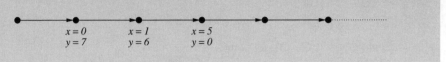

$$x = 0 \qquad x = 1 \qquad x = 5$$
$$y = 7 \qquad y = 6 \qquad y = 0$$

3.3.3   Introducing message-passing

Recall that we earlier had

$$Spec_1 \;=\; a \wedge \bigcirc b \wedge \bigcirc\bigcirc c \wedge \bigcirc\bigcirc\bigcirc (a \vee b)$$
$$Spec_2 \;=\; x \wedge \bigcirc\bigcirc y \wedge \bigcirc\bigcirc\bigcirc z$$

and that the synchronous, concurrent combination of the two elements was simply $Spec_1 \wedge Spec_2$, that is

$$(a \wedge x) \wedge \bigcirc b \wedge \bigcirc\bigcirc (c \wedge y) \wedge \bigcirc\bigcirc\bigcirc (z \wedge (a \vee b)) \,.$$

Notice how there is *no* interaction between $\{a, b, c\}$ and $\{x, y, z\}$.

If we want information from one element to affect the computation within the other, we need some form of communication. One way to achieve this is to identify the aspects in each system (hence, in each specification) that correspond to information coming into an element, or going out of it. For example, we will now ensure that a specification, $Spec_i$, contains:

- the set of propositions used within it, $Props(Spec_i)$;

- the set of propositions corresponding to *incoming* information, $In(Spec_i)$; and

- the set of propositions corresponding to *outgoing* information, $Out(Spec_i)$.

The idea here is that the propositions in $In(Spec_i)$ become true when an item of information is received, while making the propositions in $Out(Spec_i)$ true has the side-effect of passing information out of the element.

In general, when we link elements, we also link their specifications. Thus, the specification of the whole system would be

$$Spec_1 \wedge Spec_2 \wedge Comms(E_1, E_2)$$

where *Comms* links messages sent by each element to those received by the other. For example, for $a \in Out(Spec_1)$ and $x \in In(Spec_2)$,

$$a \Rightarrow \Diamond x$$

specifies that once a is made true within E_1 then x will eventually be made true within E_2.

Example 3.13 *Assume we had specifications of E_1 and E_2 as earlier:*

$$Spec_1 \;=\; a \wedge \bigcirc b \wedge \bigcirc\bigcirc c \wedge \bigcirc\bigcirc\bigcirc(a \vee b)$$

$$Spec_2 \;=\; x \wedge \bigcirc\bigcirc y \wedge \bigcirc\bigcirc\bigcirc z$$

Also, assume that $a \in Out(Spec_1)$ and $x \in In(Spec_2)$.

If we define $Comms(E_1, E_2) = \Box(a \Rightarrow \bigcirc x)$, this says that it is always the case that, if 'a' becomes true in E_1's specification, then, in the next moment[2] x' will be true in E_2's specification. Thus $Spec_1 \wedge Spec_2 \wedge Comms(E_1, E_2)$ now gives us

$$(a \wedge x) \wedge \bigcirc(b \wedge x) \wedge \bigcirc\bigcirc(c \wedge y) \wedge \bigcirc\bigcirc\bigcirc(z \wedge ((a \wedge \bigcirc x) \vee b)).$$

Example 3.14 *Again with*

$$Spec_1 \;=\; a \wedge \bigcirc b \wedge \bigcirc\bigcirc c \wedge \bigcirc\bigcirc\bigcirc(a \vee b)$$

$$Spec_2 \;=\; x \wedge \bigcirc\bigcirc y \wedge \bigcirc\bigcirc\bigcirc z$$

but now with

$$Comms(E_1, E_2) = \quad \Box(a \Rightarrow \bigcirc x) \wedge \Box(y \Rightarrow \bigcirc \neg a)$$

then $Spec_1 \wedge Spec_2 \wedge Comms(E_1, E_2)$ becomes

$$(a \wedge x) \wedge \bigcirc(b \wedge x) \wedge \bigcirc\bigcirc(c \wedge y) \wedge \bigcirc\bigcirc\bigcirc(z \wedge \neg a \wedge b)).$$

Notice, particularly, how the execution within E_2 (i.e. y becoming true) now affects the choices within E_1 (i.e. a being false and so b being chosen from $a \vee b$).

Alternative communication properties

In the above examples we had a relatively simple '*Comms*' formula whereby a change in one element affected another element in the *next* state. Of course, because we have the power of temporal logic, we can specify much more complex communication properties. We will give some examples of such properties below, but will also introduce more notation that can be useful in describing such behaviours.

Our specifications are becoming larger, so we will begin by putting them into SNF. Thus, we might consider two specifications for elements E_3 and E_4, respectively:

$$Spec_3: \qquad \Box \begin{bmatrix} \mathbf{start} & \Rightarrow & a \\ a & \Rightarrow & \bigcirc b \\ c & \Rightarrow & \bigcirc d \end{bmatrix}$$

$$Spec_4: \qquad \Box [\; x \;\Rightarrow\; \bigcirc y \;]$$

Now let us introduce a little graphical notation. Considering element E_3 on its own, we might represent it graphically as follows.

---

[2] Recall that, since the combination is synchronous, both elements share the same notion of 'next'.

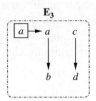

The idea here is that we know that if we have 'a,' then 'b' will occur next, and if we have 'c,' then 'd' will occur next. In addition, we know that 'a' is true at the start, represented by $\boxed{a}$.

Similarly with element E_4:

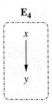

Notice that, on their own, these elements do very little. In E_3, a is initially made true and then b is made true at the next step. Since c has not yet been made true, the element E_3 does not constrain d. If c were to become true, then d would be forced to be true in the next state. Element E_4 does even less, because x has not yet been made true! If x is made true, then y must be true in the next state.

Now, we add a simple *Comms* formula:

$$Comms(E_3, E_4) = \Box(b \Rightarrow \bigcirc x)$$

This ensures that

- once b is made true in $Spec_3$ (at $i = 1$), x will be made true at the next step after that ($i = 2$), and so
- $Spec_4$ ensures that y will be made true at $i = 3$.

Now, if we instead use

$$Comms(E_3, E_4) = \Box(b \Rightarrow \Diamond x)$$

the effect on x will occur at *some* indeterminate moment in the future once b has occurred. Graphically, we might represent this as

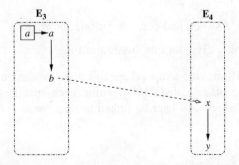

In the diagram time is taken to be increasing as we move down. Thus, a occurs before b and the message from b to x takes some time (but we are unsure exactly how long, and so the gradient of this line is indeterminate).

Now we can extend this further with communication back from E_4 to E_3. For example, if we extend $Comms(E_3, E_4)$ to

$$Comms(E_3, E_4) = \Box(b \Rightarrow \Diamond x) \land \Box(y \Rightarrow \Diamond c),$$

then making y true in E_4 will have the effect of *eventually* making c true back in element E_3. This, in turn, will ensure that d is satisfied at the next moment after this:

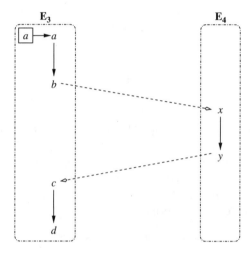

Thus, from the combination $Spec_3 \land Spec_4 \land Comms(E_3, E_4)$ we can infer that d will eventually become true, that is $\Diamond d$.

3.3.4   Specification case study

At this point we will describe a slightly larger case study. Still necessarily small, this example will use the techniques described above as well as introducing the specification of broadcast communication. In essence, this example is meant to bring together much of what we have covered in the last few sections.

The case study is very simple. It essentially concerns a *resource controller*, but is couched in terms of a 'Parent' allocating sweets to two children. So:

- the Parent has a bag of sweets;
- each child wants a sweet, and can ask for one; however,
- the Parent will only give out one sweet at a time.

In what follows we will provide temporal specifications of the Parent and both children (one called 'Jack', the other called 'Jill'). Within each specification there are several propositions, some of which will later be linked using *Comms* formulae.

Parent specification

The propositions included here are:

asked_jill – true when the Parent knows that Jill just asked for a sweet;

asked_jack – true when the Parent knows that Jack just asked for a sweet;

give_jill – true when the Parent gives a sweet to Jill; and

give_jack – true when the Parent gives a sweet to Jack.

Thus, the simple specification of *Parent* is

$$\Box \left[\begin{array}{rcl} asked\_jill & \Rightarrow & \Diamond give\_jill \\ asked\_jack & \Rightarrow & \Diamond give\_jack \\ \textbf{start} & \Rightarrow & \neg give\_jill \vee \neg give\_jack \\ \textbf{true} & \Rightarrow & \bigcirc(\neg give\_jill \vee \neg give\_jack) \end{array} \right]$$

The intuitive meanings of these formulae are simply:

- if Jill asked for a sweet, then Jill will eventually receive one;
- if Jack asked for a sweet, then Jack will eventually receive one;
- at most one child can receive a sweet at the start;
- at most one child can receive a sweet at any moment after that.

Finally, notice that

$$In(Parent) = \{asked\_jill, asked\_jack\}$$
$$Out(Parent) = \{give\_jill, give\_jack\}$$

thus Parent receives communications via *asked_jill* and *asked_jack*, and effectively sends communications through *give_jill* and *give_jack*.

Jill specification

The particular behaviour that Jill exhibits is that of a child who requests a sweet as often as possible. The specific proposition used within the Jill specification is *req_jill*, which is true when Jill asks for a sweet. Thus, the basic specification of Jill is

$$\Box \left[\begin{array}{rcl} \textbf{start} & \Rightarrow & req\_jill \\ req\_jill & \Rightarrow & \bigcirc req\_jill \end{array} \right]$$

showing that, not only does Jill ask initially but, once Jill has asked, then Jill will ask again in the next state.

Now, to implement communication between Jill and Parent we will define *Comms* (Jill, Parent) to be:

$$\Box(req\_jill \Rightarrow \Diamond asked\_jill) \wedge \Box(give\_jill \Rightarrow \Diamond got\_jill)$$

Notice that Jill actually has the potential to use another proposition, *got_jill*, but does not do so in the specification above. Jill effectively does not care about communication back from Parent.

Jack specification

Jack will have similar behaviour to Jill but will only make a request for sweets at every second moment in time, rather than at every moment. Again, the propositions used in the specification of Jack are *req_jack* and *wait*, although *got_jack* could also have been used. Note that *wait* is an *internal* proposition and so is not involved in any communication. Jack's specification is:

$$\square \left[\begin{array}{rcl} \textbf{start} & \Rightarrow & req\_jack \\ \textbf{start} & \Rightarrow & \neg wait \\ req\_jack & \Rightarrow & \bigcirc wait \\ req\_jack & \Rightarrow & \bigcirc \neg req\_jack \\ wait & \Rightarrow & \bigcirc req\_jack \\ wait & \Rightarrow & \bigcirc \neg wait \end{array} \right]$$

Again, to implement communication, we define *Comms*(Jack, Parent) to be

$$\square(req\_jack \Rightarrow \diamondsuit asked\_jack) \ \wedge \ \square(give\_jack \Rightarrow \diamondsuit got\_jack)$$

Properties of the system

Let us consider the specifications of Parent, Jack and Jill, together with the appropriate communications formulae *Comms*(Jack, Parent) and *Comms*(Jill, Parent). The pattern of these *Comms* formulae is:

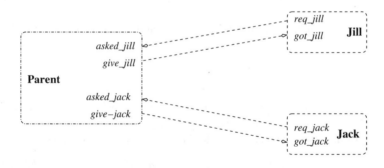

From the combined temporal specification

$$Spec_{\text{Parent}} \wedge Spec_{\text{Jill}} \wedge Spec_{\text{Jack}} \wedge Comms(\text{Jack, Parent}) \wedge Comms(\text{Jill, Parent})$$

we can then establish the following properties.

1. Jill will ask infinitely often, that is $\square\diamondsuit asked\_jill$. Note that, if the *Comms* formula ensured *instantaneous* communication then this, together with the synchronous nature of the example, could even allow us to infer $\square asked\_jill$.

2. Jack will ask infinitely often, that is $\square\diamondsuit asked\_jack$.

Recall also that Parent has a *liveness* constraint saying that, if a child has asked, some allocation will be given, as well as a *safety* constraint saying that allocations to Jack and Jill must never happen at exactly the same moment in time.

So, from (1) and (2) above, and from $Spec_{\mathsf{Parent}}$, we can infer that both $\Box \Diamond got\_jill$ and $\Box \Diamond got\_jack$. And so each child will receive sweets infinitely often, but not at the same time as we know, from $Spec_{\mathsf{Parent}}$, that

$$\bigcirc \Box (\neg give\_jack \vee \neg give\_jill) \, .$$

Jealousy and broadcast

Imagine we now change the behaviour of Jack so that this element only asks for a sweet when it sees Jill being given one. To do this, Jack must be able to see whenever the Parent allocates a sweet to Jill.

In order to specify this, we utilize a simple form of broadcast communication whereby *give_jill* in Parent's specification is linked to more than one other proposition. Specifically we modify the *Comms*(Jack, Parent) formula to now be

$$\Box (req\_jack \Rightarrow \Diamond asked\_jack)$$

$$\wedge \Box (give\_jack \Rightarrow \Diamond got\_jack)$$

$$\wedge \Box (give\_jill \Rightarrow \Diamond jealous)$$

Thus, *give_jill* becoming true affects multiple other elements, with '*jealous*' being made true in Jack's specification because of *give_jill*. Now that *jealous* is available for use in Jack's specification, we can modify Jack's specification to be simply:

$$jealous \Rightarrow \bigcirc req\_jack \, .$$

So now, Jack only asks for a sweet in response to seeing Jill receive a sweet. The revised communication pattern is as follows.

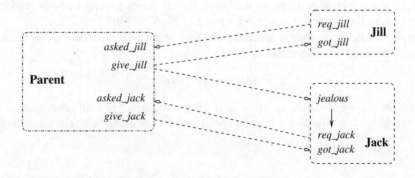

Notice how we have just changed the *Comms* formula between element specifications in order to specify a different variety of communication.

We could go on developing this example, but will finish here and turn to the question of what we might do with any temporal specification we produce.

Exercise 3.3 *Here is a temporal specification for a simple message-passing system consisting of two elements, P and Q.*

$$Spec_P: \Box \begin{bmatrix} \textbf{start} \Rightarrow a \\ a \Rightarrow \bigcirc b \\ b \Rightarrow \bigcirc c \\ d \Rightarrow \bigcirc e \end{bmatrix} \qquad Spec_Q: \Box \begin{bmatrix} x \Rightarrow \bigcirc y \\ y \Rightarrow \Diamond w \end{bmatrix}$$

(a) *What is the behaviour of $Spec_P$ on its own?*

(b) *Given $Comms(P, Q) = \Box(b \Rightarrow \Diamond x)$, what is the behaviour of*

$$Spec_P \wedge Spec_Q \wedge Comms(P, Q) ?$$

(c) *Given $Comms(P, Q) = \Box(b \Rightarrow \Diamond x) \wedge \Box(w \Rightarrow \Diamond d)$, what is the behaviour of*

$$Spec_P \wedge Spec_Q \wedge Comms(P, Q) ?$$

(d) *Why does*

$$Spec_P \wedge Spec_Q \wedge \Box(b \Rightarrow \Diamond x) \wedge \Box(w \Rightarrow \Diamond a)$$

imply $\Box \Diamond c$?

3.3.5 Refinement

There are many things that we can do with a temporal formula describing either a set of elements or some expected behaviours. Several of these will be investigated further in subsequent chapters, but two things we will consider here are: what if we want to develop another version of the specification and how will this correspond to our first version? This aspect is particularly useful in formal development methods, and can be captured under the general topic of *refinement* [229, 375, 380].

Imagine we have a temporal specification, *Spec*, for a system. Then we might establish some properties of this specification, for example

$$\vdash Spec \Rightarrow Properties.$$

Now, maybe the specification we have is missing some detail or some particular element, so we might wish to produce a revised specification, say *Spec'*. Clearly we do not want to have to prove

$$\vdash Spec' \Rightarrow Properties$$

if we can avoid it, and surely we would like all the behaviours of *Spec'* to be also behaviours of *Spec*. Thus, as is common in most formal specification techniques, we can *refine Spec* to a new specification *Spec'* as long as we can establish that

$$\vdash Spec' \Rightarrow Spec.$$

Example 3.15 *Let us consider a sample initial temporal specification and its refinement:*

$$Spec = p \wedge \Diamond q$$
$$Spec' = p \wedge \bigcirc q$$

Since $\vdash (p \wedge \bigcirc q) \Rightarrow (p \wedge \Diamond q)$, *then Spec' is indeed a refinement of Spec.*
   Consequently, if we know that $\vdash Spec \Rightarrow (\Diamond p \wedge \Diamond q)$, *which is indeed true, then it must also be the case that* $\vdash Spec' \Rightarrow (\Diamond p \wedge \Diamond q)$.

What does the above mean in terms of temporal sequences when we say '$\varphi \Rightarrow \psi$'? Recall that models for PTL are of the form $\langle \mathbb{N}, \pi \rangle$, represented pictorially as

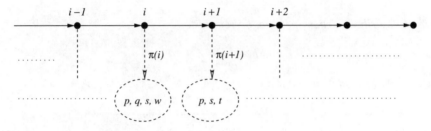

So, when we say $\varphi \Rightarrow \psi$, we mean

   for all π, if $\langle \mathbb{N}, \pi \rangle \models \varphi$, then $\langle \mathbb{N}, \pi \rangle \models \psi$.

An alternative way to say this is that

   the set of models that satisfy φ *is a **subset** of* the set of models that satisfy ψ.

So, a sequence of refinements such as:

$$Spec_1$$
$$\Uparrow$$
$$Spec_2$$
$$\Uparrow$$
$$\ldots \ldots$$
$$\ldots \ldots$$
$$\Uparrow$$
$$Spec_n$$

are successively reducing the number of possible models described. Since we identify models with execution sequences, this also means that the number of possible execution sequences is decreasing. Thus, as we refine our specifications, the number of possible execution sequences or implementations decreases.

Example 3.16 *In the previous example, we had*

$$Spec = p \wedge \Diamond q,$$

which has models such as

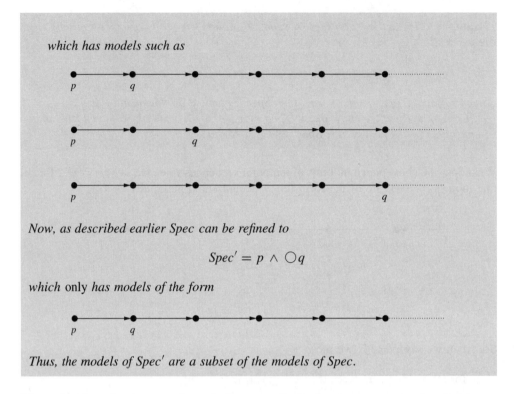

Now, as described earlier Spec can be refined to

$$Spec' = p \wedge \bigcirc q$$

which only *has models of the form*

Thus, the models of Spec' are a subset of the models of Spec.

Thus, the idea of refinement is typically to gradually become more certain of the executions we are interested in. Consequently, the specifications are refined to reduce their non-determinism. It is unlikely that the final specification reached will be purely deterministic, but it will involve many fewer possibilities than the original specification.

3.4 Advanced topics

Perhaps with more than any other chapter in this book, there number a *huge* number of topics that we could explore further. Work on temporal specification and temporal semantics has been tackled ever since Pnueli's original use of temporal logic in this area; work on logical semantics and specifications has been carried out for even longer. Thus, we can only hope to cover an arbitrary selection of the possible topics here, and then only in a shallow way. The interested reader is invited to follow up this introduction to advanced topics both by following references we give, and by examining some of the excellent books in this area, for example [210, 224, 299, 326, 339, 363].

3.4.1 Alternative models of concurrency

Earlier in this chapter we mentioned some alternative methods of concurrently organizing elements, but then chose to use synchronous concurrency. This choice was mainly taken because the synchronous concurrency model is the easiest to describe using discrete temporal logic. However, we can, with a little more work, and perhaps some more powerful temporal logics, represent other models of concurrency.

Interleaving concurrency

In the interleaving model, only one of the elements can execute at any one time, and so periods of activity for one element are interspersed (or interleaved) with periods of activity for the other. For example, if we label all steps of element E_i by the predicate '$executing(E_i)$', we might represent the specification of such a concurrent system by

$$Spec_1 \wedge Spec_2 \wedge \Box\neg(executing(E_1) \wedge executing(E_2)).$$

Thus, this ensures a pattern of execution such as

especially if we add $\Box(executing(E_1) \vee executing(E_2))$ to ensure that *exactly one* of E_1 or E_2 is executing at any moment in time.

Asynchronous execution

In the synchronous model all elements work on the same clock (and so have a common notion of 'next'), but *asynchronous* execution allows elements to have different clocks and so to work potentially at different speeds. A typical pattern of execution could be

especially... E_1, E_2 diagram

Modelling such a situation can be achieved using a temporal logic that has a *dense* model of time, for example the *Temporal Logic of the Reals (TLR)* [46, 313]. Without describing the syntax of TLR, we here give an indication of how TLR can be used to model both the semantics of individual elements and of asynchronous compositions of such elements [192].

Individual execution. The individual execution states of the program are essentially mapped down on to the Real Number line, describing the intervals over which the elements execute. Thus, the interval of time during which the element is generating one of its (local) states is characterized by the proposition *tick* being true over an interval on the Real Number line.

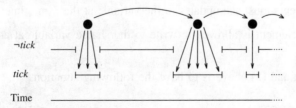

So, over the period that our individual local execution remains within one state, then *tick* will be true. When the local execution leaves this state, then *tick* becomes false for a period. And so on.

Multiple elements.   With multiple elements, we will have multiple *tick* periods. Since we are modelling asynchronous computation, these periods can all be different [192], for example.

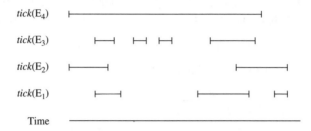

Thus, the propositions true in each element's execution are mapped down to propositions in TLR, which are true over the corresponding interval. This can, of course, become quite complex.

3.4.2   Alternative models of communication

As with concurrency, there are further options for communication that we did not explore. We will consider two of these below.

Shared variables

Here, communication between two elements is achieved by ensuring that certain variables are *shared* between the elements. Thus, each element can read from, and write to, these variables.

Consequently, for simple communication, one element just writes to a shared variable, then the other element reads the value of that variable. The usual constraints on shared variables are that:

1. both elements must see *exactly* the same values for a shared variable;

2. both elements must see updates to the variable at the same time; and

3. only one element is allowed to write to the shared variable at any one execution step.

A solution to (1) and (2) above is to have the following situation.

where we must ensure it is always the case that 'x' and 'y' match, that is x $\Leftrightarrow$ y. Thus, 'x' and 'y' represent each element's view of the shared variable.

Solving point (3) above is a little more difficult. One common approach is to add a *writing_to*(E_i, x) predicate into the specification of element E_i such that we make this true whenever the shared variable x is written by E_i. Effectively, this means that, for example, the semantics for assignment in element E_i becomes

$$[\![x:=2]\!] = \bigcirc((x = 2) \land writing\_to(E_i, x)).$$

Thus, if x in element E_3 and y in element E_4 are to refer to the same shared variable, the full specification of synchronous concurrency with these shared variables becomes

$$Spec_3 \land Spec_4$$

$$\land \quad \Box(x \Leftrightarrow y)$$

$$\land \quad \Box\neg(writing\_to(E_3, x) \land writing\_to(E_4, y))$$

Example 3.17 *Suppose we have*

$$Spec_C: \Box \left[\begin{array}{l} start \Rightarrow x = 0 \\ \exists v. (x = v) \land odd(x) \Rightarrow \bigcirc(x = (v + 1) \land writing\_to(C, x)) \end{array} \right]$$

$$Spec_D: \Box \left[\quad \exists w. (y = w) \land even(y) \Rightarrow \bigcirc(y = (w + 1) \land writing\_to(D, y)) \quad \right]$$

Then, conjoining the specifications together with formulae characterizing the shared variables, we get

$$Spec_C \land Spec_D \land \quad \Box(x \Leftrightarrow y) \land \quad \Box\neg(writing\_to(C, x) \land writing\_to(D, y))$$

This can be viewed as

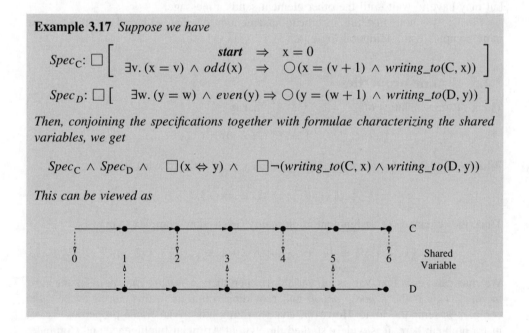

Channel communication

This form of communication essentially restricts general message passing with two elements:

- messages can only be passed down specified *channels*; and

- message send and receive actions must be *synchronised*.

Since channels will occur later. Let us explain these in more detail.

Channels are unidirectional structures that exist between two processes (they were characterized in various algebraic approaches such as CCS [376] and CSP [275]. In general only one message at a time is allowed within a channel and communication can only take place through existing channels. For example, if a channel exists between E_1 and E_2, then E_1 can send a message to E_2 down this channel:

However, if there is no channel from E_1 to E_3, then communication to E_3 is impossible. Note the distinction with general message passing, where a message can be addressed to any other element.

Synchronised Communication ensures that one element can attempt to send a message via a channel, but may have to wait until the element at the other end of the channel is ready to receive it. Similarly, an element can try to receive a message from a channel, but may have to wait until the other element sends a message.

Finally, we note that the specification/semantics of channel-based systems can be quite complex using temporal logic [45].

3.4.3 Frame properties

Think of the semantics of '$x:=0; y:=3$', that is

$$[\![x:=0; y:=3]\!] = \bigcirc((x = 0) \wedge \bigcirc(y = 3)).$$

What is the value of x at the time point when $y = 3$? That is, what is

$$[\![x:=0; y:=3; x:=x+1]\!] ?$$

Thus, the semantics of assignment is, in reality, even more complex than

$$[\![x:=v; s]\!] = \exists w. (x = w) \wedge \bigcirc(x = v(x/w) \wedge [\![s]\!]).$$

We must also say that values of variables other than x *remain the same* in the next moment. This is the *frame problem* and providing solutions to this can again be quite complex in temporal logic. However, characterizing non-trivial frame properties[3] is not just a problem here, it is widely studied throughout Artificial Intelligence and Computer Science [264, 432, 452].

3.4.4 Proving program properties

An advantage of using temporal logic for the formal description of programming languages is that a proof system for programs is automatically derivable from such a

---

[3] In other words, describing aspects that remain the same while other, more dynamic, elements change.

semantics [360]. In particular, semantic rules of the form

$$[\![g(\mathrm{S}, \mathrm{T})]\!] \equiv f([\![\mathrm{S}]\!], [\![\mathrm{T}]\!])$$

automatically produce program proof rules of the form

$$\frac{\vdash \{\mathrm{S}\}_\theta \varphi \qquad \vdash \{\mathrm{T}\}_\theta \psi}{\vdash \{g(\mathrm{S}, \mathrm{T})\}_\theta f(\varphi, \psi)}$$

Here, the temporal assertion $\vdash \{\mathrm{S}\}_\theta \varphi$ means that the temporal formula φ is a specification of the statement S in the procedural environment θ. (The use of such procedural environments will be described later.) The above proof rule states that, if we can establish φ from program S and ψ from program T, then we know that $f(\varphi, \psi)$ is true of the combination (using 'g') of S and T.

Example 3.18 *A semantic definition for the (synchronous) parallel composition of two processes could be given as*

$$[\![\mathrm{S}\|\mathrm{T}]\!] \Leftrightarrow [\![\mathrm{S}]\!] \wedge [\![\mathrm{T}]\!].$$

This leads naturally to a proof rule of the form

$$\frac{\vdash \{\mathrm{S}\}_\theta \varphi \qquad \vdash \{\mathrm{T}\}_\theta \psi}{\vdash \{\mathrm{S}\|\mathrm{T}\}_\theta \varphi \wedge \psi}.$$

In plain language, if we can prove property φ of process S and can prove property ψ of T, then the above rule allows us to infer that $\varphi \wedge \psi$ is true of the parallel program S$\|$T.

There are two general proof rules not directly derived from semantic definitions. These general proof rules are:

    Consequence Rule   *(or Weakening Rule)*

$$\frac{\vdash \{\mathrm{S}\}_\theta \varphi \qquad \vdash \varphi \Rightarrow \psi}{\vdash \{\mathrm{S}\}_\theta \psi}$$

The link between the temporal semantics of a programming language given using $[\![\mathrm{S}]\!]_\theta$ and the proof system given using assertions of the form $\vdash \{\mathrm{S}\}_\theta \varphi$ can be shown by the following proof rule:

    Semantics Rule

$$\frac{\vdash \{\mathrm{S}\}_\theta \varphi}{\vdash [\![\mathrm{S}]\!]_\theta \Rightarrow \varphi}$$

Now, the above rules together with a temporal semantics are enough to allow us to develop a viable temporal proof system for programs. While we will not consider this further here, the interested reader should consult [44, 336, 337, 360, 363].

3.4.5 Extended programming languages

The programming languages considered in this chapter have been very simple. They can be extended in many ways, leading to increasingly complex temporal semantics. The particular form of temporal assertion considered above, $\vdash \{S\}_\theta \varphi$, means that the temporal formula φ is a specification of the statement S in the *procedural environment* θ. This procedural environment is provided if our programming language allows the definition and invocation of procedures (as most imperative languages do). In general, the procedural environment, θ, is a mapping of the form:

$$\theta: \text{ Proc } \rightarrow \text{ Sem}$$

that maps procedure names to their semantic definitions. Semantic objects are here represented by temporal formulae, so θ maps procedure names to temporal formulae.[4]

Thus, a temporal semantics for a particular language can be given by defining $[\![\ldots]\!]_\theta$ for each of the possible constructs within the language. Temporal formulae are used to represent the semantics of a statement in the programming language, so each formula can be used to give a set of temporal models over which the statement is satisfied. Procedures can be recursive, so $[\![P]\!]$ might, in turn, invoke $[\![P]\!]$ again. Thus, a sample proof rule for procedure definitions is [44, 360, 363]:

Procedure Definition Rule

$$\vdash \{S\}_{\theta \dagger [P \to \xi]} \varphi(\xi)$$

$$\overline{\vdash \{(P = B_P); S\}_\theta \exists \xi. \Box (\xi \Leftrightarrow [\![B_P]\!]_{\theta \dagger [P \to \xi]}) \wedge \varphi(\xi)}$$

Unfortunately, this rule is very complicated and requires the explicit use of the semantic function $[\![\ ]\!]$. The existential quantification is needed to bind the actual semantics of the procedure body to the assumed one. We do not expect (or wish) the reader to try to understand this rule. However, just note that extending the programming language, for example by recursive procedures, can have a significant effect on the program semantics and so on the proof theory of programs.

Rules such as the above will always be complex because of the possibility of recursive procedures. Even if quantification is not used, extensions to the temporal language such as fixpoints are often needed to describe these semantics (see Section 2.8.4).

Aside. If you *do* really want some intuition about the above rule, then here it is! When we define a new procedure, called P, with a body B_P, then we need to generate a temporal formula (called ξ) describing the semantics of P and update our procedural environment, θ, to $\theta \dagger [P \to \xi]$. In other words we extend the mapping with the meaning of the new procedure P. However, finding ξ can be difficult, especially as the body of the procedure, B_P, might well contain recursive calls to P. Thus, we wish to find a ξ such that

$$\Box (\xi \Leftrightarrow [\![B_P]\!]_{\theta \dagger [P \to \xi]}) \,.$$

---

[4] For simplicity, we here allow no arguments to procedures.

This shows how we end up with a temporal fixpoint capturing the behaviour of (potentially recursive) procedures.

3.4.6   Compositionality

Non-compositional systems imply that, in general, elements cannot be specified or verified in isolation. In the case of programming language semantics, this implies that the semantics of subprograms (e.g. procedures) cannot be combined to give the semantics of the whole program.

Thus, we require a compositional semantics, and consequently a compositional proof system for a programming language. To be compositional, we require that the semantics of a particular language construct gives the behaviour of that construct in all possible contexts (that are able to be generated within the language).

Compositional reasoning about programs requires a compositional proof theory for the language the programs are written in. Without compositionality, and so without the possibility of general hierarchical decomposition of programs and systems, global specifications must be used. Unfortunately, this also implies that global verification must be used because there is no possibility of studying the subprograms (or subsystems) in isolation. So, any formal verification might well involve the behaviours of all the subsystems.

For further discussion of these aspects, specifically with respect to temporal logics, see [44, 45, 383, 385].

3.4.7   Abstractness

For a definition to be considered as the semantics of a particular language, it must correspond to an accepted natural semantics for that language, which intuitively models the language's behaviour. In particular, a formal semantics must distinguish precisely the same programs that the natural semantics does. In other words, the two semantics induce the same equivalence classes of programs. An *operational semantics* is often used for this natural semantics. This is intended to model the behaviours of the language by being close to some implementation of that language. The equivalence of these semantics is, thus, important.

One of the major problems with the temporal logics described so far has been the difficulty in producing abstract temporal semantics for various programming languages in these logics. The main difficulty is that for semantics written in 'standard' temporal logics everything must be modelled, even those things that have no observable effect. For example, the semantics of sequential composition is naively taken as

$$[\![x : = a; \ T]\!] = \bigcirc (x = a \wedge [\![T]\!]).$$

This can cause problems, since $[\![x : = 1; \ x : = 1; \ T]\!]$ and $[\![x : = 1; \ T]\!]$ are *not* now semantically equivalent. Though this property of being able to 'count' states using the next operator is useful in some applications, it is often undesirable. Because of this, several attempts have been made to produce temporal logics that are more likely to give abstract temporal semantics. Many of these logics are subtly different from the ones used above, for example two interval temporal logics [257, 449], TLR [46] and general past-time temporal logics [42, 43, 245, 347]. In addition, the work of Lamport considers this problem of abstractness in some detail [336].

3.5   Final exercises

Exercise 3.4 *Consider the following temporal specification, for an element A.*

$$Spec_A: \Box \begin{bmatrix} \textbf{\textit{start}} & \Rightarrow & p \\ \textbf{\textit{start}} & \Rightarrow & \neg q \\ \textbf{\textit{start}} & \Rightarrow & \neg send\_msg \\ p & \Rightarrow & \bigcirc \neg p \\ p & \Rightarrow & \bigcirc \neg send\_msg \\ p & \Rightarrow & \bigcirc q \\ q & \Rightarrow & \bigcirc send\_msg \end{bmatrix}$$

(a) *What is the behaviour of this specification, that is, when is send_msg made true?*

(b) *If we extend this specification to give*

$$Spec_A: \Box \begin{bmatrix} \textbf{\textit{start}} & \Rightarrow & p \\ \textbf{\textit{start}} & \Rightarrow & \neg q \\ \textbf{\textit{start}} & \Rightarrow & \neg send\_msg \\ p & \Rightarrow & \bigcirc \neg p \\ p & \Rightarrow & \bigcirc \neg send\_msg \\ p & \Rightarrow & \bigcirc q \\ q & \Rightarrow & \bigcirc p \\ q & \Rightarrow & \bigcirc send\_msg \end{bmatrix}$$

what is the behaviour of $Spec_A$ now and how often is send_msg made true?

(c) *For the version of $Spec_A$ given in (b), is it the case that $\Box \Diamond send\_msg$ is true?*

Exercise 3.5 *Assume $Spec_A$ is the extended version as in Exercise 3.4 and $Spec_B$ is:*

$$Spec_B: \Box \begin{bmatrix} rcv\_msg & \Rightarrow & \bigcirc g \\ f & \Rightarrow & \bigcirc g \\ g & \Rightarrow & \bigcirc f \end{bmatrix}$$

(a) *What is the behaviour of $Spec_A \wedge Spec_B$ if we assume that rcv_msg is always false? (Note that we are assuming* synchronous *composition of processes here.)*

(b) *What formulae might we write in $Comms(A, B)$ to represent* instantaneous *communication between send_msg and rcv_msg in the system specified by*

$$Spec_A \wedge Spec_B \wedge Comms(A, B)?$$

(c) *What might we write, in $Comms(A, B)$, to characterize communication from A to B, via send_msg and rcv_msg, that is guaranteed to arrive, but has variable delay?*

(d) What might we write in Comms(A, B) to characterize the situation where send_msg and rcv_msg both refer to the same shared variable?

(e) If we wish to specify that a message send will be followed, at some time in the future, by a message receipt, how might we modify the above specification?

3.6 Where to next?

We have here seen how our temporal logic can be used to specify a range of computational systems. This can be achieved since linear temporal logic inherently represents properties of sequences and so we can use linear temporal logic specifically to describe the properties of *execution* sequences.

Once we have a temporal logic formula characterizing some system, it is only natural that we would like to verify some property of this system. This can be achieved by *proving* that one temporal formula implies another. This leads on to a mechanism for carrying out such proofs which will be considered in the next chapter.

In later chapters, we will see that there are other things we might wish to do with a temporal specification. For example, given some program that reputedly implements our required system, we might verify our specification against the program in order to judge if this is, in fact, the case. In Chapter 5, we will see how *model checking* can be used for this. If we do not yet have a program, but only have our temporal specification, then we might want to provide an implementation quickly by *executing* the temporal specification. We consider this in Chapter 6 where we use the specification as a program. To begin with, however, we will consider a technique for proving that a PTL formula is true (i.e. valid). This then allows us to prove properties of specifications and to carry out refinement proofs if the specification is to be developed further. We consider this *deductive* approach in Chapter 4.

4

Deduction

The past is certain, the future obscure.
— Thales of Miletus.

In this Chapter, we will:

- [INTRODUCTION] examine the idea of deciding on the truth of temporal formulae using formal rules and procedures (Section 4.1);

- [TECHNIQUE] introduce the *clausal temporal resolution* method for proof in PTL (Section 4.2);

- [SYSTEM] describe the TSPASS system which is one of those implementing the clausal temporal resolution approach (Section 4.3); and

- [ADVANCED] highlight a selection of more advanced work including alternative methods and tools (Section 4.4).

Although fewer in number than in previous chapters, exercises will again be interspersed throughout the chapter with solutions provided in Appendix B.

4.1 Temporal proof

As we saw in the last chapter, we often want to check whether one temporal formula implies another. For example, if we have specified a system using the temporal formula, ψ, then we can check whether a certain temporal property, say φ, follows from the specification of the system by establishing that φ is implied by ψ, that is, whether $\vdash \psi \Rightarrow \varphi$. This can be achieved by *proving* that one temporal formula implies another.

An Introduction to Practical Formal Methods Using Temporal Logic, First Edition. Michael Fisher.
© 2011 John Wiley & Sons, Ltd. Published 2011 by John Wiley & Sons, Ltd.

There are a variety of mechanisms for deciding proof problems in temporal logic, some of which we will mention in this chapter. However, our main technique will be a variety of *clausal resolution* for PTL. Before proceeding, it is appropriate that we review the use of clausal resolution in classical logic (see also Appendix A).

Resolution was first proposed as a proof procedure for classical logics by Robinson several decades ago [437]. Resolution was claimed to be 'machine-oriented' because it was deemed particularly suitable for proofs to be performed by computer, having only one rule of inference that may have to be applied many times. Using this approach the validity of a logical formula, φ, is checked by first negating it and translating $\neg\varphi$ into a particular form, Conjunctive Normal Form (CNF), before resolution rules may be applied. A formula in CNF can be represented as

$$C_1 \wedge C_2 \wedge \ldots \wedge C_m$$

where each C_i, for $i = 1$ to m, known as a *clause*, is a disjunction of literals and each *literal* is either an atomic formula or its negation. (For further details on the translation to CNF, see Appendix A.) Typically, the CNF formula $C_1 \wedge C_2 \wedge \ldots \wedge C_m$ is represented as a set of clauses $C = \{C_1, C_2, \ldots, C_m\}$.

The resolution inference rule (see below) is then applied repeatedly to the set of clauses derived from $\neg\varphi$ and new inferences are added to the set. Pairs of clauses are resolved using the classical (propositional) resolution inference rule

$$\frac{C_1 \vee p \\ C_2 \vee \neg p}{C_1 \vee C_2}$$

where C_1 and C_2 are disjunctions of literals and p is a proposition. The intuition here is that, since one of p or $\neg p$ *must* be false, then one of C_1 or C_2 *must* be true. In the above, the terminology used is that $C_1 \vee C_2$ is known as the *resolvent*, p and $\neg p$ are called *complementary literals* and $C_1 \vee p$ and $C_2 \vee \neg p$ are called the *parents* of the resolvent.

The resolution rule is applied to pairs of clauses from C, with new resolvents being added back into C, until either an empty resolvent (denoting **false**) is derived or no further resolvents can be generated. (In the propositional logics we are interested in, this procedure is guaranteed to terminate.) Typically, given some formula, φ, that we wish to decide the validity of, we apply the resolution process to $\neg\varphi$. If a contradiction (**false**) is derived then $\neg\varphi$ is unsatisfiable and so the original formula φ must therefore be valid. This process of determining the unsatisfiability of the negation of a formula is known as *refutation*. The resolution proof procedure is *refutation complete* for classical logic because, when applied to an unsatisfiable formula, the procedure is guaranteed to (terminate and) produce **false**.

In justification of Robinson's original statement that resolution provides a 'machine oriented' approach, many theorem provers based on resolution have indeed been developed. Resolution-based provers have been *very* successful, especially within classical propositional and first-order logics. For example, Otter [370] and Vampire [436] are resolution-based provers. The former is a popular, and widely used, automated reasoning system; the latter is currently the fastest such system for analysing classical first-order logics.

4.2   Clausal temporal resolution

Since clausal resolution is a simple and adaptable proof method for classical logics, with a bank of research into heuristics and strategies, it is perhaps surprising that relatively few attempts have been made to extend this to temporal logics. However, this is what we will describe below.

There have been *some* attempts to extend resolution-based approaches to proof in discrete, linear temporal logic and these fall, in turn, into two main classes: non-clausal and clausal. The non-clausal method described in [4], and extended to first-order temporal logic in [6], requires a large number of resolution rules, making implementation of this method problematic [365]. Recall that PTL is difficult to reason about because the inter-action between the □-operator and the ○-operator encodes a form of induction. Thus, a special temporal resolution rule is needed to handle this. While there were several earlier attempts at defining *clausal* resolution for temporal logics, namely those of [103] and [489], the approach we will describe below is the most developed and has subsequently been implemented within several tools. This clausal resolution technique was originally introduced in [185] and then was subsequently refined [141, 180, 208], extended [143, 162, 318], implemented [293, 298, 356] and applied [58, 181, 211] over a number of years.

The temporal resolution method is clausal, so we must transform a PTL formula into a clausal form, specifically into SNF that we introduced in Section 2.6. The method is a refutation approach, so given any temporal formula, X, that we want to prove, we first negate X and attempt to show the *unsatisfiability* of $\neg X$. If we succeed in establishing this unsatisfiability, then X must be valid.

In order to implement this approach, we use two types of resolution rule that are applicable to clauses, one analogous to the classical resolution rule and the other a new *temporal resolution* rule. Again, due to the inductive interaction between the □ and ○, the application of the temporal resolution rule is non-trivial, requiring specialised algorithms [148, 150]. Finally, we note that having been defined, proved correct and extended to more complex logics, the temporal resolution method has been implemented and has been shown to be generally competitive [295, 296].

Overview of the temporal resolution method

Before we delve into the detail of the temporal resolution method, we will give a general overview of the processes involved. Clearly, many of the terms will be unfamiliar to the reader, but this at least serves to give a 'roadmap' of the techniques we require.

So, to check a PTL formula, say A, for unsatisfiability we perform the following steps.

1. *Transform* A into SNF, giving a set of clauses A_S.

2. Perform *step resolution* on clauses from A_S until either:

    (a) a contradiction is derived, in which case A is unsatisfiable; or

    (b) no new resolvents are generated, in which case we continue to step (3).

3. Select an eventuality from within A_S, and perform *temporal resolution* with respect to this – if any new formulae are generated, go back to step (2).

4. If all eventualities have been resolved, then A is satisfiable, otherwise go back to step (3).

4.2.1   From PTL to SNF

As in classical resolution, temporal formulae are transformed into a specific normal form before resolution rules are applied. The normal form we use is exactly the SNF we have already seen in Section 2.6; as we will show later, this forms the basis of methods for *both* deduction and execution in PTL [195]. As described in Section 2.6, formulae in SNF are of the general form

$$\Box \bigwedge_i C_i$$

where each C_i is known as a *rule*. Throughout this chapter we will also term such formulae as *PTL-clauses* (analogous to a 'clause' in classical logic). Each C_i must be one of the following forms with k_a, k_b, l_c, l_d and l all representing literals.

$$\mathbf{start} \quad \Rightarrow \quad \bigvee_c l_c \qquad \text{(an } \textit{initial} \text{ PTL-clause)}$$

$$\bigwedge_a k_a \quad \Rightarrow \quad \bigcirc \bigvee_d l_d \quad \text{(a } \textit{step} \text{ PTL-clause)}$$

$$\bigwedge_b k_b \quad \Rightarrow \quad \Diamond l \qquad \text{(a } \textit{sometime} \text{ PTL-clause)}$$

For convenience, the outer '$\Box$' and '$\wedge$' connectives are usually omitted, and the set of PTL-clauses $\{C_i\}$ is considered. But how do we generate SNF from an initial PTL formula? As we will see below, the process is analogous to the transformation from classical logic to CNF, though necessarily a little more complex.

The transformation from PTL into SNF depends on three main classes of operations: the renaming of complex subformulae; the removal of temporal operators; and various classical style rewrite operations. Given a formula, φ, to translate to SNF, we introduce $\Box(\mathbf{start} \Rightarrow \varphi)$ and begin to apply the various translation rules. As we will see, this generates SNF that preserves the unsatisfiability of the original PTL formula. In the examples that follow, we assume that all negations have been 'pushed' to the literals. For example, formulae such as $\neg\Box p$ have been transformed to $\Diamond\neg p$. This is because many of the transformations we describe below involve the manipulation of *positive* subformulae.

Subformula renaming

Renaming, as suggested in [407], is a way of preserving the structure of a formula when translating into a normal form in classical logic. In the temporal case, a complex subformula can be replaced by a *new* proposition symbol (i.e. never used before) with the truth value of the new proposition being linked to the subformula it represents at all points in time. Thus, for a positive subformula φ, embedded within a temporal context T, we define *renaming* by

Formula	Translates to	Formula
$\Box(L \Rightarrow T(\varphi))$	$\longrightarrow$	$\Box(L \Rightarrow T(x))$ $\Box(x \Rightarrow \varphi)$

Here, 'x' is a new propositional symbol.

Example 4.1 *If we have the formula* **start** $\Rightarrow \Diamond \bigcirc p$, *then we can apply the above renaming to give the two clauses:*

$$\begin{aligned} \textbf{start} &\Rightarrow& \Diamond y \\ y &\Rightarrow& \bigcirc p \end{aligned}$$

where 'y' is a new *proposition.*

As we can see, renaming typically reduces the complexity of individual rules/clauses at the expense of increasing both the number of rules/clauses and the number of propositions. There are many different variants and possible optimizations regarding this process; the example below highlights one such.

Example 4.2 *Given the formula '*start $\Rightarrow (\bigcirc \square p \wedge \Diamond \square p)$*' we might apply renaming twice to generate*

1. **start** $\Rightarrow \bigcirc w \wedge \Diamond z$
2. $w \Rightarrow \square p$
3. $z \Rightarrow \square p$

However, an obvious improvement is to 're-use' previous renamings where possible. Thus, we could *have generated*

1. **start** $\Rightarrow \bigcirc w \wedge \Diamond w$
2. $w \Rightarrow \square p$

which contains one fewer new propositions and one fewer new formulae.

Removal of temporal operators

Recall from our definition of SNF that, apart from the outer '$\square$' the only temporal operators we retain are '$\Diamond$' and '$\bigcirc$'. We actually remove all the other operators, for example U, W, and (embedded) $\square$. This is achieved essentially by using the fixpoint definitions of these operators. Let us consider $\square \varphi$. Recall from Section 2.8.4 that '$\square$' is effectively a maximal fixpoint:

$$\square \varphi \equiv \nu \xi. (\varphi \wedge \bigcirc \xi).$$

As we also mentioned, we can use quantified propositional temporal logic (QPTL) to define $\square \varphi$. Thus, with

$$\exists x. x \Rightarrow (\varphi \wedge \bigcirc x)$$

we define the new proposition x to imply $\square \varphi$. This can be rewritten to

$$\exists x. ((x \Rightarrow \varphi) \wedge (x \Rightarrow \bigcirc x)).$$

We now essentially use this approach to remove $\square \varphi$ in our translation to SNF.

Formula	Translates to	Formula
$\Box(L \Rightarrow \Box\varphi)$	$\longrightarrow$	$\Box(L \Rightarrow x)$ $\Box(x \Rightarrow \varphi)$ $\Box(x \Rightarrow \bigcirc x)$

Here, x must again be a new proposition name (i.e. not used before). Notice how x is now true at exactly the same points as $\Box\varphi$, and so can effectively be used in its place for our later refutations.

Example 4.3 *As an example, now consider $p \Rightarrow \Box q$. Using the above (simple) translation, we get (where z is a new proposition):*

$$\Box(p \Rightarrow z)$$
$$\Box(z \Rightarrow q)$$
$$\Box(z \Rightarrow \bigcirc z)$$

Thus, once z becomes true, then the last clause ensures that it continues being true at all future states. This, together with the second formula above ensures that q is true at all future states.

Aside: Replacing temporal formulae.   Using the above we only prescribe that x implies $\Box\varphi$. This is enough for us when we apply our clausal temporal resolution procedures. However, if we want a new proposition that *exactly* mimics $\Box\varphi$, then we can use the full QPTL version

$$\exists x.x \Leftrightarrow (\varphi \wedge \bigcirc x).$$

Once we translate this to SNF, we get the additional clause

$$\Box((\neg x \wedge \varphi) \Rightarrow \bigcirc \neg x).$$

Again, there is more to this. If we actually want $\neg x$ *really* to model '$\neg\Box\varphi$', then we might even choose to add

$$\Box(\neg x \Rightarrow \Diamond\neg\varphi).$$

Returning to just the use of '$\Rightarrow$', then translation rules can be devised to remove all the unwanted temporal operators, for example $\psi U \varphi$:

Formula	Translates to	Formula
$\Box(L \Rightarrow \psi U \varphi)$	$\longrightarrow$	$\Box(L \Rightarrow \Diamond\varphi)$ $\Box(L \Rightarrow y)$ $\Box(y \Rightarrow (\varphi \vee \psi))$ $\Box(y \Rightarrow (\varphi \vee \bigcirc y))$

Note. Recall the fixpoint definition of 'U':

$$\psi U \varphi \quad \equiv \quad \mu\xi.(\varphi \vee (\psi \wedge \bigcirc\xi))$$

To ensure the 'minimal' fixpoint, we incorporate '$\Diamond\varphi$' into the translation to ensure that φ can not be avoided indefinitely.

Translation of the 'W' operator is similar (but, note the lack of the '$\Diamond$' operator in the translation, removing the necessity to have φ become true eventually):

Formula	Translates to	Formula		
$\Box(L \Rightarrow \psi W \varphi)$	$\longrightarrow$	$\Box(L$	$\Rightarrow$	w)
		$\Box(\text{w}$	$\Rightarrow$	$(\varphi \vee \psi))$
		$\Box(\text{w}$	$\Rightarrow$	$(\varphi \vee \bigcirc\text{w}))$

In both of the above, y and w are new propositions.

Exercise 4.1 *Translate '$f U (\bigcirc g)$' into SNF.*

Aside. Different variants of the normal form have been suggested [187, 195, 205]. For example, where PTL is extended to allow past-time operators the normal form has **start** or $\bullet A$ (where '$\bullet$' means in the *previous moment* in time and A is a conjunction of literals) on the left-hand side of the PTL-clauses and a present-time formula or eventuality (i.e. '$\Diamond l$') on the right-hand side. Other versions allow PTL-clauses of the form **start** $\Rightarrow \Diamond l$. These are all expressively equivalent when models with finite past are considered.

Example 4.4 *If we are certain that there always is a* previous *moment in time, then a formula of the form*

$$\Box(\bullet L \Rightarrow P)$$

can be transformed to '$\Box(L \Rightarrow \bigcirc P)$'.

4.2.2 Translation into SNF

There are a number of ways to use the translation rules above, together with several more we will introduce in this section, to transform an arbitrary PTL formula into SNF. Just by applying the translations in a different order, we can get a different, but semantically equivalent, set of clauses. In Section 4.3, we will get an idea of how practical tools such as TSPASS carry out this translation. For the moment we will just give a simple approach for translating PTL to SNF [195, 208].

Given any PTL formula, A, we carry out the following sequence of transformations.

1. *Anchor A to the start, and rename.*

Formula	Translates to	Formula
A	$\longrightarrow$	$\square(\textbf{start} \Rightarrow z)$ $\square(z \Rightarrow A)$

2. *Remove any classical constructions on the right-hand side of clauses.*
   For example:

Formula	Translates to	Formula
$\square(L \Rightarrow (A \wedge B))$	$\longrightarrow$	$\square(L \Rightarrow A)$ $\square(L \Rightarrow B)$
$\square(L \Rightarrow \neg(A \wedge B))$	$\longrightarrow$	$\square(L \Rightarrow (\neg A \vee \neg B))$
$\square(L \Rightarrow \neg(A \Rightarrow B))$	$\longrightarrow$	$\square(L \Rightarrow A)$ $\square(L \Rightarrow \neg B)$

3. *Rename any embedded formulae.*
   As above, we rename any (non-literal) formula that occurs positively within a temporal context 'T':

Formula	Translates to	Formula
$\square(L \Rightarrow T(\varphi))$	$\longrightarrow$	$\square(L \Rightarrow T(x))$ $\square(x \Rightarrow \varphi)$

For example:

Formula	Translates to	Formula
$\square(a \Rightarrow \bigcirc\bigcirc b)$	$\longrightarrow$	$\square(a \Rightarrow \bigcirc y)$ $\square(y \Rightarrow \bigcirc b)$

4. *Process negations applied to temporal operators.*
   As described above, in many cases (when we want to 'keep' the subsequent temporal operators), we just 'push' negations through a formula, for example (note that, by now, only literals occur inside temporal operators)

Formula	Translates to	Formula
$\square(L \Rightarrow \neg\bigcirc p)$	$\longrightarrow$	$\square(L \Rightarrow \bigcirc\neg p)$
$\square(L \Rightarrow \neg\square q)$	$\longrightarrow$	$\square(L \Rightarrow \lozenge\neg q)$

In other cases, we remove both the negation and temporal operator together (see below).

5. *Remove unwanted temporal operators (and their negated forms).*
   For example:

Formula	Translates to	Formula
$\square(L \Rightarrow \square q)$	$\longrightarrow$	$\square(L \Rightarrow z)$ $\square(z \Rightarrow q)$ $\square(z \Rightarrow \bigcirc z)$
$\square(L \Rightarrow \neg\lozenge r)$	$\longrightarrow$	$\square(L \Rightarrow w)$ $\square(w \Rightarrow \neg r)$ $\square(w \Rightarrow \bigcirc w)$

6. *Perform some final rewriting and simplification to get clauses into the exact SNF form.*
   For example:

Formula	Translates to	Formula
$\square(L \Rightarrow c \vee \bigcirc R)$	$\longrightarrow$	$\square((\neg c \wedge L) \Rightarrow \bigcirc R)$
$\square(\textbf{true} \Rightarrow R)$	$\longrightarrow$	$\square(\textbf{start} \Rightarrow R)$ $\square(\textbf{true} \Rightarrow \bigcirc R)$
$\square(L \Rightarrow \bigcirc \textbf{false})$	$\longrightarrow$	$\square(\textbf{start} \Rightarrow \neg L)$ $\square(\textbf{true} \Rightarrow \bigcirc \neg L)$

So, these transformations are repeatedly applied until the formula is in the required form.

Note. There are several different ways to apply the above transformations, generating different, but essentially equivalent, sets of clauses. See both [195, 208] and the later section on TSPASS to see two different approaches.

The important properties of the translation to SNF are captured by the following theorem.

Theorem 4.1 (Correctness and complexity [195, 208]) *Given a* WFF *PTL formula, φ, and a function, τ, that translates* WFF *to SNF, then*

- *if $\tau(\varphi)$ is unsatisfiable, then φ is unsatisfiable, and*

- *$\tau(\varphi)$ contains at most a linear increase, with respect to φ, in both the number of propositions and the length of formulae.*

While we will not consider the details here, we note that this theorem ensures that the translation to SNF does not somehow introduce unsatisfiability where it did not exist in the original. Also, there is an extension to this theorem showing that $\tau(\varphi)$ is unsatisfiable *if, and only if, φ* is unsatisfiable [195], which is useful if we wish to build a model for a satisfiable formula (see Chapter 6). Finally, also note that the second element within the above theorem ensures that we are not adding significant complexity by following this route.

Example 4.5 *We illustrate the translation to the normal form by carrying out an example transformation. Assume we want to show*

$$(\lozenge p \wedge \square(p \Rightarrow \bigcirc p)) \Rightarrow \lozenge \square p$$

is valid. We negate, obtaining

$$(\lozenge p \wedge \square(p \Rightarrow \bigcirc p)) \wedge \square \lozenge \neg p$$

and begin to translate this into SNF. First, we anchor the formula to the beginning of time

    1. **start** $\Rightarrow$ $(\lozenge p \wedge \square(p \Rightarrow \bigcirc p) \wedge \square \lozenge \neg p)$

and then split the three conjuncts (recall that new propositions are represented in this font).

    1. **start** $\Rightarrow$ x_1
    2. x_1 $\Rightarrow$ $\lozenge p$
    3. x_1 $\Rightarrow$ $\square(p \Rightarrow \bigcirc p)$
    4. x_1 $\Rightarrow$ $\square \lozenge \neg p$

Thus, because sets of clauses are conjunctions, then we can split a conjunction in one clause to give several separate clauses.

Formulae labelled (1) and (2) are now in our normal form. We next work on formula (3), renaming the subformula $p \Rightarrow \bigcirc p$.

    5. x_1 $\Rightarrow$ $\square x_2$
    6. x_2 $\Rightarrow$ $(p \Rightarrow \bigcirc p)$

Here we have introduced a new proposition x_2 to describe the subformula $(p \Rightarrow \bigcirc p)$.

Next, we apply the $\square$ removal rules to formula (5), to give (7), (8), and (9), and rewrite formula (6) to give (10).

    7. x_1 $\Rightarrow$ x_3
    8. x_3 $\Rightarrow$ x_2
    9. x_3 $\Rightarrow$ $\bigcirc x_3$
  10. $(x_2 \wedge p)$ $\Rightarrow$ $\bigcirc p$

Then, formulae (7) and (8) are rewritten into the normal form giving (11) – (14).

  11. **start** $\Rightarrow$ $\neg x_1 \vee x_3$
  12. **true** $\Rightarrow$ $\bigcirc(\neg x_1 \vee x_3)$
  13. **start** $\Rightarrow$ $\neg x_3 \vee x_2$
  14. **true** $\Rightarrow$ $\bigcirc(\neg x_3 \vee x_2)$

Next, we work on formula (4) renaming $\lozenge \neg p$ by the new proposition symbol x_4.

15. $x_1 \Rightarrow \Box x_4$

16. $x_4 \Rightarrow \Diamond \neg p$

Then, we remove the $\Box$ operator from formula (16), as previously

17. $x_1 \Rightarrow x_5$

18. $x_5 \Rightarrow x_4$

19. $x_5 \Rightarrow \bigcirc x_5$

and finally write formulae (17) and (18) into the normal form.

20. **start** $\Rightarrow \neg x_1 \vee x_5$

21. **true** $\Rightarrow \bigcirc(\neg x_1 \vee x_5)$

22. **start** $\Rightarrow \neg x_5 \vee x_4$

23. **true** $\Rightarrow \bigcirc(\neg x_5 \vee x_4)$

The resulting normal form is as follows.

1.   **start** $\Rightarrow x_1$

2. $x_1 \Rightarrow \Diamond p$

9. $x_3 \Rightarrow \bigcirc x_3$

10. $(x_2 \wedge p) \Rightarrow \bigcirc p$

11. **start** $\Rightarrow \neg x_1 \vee x_3$

12. **true** $\Rightarrow \bigcirc(\neg x_1 \vee x_3)$

13. **start** $\Rightarrow \neg x_3 \vee x_2$

14. **true** $\Rightarrow \bigcirc(\neg x_3 \vee x_2)$

16. $x_4 \Rightarrow \Diamond \neg p$

19. $x_5 \Rightarrow \bigcirc x_5$

20. **start** $\Rightarrow \neg x_1 \vee x_5$

21. **true** $\Rightarrow \bigcirc(\neg x_1 \vee x_5)$

22. **start** $\Rightarrow \neg x_5 \vee x_4$

23. **true** $\Rightarrow \bigcirc(\neg x_5 \vee x_4)$

Once a formula has been transformed into SNF, we can apply step resolution and temporal resolution operations, together with simplification and subsumption rules; all of these will be explained in the forthcoming sections.

Exercise 4.2 *Consider the PTL axiom*

$$\vdash \bigcirc(a \Rightarrow b) \Rightarrow (\bigcirc a \Rightarrow \bigcirc b).$$

To check the validity of this, we want to negate it and translate to SNF. Negating gives

$$\bigcirc(a \Rightarrow b) \wedge \bigcirc a \wedge \bigcirc \neg b$$

Now rewrite this to SNF.

4.2.3 Merged SNF

To apply the temporal resolution rule (see Section 4.2.5), it is often convenient to combine one or more step PTL-clauses. Consequently, a variant on SNF called *merged-SNF (or SNF$_m$)* [185] is also defined. Given a set of step PTL-clauses in SNF, any step PTL-clause in SNF is also a PTL-clause in SNF$_m$. Any two-step PTL-clauses in SNF$_m$ may be combined to produce a step PTL-clause in SNF$_m$ as follows:

$$\frac{\begin{array}{rcl} L_1 & \Rightarrow & \bigcirc R_1 \\ L_2 & \Rightarrow & \bigcirc R_2 \end{array}}{(L_1 \wedge L_2) \quad \Rightarrow \quad \bigcirc(R_1 \wedge R_2)}$$

Thus, any possible conjunctive combination of SNF PTL-clauses can be represented in SNF$_m$.

Now we begin to describe the resolution and simplification operations, beginning with simplification.

4.2.4 Simplification and subsumption

The following transformation is used for PTL-clauses which imply **false** (where L is a conjunction of literals).

Formula	Translates to	Formula
$\square(L \Rightarrow \bigcirc\textbf{false})$	$\longrightarrow$	$\square(\textbf{start} \Rightarrow \neg L)$ $\square(\textbf{true} \Rightarrow \bigcirc\neg L)$

Thus, if, by satisfying L, a contradiction is produced in the next moment, then L must *never* be satisfied. The new constraints generated effectively represent $\square\neg L$ since **start** $\Rightarrow \neg L$ ensures that L is false at the initial state and **true** $\Rightarrow \bigcirc\neg L$ ensures that L is false at *any* next moment. This transformation keeps formulae in SNF and may, in turn, allow further step resolution inferences to be carried out.

Aside.   Some versions of SNF allow *global* rules such as '$L \Rightarrow R$', where L and R contain *no* temporal operators. This would allow **true** $\Rightarrow \neg L$ in the above case. We will see in Section 4.3.1 that another variety of our normal form allows just such classical formulae that are globally true. This would simply allow '$\neg L$' to be added in the above case.

Example 4.6 *Assume we have one SNF clause, namely*

   1. *push_the_button* $\Rightarrow$ $\bigcirc$**false**

We can apply the above simplification rule to replace this by two new SNF clauses:

   2. **start** $\Rightarrow$ ¬*push_the_button*
   3. **true** $\Rightarrow$ $\bigcirc$¬*push_the_button*

*Thus, if 'push_the_button' leads to a contradiction, then we can never have this, either at the start (**start** $\Rightarrow$ ¬push_the_button) or anywhere else (**true** $\Rightarrow$ $\bigcirc$¬push_the_button).*

Now, certain SNF PTL-clauses can be removed during simplification because they represent valid subformulae and therefore cannot directly contribute to the generation of a contradiction.

Formula	Translates to	Formula
$\square(\textbf{false} \Rightarrow \bigcirc R)$	$\longrightarrow$	
$\square(L \Rightarrow \bigcirc \textbf{true})$	$\longrightarrow$	

The first PTL-clause is valid because **false** can *never* be satisfied, and the second is valid as $\bigcirc\textbf{true}$ is *always* satisfied.

Finally, subsumption also forms part of the simplification process. Here, as in classical resolution, a PTL-clause may be removed from the PTL-clause-set if it is subsumed by another PTL-clause already present. Subsumption may be expressed as the following operation.

Formula	Translates to	Formula
$\square(L_1 \Rightarrow R_1)$ $\square(L_2 \Rightarrow R_2)$	$\xrightarrow{\vdash L_1 \Rightarrow L_2 \ \ \vdash R_2 \Rightarrow R_1}$	$\square(L_2 \Rightarrow R_2)$

The side conditions $\vdash L_1 \Rightarrow L_2$ and $\vdash R_2 \Rightarrow R_1$ must hold before this subsumption operation can be applied and, in this case, the PTL-clause $L_1 \Rightarrow R_1$ can be deleted without losing information.

Example 4.7 *If we have*

1. $\quad\textbf{start} \ \Rightarrow \ a$
2. $\quad\textbf{start} \ \Rightarrow \ a \vee b$
3. $\qquad\quad z \ \Rightarrow \ \bigcirc y$
4. $\ (w \wedge z) \ \Rightarrow \ \bigcirc(y \vee x)$

then this clause set simplifies to

1. $\textbf{start} \ \Rightarrow \ a$
3. $\quad\ \ z \ \Rightarrow \ \bigcirc y$

because '$\textbf{start} \Rightarrow a$*' provides more information than '*$\textbf{start} \Rightarrow (a \vee b)$*' and, if '*$w \wedge z$*' is true then z must necessarily be true, so '*$z \Rightarrow \bigcirc y$*' provides more information than '*$(w \wedge z) \Rightarrow \bigcirc(y \vee x)$*'.*

4.2.5   Step resolution

Pairs of initial or step PTL-clauses may be resolved using the following (resolution) operations (where R_1 and R_2 are disjunctions of literals, L_1 and L_2 are conjunctions of literals and p is a proposition).

$$\frac{\begin{array}{rcl}\textbf{start} &\Rightarrow& R_1 \vee p \\ \textbf{start} &\Rightarrow& R_2 \vee \neg p\end{array}}{\begin{array}{rcl}\textbf{start} &\Rightarrow& R_1 \vee R_2\end{array}} \qquad \frac{\begin{array}{rcl}L_1 &\Rightarrow& \bigcirc(R_1 \vee p) \\ L_2 &\Rightarrow& \bigcirc(R_2 \vee \neg p)\end{array}}{\begin{array}{rcl}(L_1 \wedge L_2) &\Rightarrow& \bigcirc(R_1 \vee R_2)\end{array}}$$

The first operation is called *initial* resolution; the second is termed *step* resolution.

We can see how these rules closely resemble the classical resolution rule from Section 4.1. Thus, step resolution effectively consists of the application of the standard classical resolution rule to formulae representing constraints at a particular moment in time, while initial resolution is essentially classical resolution at 'the beginning of time'.

Example 4.8 *We can now prove an instance of one of the PTL axioms using only step resolution and initial resolution, namely*

$$\vdash \bigcirc(a \Rightarrow b) \Rightarrow (\bigcirc a \Rightarrow \bigcirc b).$$

*To establish validity, we negate, transform to SNF, and apply the resolution procedure. If **false** is generated, then the negated formula is unsatisfiable, and so the original formula is valid. In Exercise 4.2 we negated this formula and generated:*

1.	**start**	$\Rightarrow$	f
2.	f	$\Rightarrow$	$\bigcirc$x
3.	**start**	$\Rightarrow$	$(\neg x \vee \neg a \vee b)$
4.	**true**	$\Rightarrow$	$\bigcirc(\neg x \vee \neg a \vee b)$
5.	f	$\Rightarrow$	$\bigcirc a$
6.	f	$\Rightarrow$	$\bigcirc \neg b$

Refutation can be achieved using just step resolution, initial resolution and simplification operations, as follows.

7.	f	$\Rightarrow$	$\bigcirc(\neg x \vee \neg a)$	[4, 6 *Step Resolution*]
8.	f	$\Rightarrow$	$\bigcirc \neg x$	[5, 7 *Step Resolution*]
9.	f	$\Rightarrow$	$\bigcirc$**false**	[2, 8 *Step Resolution*]
10.	**start**	$\Rightarrow$	$\neg$f	[9 *Simplification*]
11.	**true**	$\Rightarrow$	$\bigcirc \neg$f	[9 *Simplification*]
12.	**start**	$\Rightarrow$	**false**	[1, 10 *Initial Resolution*]

A contradiction has been obtained, meaning the negated formula is unsatisfiable and therefore the original formula is valid.

4.2.6 Temporal resolution

Now, we move on to the key inference rule in clausal temporal resolution, namely the temporal resolution rule itself. Essentially, this rule resolves together formulae containing the '$\square$' and '$\lozenge$' connectives. However, the inductive interaction between the '$\bigcirc$' and '$\square$' connectives in PTL ensures that the application of such an operation is non-trivial. Further, as the translation to SNF restricts the PTL-clauses to be of a certain form, the application of such an operation will be between a sometime PTL-clause and a *set* of step PTL-clauses that together ensure a complementary literal will *always* hold.

Intuition. Essentially, we would like to resolve clauses such as $L_1 \Rightarrow \Diamond l$ and $L_2 \Rightarrow \Box \neg l$ together via a rule such as

$$\frac{\begin{array}{c} L_1 \Rightarrow \Diamond l \\ L_2 \Rightarrow \Box \neg l \end{array}}{\neg (L_1 \wedge L_2)}$$

However, in translating to SNF, we have removed clauses such as $L_2 \Rightarrow \Box \neg l$. Thus, part of the (necessary) complexity of the temporal resolution operation comes from having to reconstruct $\Box \neg l$ from a set of step PTL clauses. Recall that the *induction axiom* in PTL is of the form

$$\vdash \Box(\varphi \Rightarrow \bigcirc \varphi) \Rightarrow (\varphi \Rightarrow \Box \varphi).$$

So, we essentially try to collect together a set of step PTL clauses

$$\begin{array}{ccc} L_0 & \Rightarrow & \bigcirc R_0 \\ \ldots & \Rightarrow & \ldots \\ L_n & \Rightarrow & \bigcirc R_n \end{array}$$

such that each R_j, when satisfied, will 'trigger' at least one L_i

$$\bigvee_j R_j \Rightarrow \bigvee_i L_i$$

and each R_j will ensure $\neg l$

$$\bigvee_j R_j \Rightarrow \neg l$$

Intuitively, this set of clauses is such that, once one is 'fired', then $\neg l$ will be required and another clause in the set will be 'fired' at the next step. Thus, together, these clauses ensure that $\neg l$ occurs forever.

So, once we have

$$\neg l \wedge \bigvee_j R_j$$

in one state then we will again have this true in the next state. Thus

$$\Box((\neg l \wedge \bigvee_j R_j) \Rightarrow \bigcirc(\neg l \wedge \bigvee_j R_j)).$$

Now, if we apply our PTL induction axiom to this formula, we get

$$(\neg l \wedge \bigvee_j R_j) \Rightarrow \Box(\neg l \wedge \bigvee_j R_j).$$

This then gives us

$$\bigvee_i L_i \Rightarrow \bigcirc \Box(\neg l \wedge \bigvee_j R_j)$$

which simplifies to

$$\bigvee_i L_i \Rightarrow \bigcirc \Box \neg l.$$

So, when we have a set of clauses with the above conditions then we know that $\bigcirc \Box \neg l$ is inferred and so we can resolve these clauses with any *sometime clause* of the form '$C \Rightarrow \Diamond l$'.

Example 4.9

1.	f	$\Rightarrow$	$\bigcirc \neg r$
2.	$(q \wedge \neg r)$	$\Rightarrow$	$\bigcirc \neg r$
3.	q	$\Rightarrow$	$\bigcirc p$
4.	p	$\Rightarrow$	$\bigcirc (f \vee q)$
5.	p	$\Rightarrow$	$\bigcirc \neg r$
6.	p	$\Rightarrow$	$\bigcirc g$
7.	$(f \wedge g)$	$\Rightarrow$	$\bigcirc (p \vee q)$

If we have $C \Rightarrow \Diamond r$, then the above set of clauses form an appropriate set to apply the temporal resolution operation to because they effectively generate $\bigcirc \Box \neg r$. To see this, consider 3 possible configurations:

config1 . . . . *f and g are true, but r is false;*
config2 . . . . *q is true, but r is false; and*
config3 . . . . *p is true, but r is false.*

Now, assume we begin in config1. *Clauses (1) and (7) show us that we either move to* config2 *or* config3. *If we are in* config2, *then clauses (2) and (3) show us that we next move to* config3. *If we are in* config3, *then clauses (4), (5), and (6) force us to move to either* config1 *or* config2. *Thus, once we are in any of these configurations, we remain within them forever. Since whichever one we are in, then r must be false, then the above clauses effectively represent*

$$((f \wedge g \wedge \neg r) \vee (q \wedge \neg r) \vee (p \wedge \neg r)) \Rightarrow \bigcirc \Box \neg r$$

which is a rewriting of the left-hand sides of the clauses (1) – (7).

Aside: SNF$_m$.   The SNF$_m$ form is useful for representing transitions between configurations as above. To see this, let us combine some of the clauses from Example 4.9 to give SNF$_m$ clauses. Merging clauses (1) and (7) gives us

$$(f \wedge g) \Rightarrow \bigcirc (\neg r \wedge (p \vee q)).$$

Merging clauses (2) and (3) gives

$$(q \wedge \neg r) \Rightarrow \bigcirc (p \wedge \neg r),$$

while merging clauses (4), (5) and (6) gives (with a little rewriting)

$$p \Rightarrow \bigcirc ((g \wedge f \wedge \neg r) \vee (g \wedge q \wedge \neg r)).$$

Notice how these three SNF$_m$ rules effectively describe transitions between configurations. This gives us a hint of how to find appropriate sets of clauses to apply the clausal temporal resolution rule to; see Section 4.2.8.

For the moment let us delay discussion of *how* to find such a set of clauses and consider how we use them. The general temporal resolution operation, written as an inference rule, becomes

$$
\begin{array}{rcl}
A & \Rightarrow & \bigcirc\square\neg l \\
C & \Rightarrow & \Diamond l \\
\hline
C & \Rightarrow & (\neg A)\mathcal{W}l
\end{array}
$$

The intuition behind the resolvent is that, once C has occurred then we are committed to eventually make l true. So, A must *not* be satisfied until l has occurred. If A does occur before l has been achieved, then we have a contradiction since l must eventually be true, yet A ensures that l will always be false.

Aside: Temporal resolvent. The generation of $C \Rightarrow (\neg A)\mathcal{U}l$ as a resolvent would also be sound. However because $(\neg A)\mathcal{U}l \equiv ((\neg A)\mathcal{W}l) \wedge \Diamond l$ the resolvent $C \Rightarrow (\neg A)\mathcal{U}l$ would be equivalent to the pair of resolvents $C \Rightarrow (\neg A)\mathcal{W}l$ and $C \Rightarrow \Diamond l$. The latter is subsumed by the sometime PTL-clause we have resolved with. So this leaves only the '$\mathcal{W}$' formula.

Example 4.10 *Consider the formulae*

$$
\begin{array}{rcl}
gambler & \Rightarrow & \bigcirc\square\neg rich \\
ambitious & \Rightarrow & \Diamond rich
\end{array}
$$

In English, we might say that a gambler *will never be* rich *in the future, while someone who is* ambitious *will eventually be* rich. *Applying the clausal temporal resolution rule to the above gives us*

$$
ambitious \;\Rightarrow\; (\neg gambler)\mathcal{W}rich.
$$

So, someone who is ambitious *should not be a* gambler *unless they become* rich.

In SNF we have no PTL-clauses of the form $A \Rightarrow \bigcirc\square\neg l$. So the full temporal resolution operation applies between a sometime PTL-clause and a *set* of SNF$_m$ PTL-clauses that together imply $A \Rightarrow \bigcirc\square\neg l$. The temporal resolution operation, in detail, is

$$
\begin{array}{rcl}
L_0 & \Rightarrow & \bigcirc R_0 \\
\ldots & \Rightarrow & \ldots \\
L_n & \Rightarrow & \bigcirc R_n \\
C & \Rightarrow & \Diamond l \\
\hline
C & \Rightarrow & \left[\bigwedge_{i=0}^{n}(\neg L_i)\right]\mathcal{W}l
\end{array}
$$

with the side conditions that, for all j such that $0 \leq j \leq n$,

$$\vdash R_j \Rightarrow \neg l; \text{ and}$$

$$\vdash R_j \Rightarrow \bigvee_{i=0}^{n} L_i.$$

Here, the side conditions are simply propositional formulae so they must hold in (classical) propositional logic. The first side condition ensures that, by satisfying any R_j then $\neg l$ will be satisfied. The second shows that once an R_j is satisfied then at least one of the left-hand sides (L_i) will also be satisfied. Hence, if any L_i is satisfied then, in the next moment, R_i is satisfied, as is $\neg l$, and as is L_k for some k and so on, so that

$$(\bigvee_{i} L_i) \Rightarrow \bigcirc \square \neg l.$$

Terminology.   The set of SNF$_m$ PTL-clauses $L_i \Rightarrow \bigcirc R_i$ that satisfy these side conditions are together known as *a loop in* $\neg l$. The disjunction of the left-hand side of this set of SNF$_m$ PTL-clauses, that is

$$\bigvee_{i} L_i$$

is known as a *loop formula* for $\neg l$.

Once we have applied the clausal temporal resolution rule, the resolvent must be translated into SNF before any further resolution steps are applied. Recall that formulae such as

$$C \Rightarrow (\bigwedge_{i=0}^{n} \neg L_i) W l$$

are transformed into '$C \Rightarrow w$', where the new proposition 'w' is constrained by

$$\square(w \quad \Rightarrow \quad (l \vee \bigwedge_{i=0}^{n} \neg L_i)$$
$$\square(w \quad \Rightarrow \quad (l \vee \bigcirc w)$$

However, a translation to the normal form is given below that avoids the renaming of the subformula

$$\bigwedge_{i=0}^{n} \neg L_i$$

where t is a new proposition symbol and $i = 0, \ldots, n$. Thus, for each of the PTL-clauses (4.1), (4.2) and (4.5) there are $n + 1$ copies, one for each L_i.

$$\textbf{start} \Rightarrow \neg C \vee l \vee \neg L_i \tag{4.1}$$

$$\textbf{true} \Rightarrow \bigcirc(\neg C \vee l \vee \neg L_i) \tag{4.2}$$

$$\textbf{start} \Rightarrow \neg C \vee l \vee t \tag{4.3}$$

$$\textbf{true} \Rightarrow \bigcirc (\neg C \vee l \vee \text{t}) \tag{4.4}$$

$$\text{t} \Rightarrow \bigcirc (l \vee \neg L_i) \tag{4.5}$$

$$\text{t} \Rightarrow \bigcirc (l \vee \text{t}) \tag{4.6}$$

We note that only the resolvents (4.1), (4.2) and (4.5) depend on the particular loop being resolved with, that is, contain a reference to L_i. The clauses (4.3), (4.4) and (4.6) do not depend on the particular loop found and are relevant for *any* resolution with $C \Rightarrow \Diamond l$. So, typically, (4.3), (4.4) and (4.6) are added earlier in the resolution process.

Example 4.11 *Consider the situation where we have two candidates for applying the temporal resolution operation:*

$$(a \vee b \vee c) \Rightarrow \bigcirc \square \neg m$$
$$d \Rightarrow \Diamond m$$

The resolvent is

$$d \Rightarrow (\neg a \wedge \neg b \wedge \neg c) W m.$$

Now, if we rewrite this resolvent, the separate SNF clauses we get are as follows:

1.	**start** $\Rightarrow$	$\neg d \vee m \vee \neg a$
2.	**start** $\Rightarrow$	$\neg d \vee m \vee \neg b$
3.	**start** $\Rightarrow$	$\neg d \vee m \vee \neg c$
4.	**true** $\Rightarrow$	$\bigcirc (\neg d \vee m \vee \neg a)$
5.	**true** $\Rightarrow$	$\bigcirc (\neg d \vee m \vee \neg b)$
6.	**true** $\Rightarrow$	$\bigcirc (\neg d \vee m \vee \neg c)$
7.	**start** $\Rightarrow$	$\neg d \vee m \vee \text{x}$
8.	**true** $\Rightarrow$	$\bigcirc (\neg d \vee m \vee \text{x})$
9.	$\text{x} \Rightarrow$	$\bigcirc (m \vee \neg a)$
10.	$\text{x} \Rightarrow$	$\bigcirc (m \vee \neg b)$
11.	$\text{x} \Rightarrow$	$\bigcirc (m \vee \neg c)$
12.	$\text{x} \Rightarrow$	$\bigcirc (m \vee \text{x})$

where, again, x *is a new proposition.*

4.2.7 Review of the resolution process

Given any temporal formula, A, to be tested for unsatisfiability, the following steps are performed [208].

1. Translate A into SNF, giving A_S.

2. Perform step resolution (including simplification and subsumption) on A_S until either

   (a) **start** $\Rightarrow$ **false** is derived – terminate noting that A is unsatisfiable; or

   (b) no new resolvents are generated – continue to step (3).

3. Select an eventuality from the right-hand side of a sometime PTL-clause within A_S, for example $\Diamond l$. Search for loop formulae for $\neg l$.

4. Construct loop resolvents for the loop formulae detected and each sometime PTL-clause with $\Diamond l$ on the right-hand side. If any new formulae (i.e. that are not subsumed by PTL-clauses already present) have been generated, go to step (2).

5. If all eventualities have been resolved, that is no new formulae have been generated for any of the eventualities, terminate declaring A satisfiable, otherwise go to step (3).

The key formal results concerning the temporal resolution process are proved in [208] including:

- *soundness*, that is that if a contradiction is generated via clausal temporal resolution, then the original PTL formula was unsatisfiable;

- *completeness*, that is that if a PTL formula is unsatisfiable, then a contradiction will be derived by applying clausal temporal resolution to it; and

- *termination*, that is that any clausal temporal resolution process will terminate, regardless of its initial PTL formula.

Exercise 4.3 *What are the resolvents, in SNF, of the result of applying the temporal resolution rule as in Example 4.10 earlier.*

Example 4.12 *Assume we wish to show that the following set of PTL-clauses (already translated into SNF) is unsatisfiable.*

1. **start** $\Rightarrow$ f
2. **start** $\Rightarrow$ a
3. **start** $\Rightarrow$ p
4. f $\Rightarrow$ $\Diamond\neg p$
5. f $\Rightarrow$ $\bigcirc a$
6. a $\Rightarrow$ $\bigcirc(b \vee x)$
7. b $\Rightarrow$ $\bigcirc a$
8. b $\Rightarrow$ $\bigcirc p$
9. a $\Rightarrow$ $\bigcirc p$
10. a $\Rightarrow$ $\bigcirc\neg x$

Step resolution occurs as follows.

11. a $\Rightarrow$ $\bigcirc b$ [6, 10 *Step Resolution*]

By merging PTL-clauses (9) and (11), and (7) and (8) into SNF_m we obtain the following loop in p (in SNF_m form):

$$a \Rightarrow \bigcirc(b \wedge p) \quad [9, 11 \; SNF_m]$$
$$b \Rightarrow \bigcirc(a \wedge p) \quad [7, 8 \; SNF_m]$$

for resolution with PTL-clause (4). (Effectively, this gives us $(a \vee b) \Rightarrow \bigcirc \square p$.) *The resolvents after temporal resolution are PTL-clauses 12 – 20 below.*

12.	**start** $\Rightarrow$	$\neg f \vee \neg p \vee \neg a$	$[4, 7, 8, 9, 11$	*Temporal Resolution*]
13.	**true** $\Rightarrow$	$\bigcirc(\neg f \vee \neg p \vee \neg a)$	$[4, 7, 8, 9, 11$	*Temporal Resolution*]
14.	**start** $\Rightarrow$	$\neg f \vee \neg p \vee \neg b$	$[4, 7, 8, 9, 11$	*Temporal Resolution*]
15.	**true** $\Rightarrow$	$\bigcirc(\neg f \vee \neg p \vee \neg b)$	$[4, 7, 8, 9, 11$	*Temporal Resolution*]
16.	**start** $\Rightarrow$	$\neg f \vee \neg p \vee t$	$[4, 7, 8, 9, 11$	*Temporal Resolution*]
17.	**true** $\Rightarrow$	$\bigcirc(\neg f \vee \neg p \vee t)$	$[4, 7, 8, 9, 11$	*Temporal Resolution*]
18.	$t \Rightarrow$	$\bigcirc(\neg p \vee \neg a)$	$[4, 7, 8, 9, 11$	*Temporal Resolution*]
19.	$t \Rightarrow$	$\bigcirc(\neg p \vee \neg b)$	$[4, 7, 8, 9, 11$	*Temporal Resolution*]
20.	$t \Rightarrow$	$\bigcirc(l \vee t)$	$[4, 7, 8, 9, 11$	*Temporal Resolution*]

where t *is a new proposition. Now, the proof concludes as follows:*

21.	**start** $\Rightarrow$	$\neg f \vee \neg a$	$[3, 12$	*Initial Resolution*]
22.	**start** $\Rightarrow$	$\neg f$	$[2, 21$	*Initial Resolution*]
23.	**start** $\Rightarrow$	**false**	$[1, 22$	*Initial Resolution*]

A contradiction has been obtained; hence the original set of PTL-clauses is unsatisfiable.

4.2.8   Loop search

The most complex part of the clausal temporal resolution approach is the search for the set of SNF-clauses to use in the application of the temporal resolution operation. Detailed explanations of the techniques developed for this search are provided in [148, 150].

There are several different approaches to finding appropriate sets of clauses and, in order to describe some of these, we will use a running example. Recall the set of SNF clauses from Example 4.9:

1.	f	$\Rightarrow$	$\bigcirc \neg r$
2.	$(q \wedge \neg r)$	$\Rightarrow$	$\bigcirc \neg r$
3.	q	$\Rightarrow$	$\bigcirc p$
4.	p	$\Rightarrow$	$\bigcirc(f \vee q)$
5.	p	$\Rightarrow$	$\bigcirc \neg r$
6.	p	$\Rightarrow$	$\bigcirc g$
7.	$(f \wedge g)$	$\Rightarrow$	$\bigcirc(p \vee q)$

and assume that we are attempting to apply temporal resolution with a clause such as

$$\textbf{true} \Rightarrow \Diamond r \,.$$

We will now describe three approaches to finding appropriate loops.

Loop search via SNF$_m$ graphs

As we saw in Example 4.9, we can take all possible combinations of the SNF clauses to give a (large) set of SNF$_m$ clauses. Taking some small examples, we see that combining clauses (1) and (7) above gives us

$$(f \wedge g) \Rightarrow \bigcirc(\neg r \wedge (p \vee q)).$$

Combining clauses (4), (5) and (6) generates

$$p \Rightarrow \bigcirc((g \wedge f \wedge \neg r) \vee (g \wedge q \wedge \neg r))$$

and the combination of clauses (2) and (3) gives us

$$(q \wedge \neg r) \Rightarrow \bigcirc(p \wedge \neg r).$$

Now, we might represent these clauses as edges on a graph as follows.

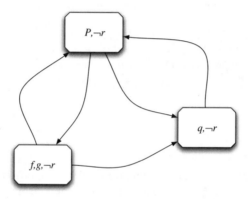

Here, each node in the graph is a conjunction of literals and each transition corresponds to some SNF$_m$ clause. So, this gives us a simple way to find appropriate loop formulae:

1. combine all SNF clauses together to get a large set of SNF$_m$ clauses;

2. represent the SNF clauses as transitions on a graph such as above;

3. search the graph for *strongly connected components* throughout which our target literal is true.

A strongly connected component is a subgraph where each transition leads to other nodes in the subgraph. In the case above, our target literal is '$\neg r$' and we see that Figure 4.1 describes a strongly connected component in which the target literal occurs at every node. Thus, we now know that

$$((f \wedge g \wedge \neg r) \vee (q \wedge \neg r) \vee (p \wedge \neg r)) \Rightarrow \bigcirc\Box\neg r$$

represents an appropriate loop to which we can apply the temporal resolution rule.

Loop search via dixon's breadth-first search

While the use of the above SNF$_m$-based search is possible, it is inefficient and unwieldy. Although searching for strongly connected components [13] will be quite quick, the number of potential SNF$_m$ clauses, and hence the graph, is *huge*. An alternative approach, and one that has led to the mechanisms used in contemporary clausal temporal resolution systems, was developed by Dixon [147, 150].

Rather than giving the full algorithm, we will simply provide some intuition of how this approach works by tackling the above problem. We begin by recognizing that, since we are looking for $\bigcirc\square\neg r$, the literal we need at every next state is $\neg r$. So, let us identify the SNF clauses within our original set that imply $\bigcirc\neg r$, namely:

$$
\begin{array}{rrcl}
1. & f & \Rightarrow & \bigcirc\neg r \\
2. & (q \wedge \neg r) & \Rightarrow & \bigcirc\neg r \\
5. & p & \Rightarrow & \bigcirc\neg r
\end{array}
$$

Thus, for $\neg r$ to be satisfied in the next state, we require

$$f \vee (q \wedge \neg r) \vee p$$

to be satisfied in the current state. Let us describe this graphically, as follows.

$$\boxed{f \vee (q \wedge \neg r) \vee p}$$

So, if we are to continue in a loop we know that we need one of f, $q \wedge \neg r$ or p to be satisfied. In addition, we still need $\neg r$ to be satisfied. So, how can we generate either $f \wedge \neg r$, $q \wedge \neg r$, or $p \wedge \neg r$? Well, we can examine SNF$_m$ combinations to see if we can achieve these. Specifically $f \wedge \neg r$, $q \wedge \neg r$ or $p \wedge \neg r$ all occur on the right-hand sides of the following combinations:

$$
\begin{array}{rrcl}
2+3: & (q \wedge \neg r) & \Rightarrow & \bigcirc(\neg r \wedge p) \\
4+5+6: & p & \Rightarrow & \bigcirc((f \wedge g \wedge \neg r) \vee (q \wedge g \wedge \neg r)) \\
1+7: & (f \wedge g) & \Rightarrow & \bigcirc(\neg r \wedge (p \vee q))
\end{array}
$$

So, we extend our graph:

$$\boxed{f \vee (q \wedge \neg r) \vee p}$$
$$\uparrow$$
$$\boxed{(f \wedge g) \vee (q \wedge \neg r) \vee p}$$

We continue now with the same procedure, checking what clauses together imply the above disjunctions while also implying $\neg r$. We again construct another node:

$$\boxed{f \vee (q \wedge \neg r) \vee p}$$
$$\uparrow$$
$$\boxed{(f \wedge g) \vee (q \wedge \neg r) \vee p}$$
$$\uparrow$$
$$\boxed{(f \wedge g) \vee (q \wedge \neg r) \vee p}$$

Once we have seen the same node twice we can stop, having detected a loop. Thus, if any of $f \wedge g$, $q \wedge \neg r$ or p occur, then $\bigcirc\square\neg r$ is guaranteed. Consequently, we can apply temporal resolution to

$$
\begin{array}{rcl}
\mathbf{true} & \Rightarrow & \Diamond r \\
((f \wedge g) \vee (q \wedge \neg r) \vee p) & \Rightarrow & \bigcirc\square\neg r
\end{array}
$$

This approach has been defined, proved correct and implemented [148, 150]. It has the advantage that we are here carrying out a *directed* search for clauses that are truly relevant, rather than looking at all combinations of clauses (as in the SNF$_m$ approach above), and that the steps at each stage are relatively straightforward. The whole graph construction process is guaranteed to terminate, and often requires only a few nodes to be constructed before an appropriate loop is found (loops are rarely very complex).

Mechanised loop search

The approach used in most contemporary clausal temporal resolution provers is essentially that of Dixon's search algorithm as above. However, an important subsequent development concerned the recognition that most of the steps in identifying nodes, generating combinations and recognizing termination could be carried out by utilizing classical automated reasoning systems. Thus, in [151], the Otter [370] system was used for this, and so clausal temporal resolution can now be implemented directly *on top* of a classical resolution system. As we will see in the next section, this is the case for TSPASS which is built on top of the SPASS automated reasoning system [456, 495, 496]. It is important to note, however, that since PTL is significantly more complex than classical propositional logic, quite a lot of extra work must potentially be carried out within the classical reasoning system.

In the next section we will look at the TSPASS system that implements clausal temporal resolution.

4.3   The TSPASS system

TSPASS [356, 357, 471] is an implementation of clausal temporal resolution based on the (classical) first-order theorem prover SPASS [495, 496, 456]. It actually implements clausal temporal resolution for *monodic* first-order temporal logic, but we will here deal only with the PTL aspects. (Recall that a monodic temporal formula is one whose temporal subformulae have, at most, one free variable.) TSPASS implements a number of simplifications and enhancements over the clausal temporal resolution approach described in Section 4.2, as follows.

- It uses the *simplified* clausal temporal resolution calculus [141].

- It is based on *ordered* resolution [33].

- It uses SPASS underneath, not only to carry out step resolution and simplification operations, but also to automatically search for loops that can be used in the temporal resolution operation (as described above).

We will explain these aspects in subsequent sections.

TSPASS was implemented by Michel Ludwig and is currently available via

> http://www.csc.liv.ac.uk/~michel/software/tspass

as well as at

> http://www.csc.liv.ac.uk/~michael/TLBook/TSPASS-System

4.3.1 Simplified clausal temporal resolution

The basic ideas behind clausal temporal resolution have been explained above. The key inference rule is the clausal temporal resolution rule which, as we have seen, generates quite a complex resolvent. However, it turns out that, if we take a simplified original problem, then we do not need *such* a complex clausal temporal resolution rule. In particular, in [141], it was shown that if an SNF problem

1. has only *one* eventuality (the *single eventuality* case), and

2. the sometime SNF-clause that contains that eventuality only has '**true**' on its left-hand side (the *unconditional eventuality* case),

then we only need the simplified clausal temporal resolution rule:

$$
\begin{aligned}
L_0 &\Rightarrow \bigcirc R_0 \\
\dots &\Rightarrow \dots \\
L_n &\Rightarrow \bigcirc R_n \\
\mathbf{true} &\Rightarrow \Diamond l \\
\hline
\mathbf{true} &\Rightarrow \bigwedge_{i=0}^{n} \neg L_i
\end{aligned}
$$

Notice, here, that the resolvent is actually just a non-temporal formula which we would previously have rewritten into SNF.

Example 4.13 *Let us begin with the simple set of SNF clauses:*

1. $a \Rightarrow \bigcirc b$
2. $a \Rightarrow \bigcirc \neg m$
3. $b \Rightarrow \bigcirc a$
4. $b \Rightarrow \bigcirc \neg m$
5. $\mathbf{true} \Rightarrow \Diamond m$

We can see that one loop is caused by '$a \lor b$', that is:

6. $(a \lor b) \Rightarrow \bigcirc \Box \neg m$.

Applying the simplified temporal resolution rule between (5) and (6) now gives us

7. $\mathbf{true} \Rightarrow \neg a \lor \neg b$.

A refined normal form

The fact that we deal above with only sometime-clauses of the form '$\mathbf{true} \Rightarrow \Diamond \dots$' and generate resolvents that are purely classical (i.e. not temporal) leads us to use an alternative (but equivalent) formulation of the normal form. This is used not only in more recent papers on clausal temporal resolution, but has been found to be particularly useful in developing clausal resolution for FOTL [143, 318] and is the basis of modern clausal resolution provers for PTL such as TSPASS. Thus, a *temporal problem* [141] comprises four sets of formulae, $\mathcal{I}$, $\mathcal{U}$, $\mathcal{E}$, and $\mathcal{S}$, where

$\mathcal{I}$ represents the *initial part* of the problem, and contains non-temporal formulae that should be satisfied at the first moment in time – thus where previously we had '**start** $\Rightarrow p$' we would just add 'p' to $\mathcal{I}$.

$\mathcal{U}$ represents the *universal part* of the problem, and contains non-temporal formulae that are *universally* true – if ever we want to say that some (non-temporal) property is globally true, for example '**true** $\Rightarrow (q \vee r)$', then we would previously have had both '**start** $\Rightarrow (q \vee r)$' and '**true** $\Rightarrow \bigcirc (q \vee r)$', but now we just add '$q \vee r$' to $\mathcal{U}$.

$\mathcal{E}$ is the *sometime part* of the problem, containing eventualities that must be satisfied infinitely often – since our sometime-clauses will now be of the form '**true** $\Rightarrow \lozenge s$' we just store '$\lozenge s$' in $\mathcal{E}$.

$\mathcal{S}$ is the *step part* of the problem, containing step clauses – each element of $\mathcal{S}$ is an SNF step clause, for example '$(a \wedge b) \Rightarrow \bigcirc (c \vee d)$'.

So, the full temporal problem is characterized by the formula

$$\bigwedge \mathcal{I} \wedge \square \bigwedge \mathcal{U} \wedge \square \bigwedge \mathcal{S} \wedge \square \bigwedge \mathcal{E}$$

where '$\bigwedge$' just means we conjoin together all elements of the subsequent set.

Example 4.14 *Imagine that we have transformed a PTL formula into SNF giving the following clauses (actually a simplified version of Example 4.5 with all the conditional sometime clauses changed to unconditional ones):*

1.	**start**	$\Rightarrow$	x_1
2.	**true**	$\Rightarrow$	$\lozenge p$
3.	x_3	$\Rightarrow$	$\bigcirc x_3$
4.	$(x_2 \wedge p)$	$\Rightarrow$	$\bigcirc p$
5.	**start**	$\Rightarrow$	$\neg x_1 \vee x_3$
6.	**true**	$\Rightarrow$	$\bigcirc(\neg x_1 \vee x_3)$
7.	**start**	$\Rightarrow$	$\neg x_3 \vee x_2$
8.	**true**	$\Rightarrow$	$\bigcirc(\neg x_3 \vee x_2)$
9.	**true**	$\Rightarrow$	$\lozenge \neg p$
10.	x_5	$\Rightarrow$	$\bigcirc x_5$
11.	**start**	$\Rightarrow$	$\neg x_1 \vee x_5$
12.	**true**	$\Rightarrow$	$\bigcirc(\neg x_1 \vee x_5)$
13.	**start**	$\Rightarrow$	$\neg x_5 \vee x_4$
14.	**true**	$\Rightarrow$	$\bigcirc(\neg x_5 \vee x_4)$

This might have alternatively been represented as the temporal problem:

$\mathcal{I} = \{x_1\}$

$\mathcal{U} = \{\neg x_1 \vee x_3, \neg x_3 \vee x_2, \neg x_1 \vee x_5, \neg x_5 \vee x_4\}$

> $\mathcal{E} = \{\Diamond \neg p, \Diamond p\}$
> $\mathcal{S} = \{x_3 \Rightarrow \bigcirc x_3, (x_2 \wedge p) \Rightarrow \bigcirc p, x_5 \Rightarrow \bigcirc x_5\}$
>
> *The two representations are equivalent, but the latter is clearly more concise.*

If we look at the new clausal temporal resolution rule introduced above, we can see why this problem form is convenient. In applying clausal temporal resolution, we just take several elements from $\mathcal{S}$, one element from $\mathcal{E}$ (there may indeed only be one) and generate a new element in $\mathcal{U}$. With this style of calculus we are required to carry out quite a lot of proof within the non-temporal sets $\mathcal{I}$ and $\mathcal{U}$, which explains why modern clausal temporal resolution provers are based on classical tools, such as Otter [151, 370], Vampire [436] and SPASS [495, 496, 456].

Now we turn back to the original simplified form. As stated above, if our problem has only *one* eventuality (the *single eventuality* case) and the sometime SNF-clause that contains that eventuality only has '**true**' on its left-hand side (the *unconditional eventuality* case), then we can use a simplified clausal resolution rule. Happily, we can transform *any* SNF problem (and, thus, any PTL problem) into this unconditional, single eventuality case [141]. Let us see how.

From conditional to unconditional eventualities

Let us first consider how we might transform a sometime clause of the form '$p \Rightarrow \Diamond q$' to a set of clauses in which the only sometime clause is of the form '**true** $\Rightarrow \Diamond \ldots$'. Let us recall what $p \Rightarrow \Diamond q$ means, intuitively. If p never becomes true, then there is nothing to satisfy. However, whenever p becomes true then we can see a sequence of states in which q is false before we see a state where q is true (and we will definitely see such a state). So, consider the following SNF clauses incorporating the *new* proposition *wait_for_q*.

$$(p \wedge \neg q) \Rightarrow wait\_for\_q$$
$$wait\_for\_q \Rightarrow \bigcirc (q \vee wait\_for\_q)$$
$$\textbf{true} \Rightarrow \Diamond \neg wait\_for\_q$$

It turns out that the above clauses are satisfied in exactly the same situations that $p \Rightarrow \Diamond q$ is satisfied at, and so we can replace $p \Rightarrow \Diamond q$ by the above [141]. Further intuition is as follows. The above states that *wait_for_q* must be *false* infinitely often. To ensure *wait_for_q* is false then, either p has not occurred or, if it has, q has subsequently been satisfied. The only way for $\square(\textbf{true} \Rightarrow \Diamond \neg wait\_for\_q)$ to fail is if, after some point, *wait_for_q* is always true. But for *wait_for_q* to be true, we must have had a p but not yet reached a q. Since, we originally had $\square(p \Rightarrow \Diamond q)$, then this situation cannot occur and so *wait_for_q* cannot go on being true forever.

From many eventualities to one

Now, how do we go from a problem with many eventualities to a problem with only one? The basic idea is that if we have several eventualities, for example $\Diamond a$, $\Diamond b$, and

$\Diamond c$, then there will be a point where a, b and c have *all* been satisfied. We just need to make sure that this point is eventually reached. If we require that $\Diamond a$, $\Diamond b$, and $\Diamond c$ keep on occurring, for example because we have **true** $\Rightarrow \Diamond a$, **true** $\Rightarrow \Diamond b$, and **true** $\Rightarrow \Diamond c$ then our 'success point' must keep on occurring (infinitely often).

Example 4.15 *Let us assume that we have the step clauses*

$$\text{true} \Rightarrow \Diamond a$$
$$\text{true} \Rightarrow \Diamond b$$
$$\text{true} \Rightarrow \Diamond c$$

and let us define a new *proposition that describes a point where a, b and c have occurred. Let us call this proposition 'success', and consider the following clauses (not yet in our required form).*

$$\textbf{start} \quad \Rightarrow \quad success$$
$$success \quad \Rightarrow \quad \Diamond \bigcirc ((\neg success \, \mathsf{S} \, a) \wedge (\neg success \, \mathsf{S} \, b) \wedge (\neg success \, \mathsf{S} \, c) \wedge success)$$

Informally, $f \, \mathsf{S} \, g$ means that g occurred sometime in the past and at all moments between that point and now (but not including either one) then f must hold. Thus, our above formula ensures that there is some point in the future where

- *success is satisfied, yet*

- *before now, success has not been satisfied since a became true,*

- *before now, success has not been satisfied since b became true, and*

- *before now, success has not been satisfied since c became true.*

Thus, 'success', marks a point when we know that all of 'a', 'b' and 'c' have occurred (the semantics of 'S' forces this) and, because we know we must again find another such point in the future, then 'a', 'b' and 'c' are guaranteed to occur infinitely often in the future. For example:

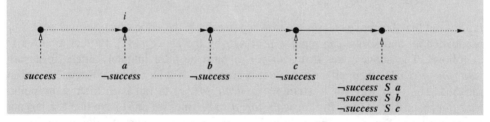

Once we have introduced formulae such as in the example above, we can again translate them into our normal form. However, there is now only *one* eventuality remaining.

Aside: Translating: 'S'.   The fact that we now have past-time operators in the example above might appear problematic, but such operators are easily translated into the normal form as follows. Consider '$x \Rightarrow (p\mathcal{S}\,q)$'. We translate this into the following set of SNF clauses (again requiring a little further rewriting), introducing a new proposition, in this case 's'.

$$
\begin{aligned}
\textbf{start} &\Rightarrow \neg\mathsf{s} \\
x &\Rightarrow \mathsf{s} \\
((q \lor (p \land \mathsf{s})) &\Rightarrow \bigcirc\mathsf{s} \\
\neg((q \lor (p \land \mathsf{s})) &\Rightarrow \bigcirc\neg\mathsf{s}
\end{aligned}
$$

We can, in addition, reduce multiple sometime clauses down to one such clause, even if the clauses themselves are conditional. The following is an example of this. (If the above was difficult to follow, this example will provide a little more intuition.)

Example 4.16 *Let us begin with '$\textbf{start} \Rightarrow (\Diamond a \land \Diamond b \land \Diamond c)$' and reduce to one eventuality. Similar to above, we can first introduce 'success' and generate*

$$
\begin{aligned}
\mathit{success} &\Rightarrow \neg\mathit{success}\,\mathcal{S}\,a \\
\mathit{success} &\Rightarrow \neg\mathit{success}\,\mathcal{S}\,b \\
\mathit{success} &\Rightarrow \neg\mathit{success}\,\mathcal{S}\,c \\
\textbf{start} &\Rightarrow \Diamond\mathit{success}
\end{aligned}
$$

Here, $\neg\mathit{success}\,\mathcal{S}\,b$ means that 'b' has occurred in the past and, since that point, success has always been false. Thus the above clauses ensure that, once success becomes true, it marks the first time that a, b, and c have all been true in the past. So, now all we do is say that, at some point in the future, success must become true. Thus, we can infer that a, b, and c have all become true by then.

4.3.2   Ordered resolution

Recall the basic (classical) resolution rule:

$$
\frac{\begin{array}{c} C_1 \lor p \\ C_2 \lor \neg p \end{array}}{C_1 \lor C_2}
$$

There are often very many clauses to which this resolution rule can be applied. How shall we decide which clauses to tackle first? There might even be choices of literals to resolve on. Traditionally, resolution-based systems had quite inflexible heuristics for choosing the candidate clauses and literals (see Appendix A). The introduction of *ordered* resolution [33, 178] provided a better basis for deciding on clauses and literals, and strengthened termination properties in first-order versions of resolution. Essentially, an ordering is provided on literals which aids the choice of clauses to be resolved. Technically, TSPASS uses ordered resolution with *selection*, where this selection is a complementary mechanism for directing the search for a refutation [356]. Ordered resolution with selection is now the predominant approach used in first-order resolution-based systems (such as SPASS).

4.3.3 Loop search using SPASS

We can use SPASS to carry out *loop search* as well as more obvious resolution operations. As described earlier, and in [151], classical automated reasoning can be used for this purpose. The basic idea is as follows.

At each step in the loop search algorithm, we can generate classical (often propositional) logic clauses that can then be fed to our classical prover. Recall that in the loop search algorithm (for example for $\bigcirc \Box \neg r$) we begin by searching for clauses that imply $\bigcirc \neg r$. So, we can translate all the SNF step clauses, for example by

$$
\begin{array}{llll}
f \Rightarrow \bigcirc \neg r & \text{becomes} & f \Rightarrow \neg next\_r \\
(q \wedge \neg r) \Rightarrow \bigcirc \neg r & \text{becomes} & (q \wedge \neg r) \Rightarrow \neg next\_r \\
p \Rightarrow \bigcirc \neg r & \text{becomes} & \Rightarrow \neg next\_r
\end{array}
$$

Then we add the clause *found* $\Rightarrow$ *next_r* and pass all of these to our classical prover. If we can derive a clause of the form '$P \vee \neg found$' containing no '*next_r*' literals, then the SNF clause '$P \Rightarrow \bigcirc \neg r$' exists. And so on. TSPASS works in a more sophisticated way, but still essentially translates the loop search process into a refutation process within SPASS.

4.3.4 TSPASS in action: translation

We will look first at one component of TSPASS, namely fotl-translate, which translates arbitrary PTL (and monodic FOTL) formulae into SNF (actually into the temporal problem form above).

Let us begin with the formula given in Example 4.3, namely '$p \Rightarrow \Box q$'. Now, applying fotl-translate to 'p => (always q)' gives the output

```
begin_problem(TranslatedInput).
list_of_descriptions.
name({*Translated Input*}).
author({**}).
status(unknown).
description({**}).
end_of_list.

list_of_symbols.
predicates[(_P,0),(_R,0),(p,0),(q,0)].
end_of_list.

list_of_formulae(axioms).
formula(or(not(p),_P)).
formula(always(implies(_P,_R))).
formula(always(implies(_R,next(_R)))).
formula(always(implies(_R,q))).
end_of_list.

end_problem.
```

Some explanation of this is as follows.

- We begin with some initial information which is not important to us here. The interesting data begins at `list_of_symbols`, which defines the propositions we will see in the subsequent clauses. As well as p and q, both with arity 0, we see two new propositions, _P and _R.

- The translated SNF clauses are given in `list_of_formulae(axioms)`. Here, these four formulae correspond to

$$\begin{aligned} \textbf{start} &\Rightarrow \neg p \vee \_P \\ \_P &\Rightarrow \_R \\ \_R &\Rightarrow \bigcirc \_R \\ \_R &\Rightarrow q \end{aligned}$$

Although, in this case, the use of _P is slightly redundant, we can see how these SNF clauses correspond to our translation from Example 4.3 – just consider _R as z in Example 4.3 earlier.

Example 4.17 *Just to confirm our understanding of the temporal problem form that TSPASS utilizes, let us input $\Box \Diamond e$ to* `fotl-translate`*. Happily, this gives us*

```
list_of_formulae(axioms).
formula(always(sometime(e))).
end_of_list.
```

Next we translate an '*U*' formula, by using 'm =>(p until q)' as input to

```
fotl-translate  --extendedstepclauses
```

Note that the '--extendedstepclauses' flag here allows step-clauses to contain more than one literal on each side of the implication (the default is just one literal on each side). Again, after some pre-amble, we get

```
list_of_symbols.
predicates[(_P,0),(_S,0),(_waitforq,0),(m,0),(p,0),(q,0)].
end_of_list.
```

defining three new predicates beyond m, p and q, namely _P, _S, and _waitforq. And then the clauses:

```
list_of_formulae(axioms).
formula(or(not(m),_P)).
formula(always(implies(_P,or(p,q)))).
formula(always(implies(_P,or(_S,q)))).
formula(always(implies(_S,next(or(p,q))))).
formula(always(implies(_S,next(or(_S,q))))).
formula(always(implies(and(_P,not(q)),_waitforq))).
```

```
formula(always(implies(_waitforq,next(or(_waitforq,q))))).
formula(always(sometime(not(_waitforq)))).
end_of_list.
```

We can write the output here in our more recognizable form, as follows (we also indicate which sets within the 'temporal problem' form we expect each clause to appear in).

$$
\begin{array}{rcl}
\textbf{start} & \Rightarrow & (\neg m \vee \_P) \qquad\qquad\; [\mathcal{I}]\\
\_P & \Rightarrow & p \vee q \qquad\qquad\qquad\; [\mathcal{U}]\\
\_P & \Rightarrow & \_S \vee q \qquad\qquad\qquad [\mathcal{U}]\\
\_S & \Rightarrow & \bigcirc(p \vee q) \qquad\qquad [\mathcal{S}]\\
\_S & \Rightarrow & \bigcirc(\_S \vee q) \qquad\quad\; [\mathcal{S}]\\
(\_P \wedge \neg q) & \Rightarrow & \_\mathtt{waitforq} \qquad\qquad [\mathcal{U}]\\
\_\mathtt{waitforq} & \Rightarrow & \bigcirc(\_\mathtt{waitforq} \vee q) \;\; [\mathcal{S}]\\
\textbf{true} & \Rightarrow & \Diamond\neg\_\mathtt{waitforq} \qquad\;\; [\mathcal{E}]
\end{array}
$$

> **Exercise 4.4** *Try translating 'start* $\Rightarrow (a\, \mathcal{U}\, b)$*'. Can you explain each of the formulae in the resulting temporal problem?*

Our next example shows how a non-global eventuality is transformed into a global one. Consider 't = > sometime s' and again apply fotl-translate, giving:

```
list_of_symbols.
predicates[(_P,0),(_waitfors,0),(s,0),(t,0)].
end_of_list.

list_of_formulae(axioms).
formula(or(not(t),_P)).
formula(always(implies(and(_P,not(s)),_waitfors))).
formula(always(implies(_waitfors,next(or(_waitfors,s))))).
formula(always(sometime(not(_waitfors)))).
end_of_list.
```

This corresponds to the clause set

$$
\begin{array}{rcl}
\textbf{start} & \Rightarrow & \neg t \vee \_P\\
(\_P \wedge \neg s) & \Rightarrow & \_\mathtt{waitfors}\\
\_\mathtt{waitfors} & \Rightarrow & \bigcirc(\_\mathtt{waitfors} \vee s)\\
\textbf{true} & \Rightarrow & \Diamond\neg\_\mathtt{waitfors}
\end{array}
$$

Notice how _waitfors is generated even in this relatively simple example.

Our final translation example is more complex. We begin by taking a formula involving two eventualities ($\Diamond f$ and $\Diamond g$) and produce a new set of clauses with only one (global) eventuality. So, beginning with $\square(\Diamond f \wedge \Diamond g)$ and running

```
fotl-translate --extendedstepclauses--singleeventuality
```

on input 'always ((sometime f) & (sometime g))' we get a set of additional propositions:

$$\left\{ \begin{array}{l} \texttt{\_P, \_P1, \_P2, \_P3, \_P4, \_P5, \_P6, \_P7, \_U, \_1, \_since\_waitforf,} \\ \texttt{\_since\_waitforg\_1, \_waitfor\_P6\_2, \_waitforf, \_waitforg\_1} \end{array} \right\}$$

Together with the (quite complex) set of clauses:

```
list_of_formulae(axioms).
formula(always(and(_P,_P1))).
formula(always(implies(and(_P,not(f)),_waitforf))).
formula(always(implies(_waitforf,next(or(_waitforf,f))))).
formula(always(implies(and(_P1,not(g)),_waitforg_1))).
formula(always(implies(_waitforg_1,next(or(_waitforg_1,g))))).
formula(not(_since_waitforf)).
formula(not(_since_waitforg_1)).
formula(_1).
formula(always(or(and(_waitforf,
        or(_1,not(_since_waitforf))),_P2))).
formula(always(or(or(not(_waitforf),
        and(not(_1),_since_waitforf)),_P3))).
formula(always(or(and(_waitforg_1,
        or(_1,not(_since_waitforg_1))),_P4))).
formula(always(or(or(not(_waitforg_1),
        and(not(_1),_since_waitforg_1)), _P5))).
formula(always(implies(_1,_P7))).
formula(always(implies(_P2,next(_since_waitforf)))).
formula(always(implies(_P3,next(not(_since_waitforf))))).
formula(always(implies(_P4,next(_since_waitforg_1)))).
formula(always(implies(_P5,next(not(_since_waitforg_1))))).
formula(always(implies(_P6,next(_U)))).
formula(always(implies(_U,and(and(_since_waitforg_1,_since_
        waitforf),_1)))).
formula(always(implies(and(_P7,not(_P6)),_waitfor_P6_2))).
formula(always(implies(_waitfor_P6_2,
        next(or(_waitfor_P6_2,_P6))))).
formula(always(sometime(not(_waitfor_P6_2)))).
end_of_list.
```

or, in more familiar form:

$\mathcal{I} = \{\neg\_\texttt{since\_waitforf}, \neg\_\texttt{since\_waitforg\_1}, \_1\}$

$\mathcal{E} = \{\Diamond\neg\_\texttt{waitfor\_P6\_2}\}$

$\mathcal{S} =$

$$\left\{ \begin{array}{rcl} \texttt{\_waitforf} & \Rightarrow & \bigcirc(\texttt{\_waitforf} \lor f) \\ \texttt{\_waitforg\_1} & \Rightarrow & \bigcirc(\texttt{\_waitforg\_1} \lor g) \\ \texttt{\_P2} & \Rightarrow & \bigcirc\_\texttt{since\_waitforf} \\ \texttt{\_P3} & \Rightarrow & \bigcirc\neg\_\texttt{since\_waitforf} \\ \texttt{\_P4} & \Rightarrow & \bigcirc\_\texttt{since\_waitforg\_1} \\ \texttt{\_P5} & \Rightarrow & \bigcirc\neg\_\texttt{since\_waitforg\_1} \\ \texttt{\_P6} & \Rightarrow & \bigcirc\_\texttt{U} \\ \texttt{\_waitfor\_P6\_2} & \Rightarrow & \bigcirc(\texttt{\_waitfor\_P6\_2} \lor \texttt{\_P6}) \end{array} \right\}$$

$\mathcal{U} =$

$$\left\{ \begin{array}{c}
(\_P \land \_P1) \\
(\_P \land \neg f) \Rightarrow \_waitforf \\
(\_P1 \land \neg g) \Rightarrow \_waitforg\_1 \\
\_waitforf \land ((\_1 \lor \neg\_since\_waitforf) \lor \_P2) \\
((\neg\_waitforf \lor (\neg\_1 \land \_since\_waitforf)) \lor \_P3) \\
(\_waitforg\_1 \land (\_1 \lor \neg\_since\_waitforg\_1)) \lor \_P4 \\
(\_waitforg\_1 \lor (\neg\_1 \land \_since\_waitforg\_1) \lor \_P5) \\
\_1 \Rightarrow \_P7 \\
\_U \Rightarrow (\_since\_waitforg\_1 \land \_since\_waitforf \land \_1) \\
(\_P7 \land \neg\_P6) \Rightarrow \_waitfor\_P6\_2
\end{array} \right\}$$

It is not vital to understand all the detail of the above. However, this example does exhibit how quickly the translated version becomes difficult to understand. In spite of this, if you go through the above in detail, following the translations we have mentioned earlier in this chapter, the set of clauses produced does, indeed, make sense.

Exercise 4.5 *Recall Example 4.5 earlier, namely the translation of*

$$\Diamond p \land \Box(p \Rightarrow \bigcirc p) \land \Box \Diamond \neg p$$

into SNF. Now apply fotl-translate *to this and explain how the output corresponds to the SNF clauses given in Example 4.5.*

4.3.5 TSPASS in action: step resolution

Recall Exercise 4.2 where we attempted to show the validity of the PTL axiom

$$\vdash \bigcirc(a \Rightarrow b) \Rightarrow (\bigcirc a \Rightarrow \bigcirc b)$$

by negating, translating to SNF, and applying step resolution. Let us now use TSPASS to achieve this. Negating the above gives

$$\bigcirc(a \Rightarrow b) \land \bigcirc a \land \bigcirc \neg b.$$

So, when we input

```
always (next (a => b)   &  (next a)   &  (next   b))
```

to fotl-translate --extendedstepclauses we get new propositions _P, _P1, and _P2, and the temporal problem

```
list_of_formulae(axioms).
formula(always(and(_P,and(_P1,_P2)))).
formula(always(implies(_P,next(or(not(a),b))))).
formula(always(implies(_P1,next(a)))).
formula(always(implies(_P2,next(not(b))))).
end_of_list.
```

Aside: grouping step clauses.   Note that adding '--regroupnext' as an additional flag to the fotl-translate procedure produces the following set of clauses (the flag tries to group several 'next' operators together into one):

```
list_of_formulae(axioms).
formula(always(implies(true,next(_U)))).
formula(always(implies(_U,and(or(not(a),b),and(a,not(b)))))).
end_of_list.
```

Now if we simply feed the above (ungrouped) temporal problem to TSPASS, we get (among a *lot* of other detail)

```
. . . . .
TSPASS 0.92
SPASS beiseite: Unsatisfiable.
Problem: translate_step_axiom.out
TSPASS derived 5, backtracked 0, and kept 10 clauses.
Number of input clauses: 6
Number of eventualities: 0
Total number of generated clauses: 11
. . . . . .
```

So, the original set of clauses is unsatisfiable. Slightly more informative output can be generated using the '-DocProof' flag. Here, output includes

```
Here is a proof with depth 5, length 11 :
1[0:Inp:LS] ||    -> _P(U)*.
2[0:Inp:LS] ||    -> _P1(U)*.
3[0:Inp:LS] ||    -> _P2(U)*.
4[0:Inp:LS] || _P1(U) -> a(temp_succ(U))*.
5[0:Inp] || _P2(U) b(temp_succ(U))* ->.
6[0:Inp:LS] || _P(U) a(temp_succ(U))* -> b(temp_succ(U)).
7[0:Res:4.1,6.1:LS] || _P1(U) _P(U) -> b(temp_succ(U))*.
8[0:Res:7.2,5.1] || _P1(U) _P(U)* _P2(U) ->.
9[0:Res:1.0,8.1] || _P1(U)* _P2(U) ->.
10[0:Res:2.0,9.0:LS] || _P2(U)* ->.
11[0:Res:3.0,10.0] ||    ->.
```

This proof can be described in a more recognizable, and considerably clearer, format as follows.

1.   **true** $\Rightarrow$ _P
2.   **true** $\Rightarrow$ _P1
3.   **true** $\Rightarrow$ _P2
4.   _P1 $\Rightarrow$ $\bigcirc a$
5.   _P2 $\Rightarrow$ $\bigcirc \neg b$
6.   _P $\Rightarrow$ $\bigcirc(\neg a \vee b)$
7. (_P1 $\wedge$ _P) $\Rightarrow$ $\bigcirc b$ [4, 6   Step Resolution]

8. **true** $\Rightarrow$ $\neg\_P1 \vee \neg\_P \vee \neg\_P2$ [5, 7 Step Resolution; Simplification]
9. **true** $\Rightarrow$ $\neg\_P1 \vee \neg\_P2$ [1, 8 Resolution]
10. **true** $\Rightarrow$ $\neg\_P2$ [2, 9 Resolution]
11. **true** $\Rightarrow$ **false**   [3, 10 Resolution]

Aside: Running TSPASS.   Rather than running `fotl-translate` and TSPASS separately, a shell script, `run-tspass.sh`, runs both while passing appropriate flags on to each one.

4.3.6   TSPASS in action: temporal resolution

Now we will move on to some examples involving temporal resolution, and therefore some loop detection. Consider the following input to TSPASS describing an unsatisfiable set of clauses.

```
always(a => next a)
& always(a => next ~m)
    & always(sometime m)
        & a
```

Running TSPASS does indeed indicate unsatisfiability. Again, more informative output can be provided using the '-DocProof' flag. Here, the output includes:

```
Here is a proof with depth 1, length 3 :
1[0:Inp:LS] ||  -> a(temp_zero)*.
15[0:LoopSearch::LS] || a(U)* ->.
17[0:Res:1.0,15.0] ||  ->.
```

Thus, the proof involves

1. '*a*' is true at the start, thus `a(temp_zero)`

15. an '*a*' anywhere leads to a loop in contradiction to our global eventuality, so we must add '$\neg a$' to $\mathcal{U}$, and

17. straightforward resolution between (1) and (15) gives **false**.

Bigger examples with conjunctive loops, for example, $(a \wedge b) \Rightarrow \bigcirc \Box \neg m$, are not so informative since TSPASS typically replaces '$a \wedge b$' by a new proposition and then, in the proof produced, we just see a loop in that new proposition. Similarly, step clauses such as '$x \Rightarrow \bigcirc (b \vee c)$' are typically renamed to '$x \Rightarrow \bigcirc y$' with '$y \Rightarrow (b \vee c)$' being added to $\mathcal{U}$. So, it is sometimes hard to see exactly what is happening with the original propositions.

Thus, trying the negation of our induction axiom, that is,

$$p \wedge \Box(p \Rightarrow \bigcirc p) \wedge \Diamond \neg p$$

within TSPASS simply gives a loop in p, then a straightforward contradiction with '**start** $\Rightarrow p$'.

Exercise 4.6 *Try proving*

$$(\Diamond p \wedge \Box(p \Rightarrow \bigcirc \Diamond p)) \Rightarrow \Box \Diamond p$$

that is negate, giving

$$\Diamond p \wedge \Box(p \Rightarrow \bigcirc \Diamond p) \wedge \Diamond \Box \neg p$$

and input this to TSPASS.

More details about TSPASS together with bigger examples can all be found at the TSPASS website

A final comment about TSPASS is that it is both fast[1] and reliable, and so provides an appropriate tool both for learning about temporal proof and for tackling more sophisticated applications.

4.3.7 Model construction in TSPASS

The idea with refutation-based systems is that we are usually hoping to find a contradiction. In such a case, this means that the clauses are unsatisfiable and, typically, that the unnegated original formula is valid. So, often, a contradiction is what we desire and, in such a case, TSPASS will provide a 'proof' showing how the contradiction was derived. However, what if the expected contradiction is *not* found? What shall we do then? Fortunately, TSPASS also provides us with a way to visualize a satisfiable set of clauses.

Again, rather than explaining the underlying theory, let us look at an example.

At one point during the writing of this book, we wanted to show how TSPASS established that the formula

$$(\Diamond p \wedge \Box(p \Rightarrow \Diamond p)) \Rightarrow \Box \Diamond p$$

is valid. This was to be an example of the typical PTL induction schema: p would eventually be true; whenever p is satisfied then p would eventually be true again; and so we know p will be true infinitely often. So, we input

```
(sometime p)  & always(p => sometime p)  & (sometime always ~p)
```

to TSPASS, expecting a contradiction. However, TSPASS reported that the formula is satisfiable!

Fortunately, TSPASS has the ability to construct (finite) models of satisfiable PTL formulae. So, calling TSPASS with the flag '--ModelConstruction', the following output was produced describing a simple reduced model for the input formula:

```
Temporal model constructed:
------------- 0
{}
------------- 1
```

---

[1] Of course, we mean 'fast' for a temporal prover!

```
{p}
------------- 2
{ }
----------> 2
```

In the above, p can be false in state 0, true in state 1 and then false again in state 2. After that point the transitions loop on state 2. This tells us that p just occurring once (e.g. in state 1) is enough to satisfy the formula. When we look back at the formula this becomes clear.

$$(\Diamond p \wedge \Box(p \Rightarrow \Diamond p)) \Rightarrow \Box\Diamond p$$

is *not* valid after all. The subformula '$p \Rightarrow \Diamond p$' is trivially true, so we are left with '$\Diamond p \Rightarrow \Box\Diamond p$' which is certainly not valid!

Our mistake was that we missed out a '$\bigcirc$' operator. The real formula we *meant* was

$$(\Diamond p \wedge \Box(p \Rightarrow \bigcirc\Diamond p)) \Rightarrow \Box\Diamond p$$

Thus the *model construction* aspect of TSPASS is very useful. As we will see in Chapter 6, constructing a model from a temporal formula is very productive even beyond the clausal resolution approach. Indeed, in that chapter, we will look at how such model construction can be seen as *execution* of the base formula.

4.4 Advanced topics

4.4.1 Clausal temporal resolution: extensions

The basic clausal temporal resolution technique has been developed and extended over a number of years. Three particular directions have been explored.

- *Branching-time temporal logics*
  Bolotov *et al.* [71, 72, 73, 74, 75] have developed and extended varieties of clausal temporal resolution specifically for branching temporal logics, such as CTL. Recently, Zhang *et al.* [507] have provided an implementation of this approach.

- *Modal logics*
  The basic idea of translating to an SNF-like normal form and then applying simple resolution rules has been extended to various modal logics [389] and even to combined temporal and modal logics [155, 153, 297, 157]. Typically, the normal form is extended, for example with clauses concerning a particular modal operator, M:

$$\bigwedge_j l_j \;\Rightarrow\; \mathsf{M}\bigvee_i r_i$$

where l_j and r_i are literals. Then various resolution rules are added corresponding to the properties of the modal operator. Most simply we might have:

$$
\begin{array}{rcl}
L_1 & \Rightarrow & \mathsf{M}(R_1 \vee p) \\
L_2 & \Rightarrow & \mathsf{M}(R_2 \vee \neg p) \\
\hline
(L_1 \wedge L_2) & \Rightarrow & \mathsf{M}(R_1 \vee R_2)
\end{array}
$$

but in some cases we might also have rules such as

$$\frac{\begin{array}{rcl} L_1 & \Rightarrow & \mathsf{M}(R_1 \lor p) \\ L_2 & \Rightarrow & (R_2 \lor \neg p) \end{array}}{(L_1 \land L_2) \quad \Rightarrow \quad \mathsf{M}(R_1 \lor R_2)}$$

for example if the modal operator 'M' is reflexive. And so on. The SNF style turns out to be quite flexible and modular, allowing clauses from several different logics to be collected together.

- *First-order temporal logics*
  The clausal resolution approach has been extended to first-order temporal logics, specifically the *monodic* classes introduced by Hodkinson *et al.* [280]. Applying clausal temporal resolution to such logics is significantly more difficult than PTL, but has now reached a stage where implementations can be produced [187, 143]. The current state-of-the-art here is the TSPASS system [356, 357, 471], based on the *fine-grained* temporal resolution approach described in [318].

4.4.2 Clausal temporal resolution: implementations and refinements

The resolution method has been implemented a number of times [147–151, 154,] culminating in the TRP++ [293], TeMP [298], and now TSPASS [356, 357, 471] systems. Yet temporal logics remain complex and so deductive tools are necessarily slow in some cases. This has led to a number of refinements, both of the theory and the implementation techniques, which aim to speed the process up in certain cases. Thus, as we saw for TSPASS, the resolution process itself has been simplified [141].

In addition, various strategies have been developed to refine the process [152, 156, 179, 198]. These strategies are particularly targetted at making loop finding quicker. The basic idea, particularly within [179], is that since

$$\Diamond p \equiv p \lor \bigcirc \Diamond p$$

then, if we can analyse the clause set and work out a limit by which 'p' must occur (if it is to occur at all!) then we can effectively replace '$\Diamond p$' by

$$p \lor \bigcirc p \lor \bigcirc \bigcirc p \lor \ldots \bigcirc^n p$$

in the refutation.

Finally, a refined class of temporal logics is being developed which has *much* better complexity results and promises significant improvements for the future [158]. The basic idea here is that, in practical problems, we often identify a set of propositions, only one of which can occur at any moment in time. Thus, if we consider location information of a person in a house, then exactly one of

$$\{in\_kitchen, in\_hall, in\_bathroom, in\_bedroom, in\_living\_room\}$$

is true at any temporal state. We can often identify many such sets. Happily, the complexity of deciding a temporal logic based on such sets is *much* improved (sometimes even polynomial) both in the propositional [161] and first-order cases [162]. (See Section 4.4.7 for more on this.)

4.4.3 Temporal tableaux

One of the earliest approaches to deciding the truth of temporal formulae was semantic tableaux [500]. As in the classical case, this approach analyses the formula under consideration and builds a structure representing all the possibilities. In the case of temporal logic, this structure is a graph rather than a tree. As in the classical case, nodes are pruned under certain circumstances (e.g. contradictory information). However, in the temporal case, whole subgraphs may be removed. The structures constructed are closely related to the Büchi Automata as we have seen already. As well as work on tableau methods [501], a number of implementations have been produced, from early implementations, such as DP [245], to more sophisticated *one-pass* approaches [450] within the Logics Workbench [38, 352] and implementations of Wolper's approach such as [243, 355]. Tableau-based techniques are popular [228, 477] and modern implementations remain competitive.

4.4.4 Non-clausal temporal resolution

As mentioned earlier, it was the *non-clausal* proof method that was developed quite a few years ago. Specifically, Abadi and Manna describe non-clausal resolution rules for PTL [4], and then extended this to first-order temporal logic [6].

4.4.5 Axiomatic approaches

Given that there are a number of complete axiomatizations for PTL, for example [223], then an obvious approach is to attempt direct proofs using axioms and inference rules. This has been tried, as have other approaches such as Translation Methods [392, 394], Sequent Systems (also with ω rules) [408, 409] and Gentzen Systems [401], but to little practical effect. Since the mechanization of PTL using axioms and inference rules does not help us with first-order temporal logics and because these approaches are more complex to implement than automata, resolution or tableau-based algorithms, they remain relatively unpopular. However, a promising development is the introduction of a Natural Deduction calculus, and associated proof search techniques, for PTL [77, 78].

4.4.6 Larger temporal environments

The *Stanford Temporal Prover* (STeP system) [66], based on ideas presented in [363, 364], and providing both model checking and deductive methods for PTL-like logics, is a complex environment that has been used in order to assist the verification of concurrent and reactive systems based on temporal specifications. STeP was developed

> "to support the computer-aided formal verification of reactive, real-time and hybrid systems based on their temporal specification. Unlike most systems for temporal verification, STeP is not restricted to finite-state systems, but combines model checking with deductive methods to allow the verification

of a broad class of systems, including parameterized (N-component) circuit designs, parameterized (N-process) programs, and programs with infinite data domains."

The STeP system is available at `http://www-step.stanford.edu`.

Another well-developed alternative is the TLA approach [339]. As mentioned earlier in this book, TLA uses a simpler format of temporal logic, which then allows the introduction of stronger first-order features. Specification and verification based on TLA [108] are popular, and several tools are available, including

- SANY – a syntax analyser for TLA specifications,

- TLC – a model-checker for TLA specifications, and

- TLATeX – a LaTeX typesetter for TLA.

Current status of these tools is available at

`http://research.microsoft.com/en-us/um/people/lamport/tla/tools.html`

4.4.7   Deductive verification

Is there a future for deductive temporal verification? [158]. Seemingly the complexity of proof, even in PTL, is too much for practical verification to be envisaged. Indeed, *model-checking* is by far the most popular approach to verification. But why? Part of this popularity is due to the large amount of research and development into implementation techniques that has been carried out. However, an obvious problem in the underlying computational complexity of proof in PTL is PSPACE, while model checking can sometimes be tractable (if we check simple classes of properties). How can deductive verification hope to compete?

This begs the question: *can we improve complexity of temporal deduction?* An answer is: *yes!* As mentioned above, we can improve the complexity of the decision procedure for PTL if we can identify propositions which *cannot* be true together. Usually within PTL we have a set of propositions and, at each moment in time, any subset of these propositions can be true. It turns out that, if we partition our propositions into

- a *constrained* set, X, from which exactly one proposition is satisfied at any moment in time, together with

- an *unconstrained* set, U, acting as normal,

then, for this logic, the decision problem is *polynomial* in size of X and exponential in size of U [161]. This also extends to FOTL [162]! Thus, if we can find reasonably sized X sets, then the complexity of proof can become more manageable. The key problem is now: *engineering the use of such X sets*. At present the partition of the problem into constrained and unconstrained sets must be carried out by the specifier.

Example 4.18 *Consider the simple finite state machine:*

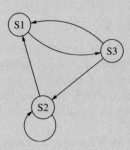

We could represent this in PTL by:

1. $In_{S1} \Rightarrow \bigcirc In_{S3}$
2. $In_{S2} \Rightarrow \bigcirc(In_{S1} \vee In_{S2})$
3. $In_{S3} \Rightarrow \bigcirc(In_{S1} \vee In_{S2})$
4. $In_{S1} \Rightarrow (\neg In_{S2} \wedge \neg In_{S3})$
5. $In_{S2} \Rightarrow (\neg In_{S1} \wedge \neg In_{S3})$
6. $In_{S3} \Rightarrow (\neg In_{S1} \wedge \neg In_{S2})$

Now, if we define $X = \{In_{S1}, In_{S2}, In_{S3}\}$, where no two propositions within X can be true at the same time, then actually we can dispense with formulae 4, 5 and 6 above!

So this form of temporal logic certainly helps with succinctness, but it also helps with deductive verification.

Some intuition – composing finite state machines: One of the central aspects of algorithmic verification techniques such as model checking is the composition of automata. This often produces an exponentially larger product automaton – this is termed the *state explosion problem*. Let us see what happens when we combine descriptions of two finite state machines in our constrained temporal logic.

Consider the two finite state machines [158]:

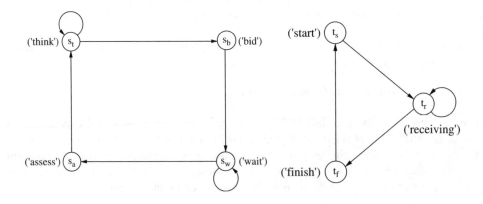

We can give temporal formulae for each of the state machines, for example, as above, and can even include (SNF) clauses ensuring that certain states are eventually reached. Then, when we take the synchronous product of the two state machines we simply take the union of the sets of clauses:

1.	$\textbf{start} \Rightarrow s_t$	2.	$s_t \Rightarrow \bigcirc (s_t \lor s_b)$
3.	$s_b \Rightarrow \bigcirc s_w$	4.	$s_a \Rightarrow \bigcirc s_t$
5.	$s_w \Rightarrow \bigcirc (s_w \lor s_a)$	6.	$\textbf{true} \Rightarrow \Diamond \neg s_t$
7.	$\textbf{start} \Rightarrow t_s$	8.	$t_s \Rightarrow \bigcirc t_r$
9.	$t_r \Rightarrow \bigcirc (t_r \lor t_f)$	10.	$t_f \Rightarrow \bigcirc t_s$

Then the decision problem for such sets of clauses is still low in comparison with unconstrained problems [158]! As you can see, however, such examples start to blur the distinction between deductive verification and the algorithmic verification approaches that we will examine next.

4.5   Final exercises

Exercise 4.7 *Use* `fotl-translate` *to translate* $\textbf{start} \Rightarrow ((\Diamond g) \land (\Diamond h))$. *Explain the resulting clauses.*

Exercise 4.8 *Recall the following set of formulae from Chapter 1.*

  (a) $\Box(try\_to\_print \Rightarrow \Diamond \neg try\_to\_print)$

  (b) $\Box(try\_to\_print \Rightarrow \bigcirc(printed \lor try\_to\_print))$

  (c) $\Box(printed \Rightarrow \bigcirc \Box \neg try\_to\_print)$

  (d) $\textbf{start} \Rightarrow try\_to\_print$

We there suggested that these formulae, together with

$$\Box \Diamond try\_to\_print .$$

is inconsistent.

  Establish this by using clausal temporal resolution, either by hand or by invoking TSPASS.

Exercise 4.9 *Recall the example of renaming from Example 4.2, namely the formula* '$\textbf{start} \Rightarrow (\bigcirc \Box p \land \Diamond \Box p)$'. *Input this into* `fotl-translate` *and see if '$\Box$' is renamed using two different new propositions or one.*

5

Model checking

The future, according to some scientists, will be exactly like the past, only far more expensive.
   – John Sladek

In this Chapter, we will:

- [INTRODUCTION] develop the idea of checking the satisfiability of a temporal formula over an appropriate structure (Section 5.1);

- [TECHNIQUE] use *Büchi Automata* to understand and analyse automated satisfiability checking of temporal formulae, also called *model checking* (Section 5.2);

- [SYSTEM] examine the `Spin` system which implements automata-theoretic model checking over execution structures generated from `Promela` programs (Section 5.3); and

- [ADVANCED] highlight a selection of more advanced work including alternative model-checkers, different logics and extended techniques (Section 5.4).

Again some exercises will be interspersed throughout the chapter with solutions provided in Appendix B.

5.1 Algorithmic verification

Imagine that we have a temporal formula, φ, that is used to specify some property that we wish to check of a system or component. Now, we have to check this against a

---

An Introduction to Practical Formal Methods Using Temporal Logic, First Edition. Michael Fisher.
© 2011 John Wiley & Sons, Ltd. Published 2011 by John Wiley & Sons, Ltd.

description of the component, for example a program. As we have seen, one (*deductive*) approach is to have another temporal formula, say Γ, that *exactly* specifies the component (e.g. this formula could be derived via a temporal semantics). Thus, Γ must characterize all the (possibly infinite number of) models (or executions) of the component. To check that the required property holds, we must prove

$$\vdash (\Gamma \Rightarrow \varphi).$$

As we saw earlier, this means establishing that

the set of models that satisfy Γ is a *subset* of the set of models that satisfy φ.

If, rather than logically specifying the component, we have a small (e.g. finite) set of execution sequences (captured by the set Σ) that are the only possible executions of the component, then it can make sense to check all these individually. Thus, in this case, we only need to check $\Sigma \models \varphi$ or

$$\forall \sigma \in \Sigma. \ \langle \sigma, 0 \rangle \models \varphi.$$

This *algorithmic* approach to verification is called *model checking* [118], and is described pictorially below.

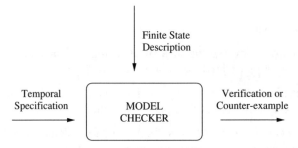

The key element here is that, if there are execution sequences of the particular component/ program/system being described that *do not* satisfy the required temporal formula, then at least one such 'failing' sequence will be returned as a counter-example. If no such counter-examples are produced then this means that all executions of the component/system/program satisfy the prescribed temporal formula.

Model checking has become *very* popular and there are a number of ways of describing this technique. For us to explain how model checking works, we begin by recalling the semantics of PTL.

5.1.1 Checking models manually

As outlined above, in order to verify that φ is satisfied on all executions of a system, captured through the set Σ, we need to establish

$$\forall \sigma \in \Sigma. \ \langle \sigma, 0 \rangle \models \varphi.$$

At first sight this seems relatively simple in that, for each sequence σ, we only need to follow the recursive definition of '$\models$' gradually taking φ apart to establish $\langle \sigma, 0 \rangle \models \varphi$.

Review from section 2.3 – definition of '$\models$':

$$\langle \sigma, i \rangle \models p \qquad \text{iff} \quad p \in \pi(i) \qquad \text{(for p} \in \text{PROP)}$$

$$\langle \sigma, i \rangle \models \neg\varphi \qquad \text{iff} \quad \text{it is not the case that } \langle \sigma, i \rangle \models \varphi$$

$$\langle \sigma, i \rangle \models \varphi \wedge \psi \quad \text{iff} \quad \langle \sigma, i \rangle \models \varphi \text{ and } \langle \sigma, i \rangle \models \psi$$

$$\langle \sigma, i \rangle \models \varphi \vee \psi \quad \text{iff} \quad \langle \sigma, i \rangle \models \varphi \text{ or } \langle \sigma, i \rangle \models \psi$$

$$\langle \sigma, i \rangle \models \textbf{start} \quad \text{iff} \quad (i = 0)$$

$$\langle \sigma, i \rangle \models \bigcirc\varphi \qquad \text{iff} \quad \langle \sigma, i + 1 \rangle \models \varphi$$

$$\langle \sigma, i \rangle \models \Diamond\varphi \qquad \text{iff} \quad \text{there exists } j \text{ such that } (j \geq i) \text{ and } \langle \sigma, j \rangle \models \varphi$$

$$\langle \sigma, i \rangle \models \Box\varphi \qquad \text{iff} \quad \text{for all } j, \text{ if } (j \geq i), \text{ then } \langle \sigma, j \rangle \models \varphi$$

$$\langle \sigma, i \rangle \models \varphi U \psi \quad \text{iff} \quad \text{there exists } j \text{ such that } (j \geq i) \text{ and } \langle \sigma, j \rangle \models \psi \text{ and}$$

$$\text{for all } k, \text{ if } (j > k \geq i), \text{ then } \langle \sigma, k \rangle \models \varphi$$

To see how such *manual* model checking is carried out, let us consider some simple examples.

Example 5.1 *Let us assume that we have an execution sequence* $\sigma = \langle \mathbb{N}, \pi \rangle$ *defining*

Here, c is true at the initial state (state '0'), b is true at state '1', and a is true at state '2'.

Now, we will verify '$c \wedge \bigcirc b \wedge \bigcirc \bigcirc a$' by checking $\langle \sigma, 0 \rangle \models c \wedge \bigcirc b \wedge \bigcirc \bigcirc a$.

From the definition of the satisfiability of conjunction, this reduces, still further, to checking all of

$$\langle \sigma, 0 \rangle \models c,$$
$$\langle \sigma, 0 \rangle \models \bigcirc b, \quad and$$
$$\langle \sigma, 0 \rangle \models \bigcirc \bigcirc a,$$

which further reduces to checking

$$\langle \sigma, 0 \rangle \models c,$$
$$\langle \sigma, 1 \rangle \models b, \quad and$$
$$\langle \sigma, 2 \rangle \models a.$$

Since 'a', 'b' and 'c' are all atomic propositions then this reduces to checking

$$c \in \pi(0), \quad and$$
$$b \in \pi(1), \quad and$$
$$a \in \pi(2).$$

All three of which are easily assessed and found to be true.

Example 5.2 *Let us assume that we have the same execution sequence as above, that is*

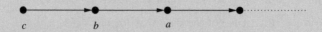

But now we will check $\langle \sigma, 0 \rangle \models c \wedge \bigcirc c \wedge \bigcirc\bigcirc \neg a$.
 Again, by the definition of the satisfiability of conjunction, this reduces to checking

$\langle \sigma, 0 \rangle \models c$,
$\langle \sigma, 0 \rangle \models \bigcirc c$, *and*
$\langle \sigma, 0 \rangle \models \bigcirc\bigcirc \neg a$.

Now, however, these reduce further to

$c \in \pi(0)$, $c \in \pi(1)$ *and* $a \notin \pi(2)$

of which only the first is true. Thus, we find that $\langle \sigma, 0 \rangle \not\models c \wedge \bigcirc c \wedge \bigcirc\bigcirc \neg a$.

Example 5.3 *Again, let us assume the same execution sequence:*

But now we will check $\langle \sigma, 0 \rangle \models \Diamond a$. *We could expand this to*

there exists a j such that $j \geq 0$ and $\langle \sigma, j \rangle \models a$

and choose $j = 2$. *However, a common alternative is to use the equivalence*

$$\Diamond\varphi \Leftrightarrow \varphi \vee \bigcirc\Diamond\varphi$$

Instantiating this equivalence with $\varphi = a$ *gives us a sequence of checks to make, namely*

1. $\langle \sigma, 0 \rangle \models a$,   *or*

2. $\langle \sigma, 0 \rangle \models \bigcirc a$,   *or*

3. $\langle \sigma, 0 \rangle \models \bigcirc\bigcirc a$,   *or* ...

of which the third is true.

Clearly, expanding '$\Diamond\varphi$' in the above way can cause problems on an infinite sequence if we do *not* find a state satisfying 'φ'. This is the problem we examine next.

Exercise 5.1 *Describe how you might check, using the relation '$\models$', whether the formula* $\Box(a \Rightarrow \bigcirc b)$ *is true on a model defined by*

$$\pi(0) \;=\; \{c\}$$

$$\begin{aligned}
\pi(1) &= \{a, c\} \\
\pi(2) &= \{b, c\} \\
\pi(3) &= \{c\} \\
\pi(4) &= \{a, b\} \\
\forall i \in \mathbb{N}. \ (i > 4) &\Rightarrow (\pi(i) = \{b\})
\end{aligned}$$

What answer should you get?

5.1.2   Manually checking infinite behaviour

Now let us consider a more complex model/execution:

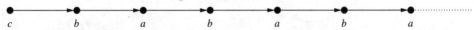

Checking formulae such as

$$c \wedge \bigcirc b \wedge \bigcirc \bigcirc a \wedge \bigcirc \bigcirc \bigcirc b$$

by following the PTL semantics is fairly straightforward. However, how do we check *infinite* behaviour, for example $\Box \varphi$, without going on forever? In order to do this, we need a better (and finite) way of representing infinite models/executions.

Representing infinite behaviour

An important property of PTL (called the *finite model property*) is that any satisfiable formula is satisfiable in a finite model/execution and so can be described in a finite structure. Generally such a description is given using some variety of graph structure. Thus the sequence above, that is

might be represented as

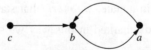

Now, we have to define the checking of formulae describing infinite behaviours on such graphs (checking finite behaviours is straightforward). To do this we provide below simple algorithms for checking the key classes of formulae (i.e. '$\Diamond \ldots$' and '$\Box \ldots$').

Checking 'always' formulae.   Recall that a fundamental property of '$\Box$' is

$$\Box \varphi \Leftrightarrow (\varphi \wedge \bigcirc \Box \varphi).$$

So, to check whether $\Box \varphi$ is satisfied at a particular state in a model, we

1. check that φ is satisfied at the current state, and

2. check that $\Box \varphi$ is satisfied in the next state.

However, if we are checking a graph structure of the form above, then we must take care not to keep on checking the same states repeatedly. So

1. whenever φ is checked in a state, we *mark* that state, and

2. if the *next* state is already marked in this way we stop, knowing that $\Box\varphi$ is *true* on this sequence.

Exercise 5.2 *Use the above method to check $\Box b$ on each of these graphs separately.*

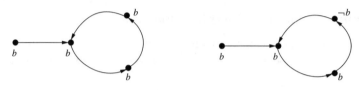

Exercise 5.3 *In the above algorithm, if the* next *state is already marked, we stop knowing that $\Box\varphi$ is true on this sequence. Why?*

Checking 'sometime' formulae.   Again, we use the fact that

$$\Diamond\varphi \Leftrightarrow (\varphi \vee \bigcirc\Diamond\varphi).$$

However, we must note that $\Diamond\varphi$ can not be 'unwound' indefinitely. So, to check $\Diamond\varphi$ at a particular state in a model, we

1. check that φ is satisfied in the current state – if it is, then $\Diamond\varphi$ is satisfied; if it is not then

2. check that $\Diamond\varphi$ is satisfied at the next state.

Again, if we are checking a cyclic graph structure then

1. whenever φ is checked in a state, we *mark* that state, and

2. if the *next* state is already marked in this way we stop, knowing that $\Diamond\varphi$ is *false* on this sequence.

Exercise 5.4 *Now use the above method to check $\Diamond b$ on each of these graphs separately.*

Exercise 5.5 *The algorithm above states that, if the* next *state is already marked in this way, we stop, knowing that $\Diamond\varphi$ is false on this sequence. Why?*

This provides an approach to the checking of temporal formulae over graph structures of the above form. If we have only a small number of different forms of temporal formulae (as in SNF) we might use this approach [186]. However, a more general way of describing and implementing model checking is based on Büchi Automata, as we will see in the next section.

5.2   Automata-theoretic model checking

Recall that, if we have a set of execution sequences (say Σ), then model checking, that is establishing

$$\Sigma \models \varphi$$

corresponds to checking that

$$\forall \sigma \in \Sigma. \ \langle \sigma, 0 \rangle \models \varphi.$$

In turn, $\langle \sigma, 0 \rangle \models \varphi$ corresponds to checking that the formula φ is satisfied on the sequence σ. However, since φ itself characterizes a set of sequences/models, then we are really checking

$$\Sigma \subseteq sequences\_satisfying\,(\varphi)\,.$$

But how can we capture the '$sequences\_satisfying\,(\varphi)$' and how can we (sensibly) test the above subset relation? One answer is to use *Büchi Automata* [95] to represent *both* sets of execution sequences and sets of temporal sequences, and then use automata-theoretic techniques in order to check the required subset relation [454, 487].

As we have already seen in Section 2.7, a Büchi Automaton, B_φ, can be used to represent the possible models of the formula φ. However, as we will see later, a Büchi Automaton, B_S, can also be used to represent all the possible execution sequences of the program or system, S. Now, both the program/system modelled and the required specification are represented using similar structures.

Pictorially, we have

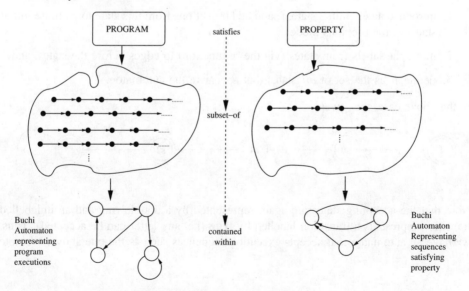

Recall that, for verification, we require that any execution of the program (captured by B_S) satisfies our required property (captured by B_φ). So, we require that the set of sequences accepted by B_S is a subset of the set of sequences accepted by B_φ, that is:

$$accepted\_by(B_S) \subseteq accepted\_by(B_\varphi)$$

or, alternatively,

$$accepted\_by(B_S) \cap \neg accepted\_by(B_\varphi) = \emptyset.$$

We can simplify this further since

$$\neg accepted\_by(B_\varphi) = accepted\_by(B_{\neg\varphi})$$

and so we are left needing to establish

$$accepted\_by(B_S) \cap accepted\_by(B_{\neg\varphi}) = \emptyset.$$

However, even testing such an intersection can be difficult and we must refine this approach still further. However, before that we will look at how Büchi Automata, such as B_S, can be generated from the system under consideration.

5.2.1   Deriving Büchi automata from a system – examples

Consider the following execution sequence from a program or system.

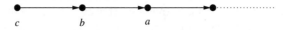

In order to devise a Büchi automaton that will capture sequences similar to that above, we can use a similar structure and then

1. introduce a new initial state, i, and add transitions from this state to all those initial states in the previous model,

2. move the labels from states (via the π function) to edges before the states, and

3. define F as the set of *all* states that appear in the automaton.

In the above case, this gives

Note that the accepting states are again represented by a double ring and an unlabelled transition represents a transition labelled by **true** (i.e. any letter can be accepted). Thus, we have an automaton that accepts execution sequences such as the one above. We note that all states in such an automaton are accepting, signifying that all states are part of legitimate execution sequences.

Consider a second execution sequence, namely

Such an execution sequence can be represented as a Büchi Automaton as follows:

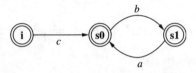

Thus, for every potential execution of the program or system in question we can build such an automaton.

Aside: Automata-theoretic view of '*Comms*' formula.   If we recall from Chapter 3, we typically had temporal specifications of several components, say $Spec_A$ and $Spec_B$, together with some *comms* formula $Comms(A, B)$. Now, assume that $Spec_A$ involves proposition x, while $Spec_B$ involves proposition y. We can build automata for $Spec_A$ and $Spec_B$, call them B_A and B_B, respectively, which have appropriate transitions in, for example

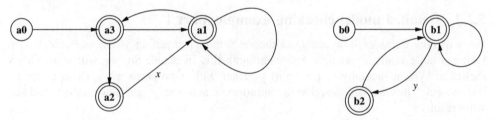

Notice that we omit many transition labels here. Notice also that, if we take the product of these two automata, then there is little interaction between them. This corresponds to the pure conjunction $Spec_A \wedge Spec_B$. But, if we add a $Comms(A, B)$ formula, such as $x \Rightarrow \Diamond y$, then $Spec_A \wedge Spec_B \wedge Comms(A, B)$ becomes *much* more interesting. Similarly, in the automata view, we now have a third automaton, B_{Comms}, corresponding to $comms(A, B)$ and interacting with both $Spec_A$, via the x label, and $Spec_B$, via the y label.

Exercise 5.6 *Generate a Büchi automaton for the Comms(A, B) formula $x \Rightarrow \Diamond^+ y$ (not $\Box(x \Rightarrow \Diamond^+ y)$) and show how (at least in general) the product construction between B_A, B_B, and B_{Comms} proceeds.*

5.2.2   Using automata theory

We have already seen how to generate automata from temporal formulae, so now we can proceed to explore more automata-theoretic aspects that we will require in model checking.

Recall that, if

$$accepted\_by(B_S) \cap accepted\_by(B_{\neg\varphi}) = \emptyset,$$

then φ is satisfied on all executions of system S. But an important aspect of such automata is that[1]

$$accepted\_by(B_S) \cap accepted\_by(B_{\neg\varphi}) = accepted\_by(B_S \times B_{\neg\varphi})$$

and so we actually only need to check

$$accepted\_by(B_S \times B_{\neg\varphi}) = \emptyset.$$

The intuitive reason for this is that the automaton $B_S \times B_{\neg\varphi}$ accepts counter-examples to $S \models \varphi$. Indeed, many model-checkers (including Spin; see Section 5.3) use this approach of effectively constructing $B_S \times B_{\neg\varphi}$. For any two automata, A_1 and A_2, if $A_1 \times A_2$ accepts a particular sequence, σ, then both A_1 and A_2 must separately accept σ. So, if we find that $B_S \times B_{\neg\varphi}$ accepts a sequence, then, because of B_S, this must be a legitimate execution sequence of the system S and, because of $B_{\neg\varphi}$, this sequence does *not* satisfy φ.

5.2.3   Detailed model checking example, part 1

We will now consider how automata-theoretic model checking works, specifically by taking a more realistic example and examining this in detail. So, we will start with a particular system (actually, a program) together with a property we wish to check of this system. Then we will describe the appropriate automata, take their product, and see what results.

Let us begin with the following program (given in a C-like language)

```
int x = random(1,4);    /* x is a random Integer of 1, 2, 3 or 4 */

while (x != 2)
do
   if                      /* switch, or multi-way if, statement */
      :: (x < 2)  ->   x:=x+1;
      :: (x > 2)  ->   x:=x-1;
   fi
od
```

---

[1] Note that the difficulties in using Büchi Automata products (see Section 2.7.3) do not occur since *all* states in B_S are accepting.

This will represent the system we wish to analyse. Some possible (temporal) executions of this program are as follows (with assignments on the transitions)

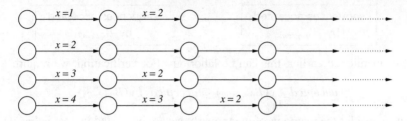

Now we wish to check the property $\Diamond(x = 2)$ on all executions of the program. Some possible sequences that satisfy this property are:

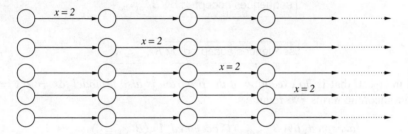

Recall that, in order to check the property, we must ensure that

$$[\text{executions of program}] \subseteq [\text{models of property}]$$

However, this can be very laborious. So, we construct a Büchi Automaton representing all executions of the program (let us call this $B_{program}$):

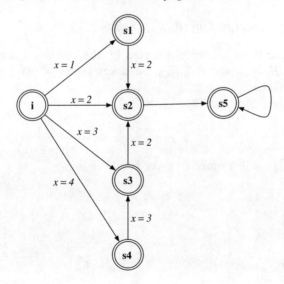

Now, we *could* generate another Büchi Automaton representing all models of the formula to be checked, that is $B_{\Diamond(x=2)}$, then we would need to show

$$\boxed{\begin{array}{c}\text{sequences accepted} \\ \text{by } B_{program}\end{array}} \quad \subseteq \quad \boxed{\begin{array}{c}\text{sequences accepted} \\ \text{by } B_{\Diamond(x=2)}\end{array}}$$

But, as we mentioned earlier, this can be laborious. For verification, we require

$$accepted\_by(B_{program}) \subseteq accepted\_by(B_{\Diamond(x=2)}).$$

Instead we consider the *negation* of the property we are interested in, that is '$\neg\Diamond(x=2)$' and work with the automaton generated from that, that is $B_{\neg\Diamond(x=2)}$.

Now, we want to check that the sets

$$\boxed{\text{sequences accepted by } B_{program}}$$

and

$$\boxed{\text{sequences accepted by } B_{\neg\Diamond(x=2)}}$$

do **not** intersect, that is *no execution of the program is also a model for* $\neg\Diamond(x=2)$. In automata-theoretic terms, we require:

$$accepted\_by(B_{program}) \cap accepted\_by(B_{\neg\Diamond(x=2)}) = \emptyset.$$

If

$$accepted\_by(B_{program}) \cap accepted\_by(B_{\neg\Diamond(x=2)})$$

is **not** empty, then that means there is a sequence that is in *both* sets, that is there is a sequence that is both a legal execution of the program and that satisfies '$\neg\Diamond(x=2)$'.

Taking intersections is not so convenient, so we go further, changing the above to a check that

$$accepted\_by(B_{program} \times B_{\neg\Diamond(x=2)}) = \emptyset.$$

In other words there is no sequence accepted by the combined automaton generated from the automata $B_{program}$ and $B_{\neg\Diamond(x=2)}$. So, a key aspect of many model-checkers is constructing the product

$$B_{program} \times B_{\neg\Diamond(x=2)}.$$

Example 5.4 *For the example above the negation of the property we are interested in* $(\Diamond(x=2))$ *is* '$\Box(x \neq 2)$', *which can give rise to* $B_{\Box(x\neq2)}$:

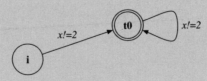

Product construction.   Taking the product of $B_{program}$ and $B_{\Box(x \neq 2)}$ we now get:

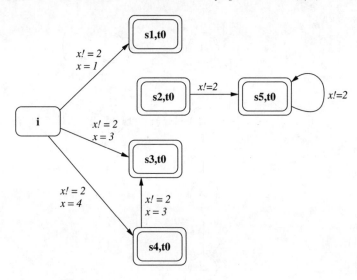

Note that we have here deleted edges that contain contradictory labels, specifically the transitions into **(s2,t0)**, which would have been labelled by both '$x = 2$' and '$x! = 2$'.

Note also that there are **no** infinite sequences, starting from an initial state, and passing through accepting states infinitely often. Therefore this automaton is *empty*, that is it accepts no sequences. This means that no possible execution of the program satisfies the negation of our required property and thus all legal executions (as captured within the Büchi Automaton) satisfy our required property. Consequently, a model-checker will return no errors in such a case.

So, now, let us consider what happens when we examine an erroneous variation of the program.

5.2.4   Detailed model checking example, part 2

Let us modify our program a little to be:

```
int x = random(1,4);    /* x is a random Integer of 1, 2, 3 or 4 */

while (x > 2)
do
    x:=x-1;
od
```

This clearly has a problem if we expect all executions to eventually ensure '$x = 2$', since if the random number generator assigns '1' to the variable, then this program does not modify the variable's value further. In particular, the variable 'x' never reaches the value '2'. So, let us see how our automata-theoretic model checking proceeds with this modified program.

The Büchi Automaton representing all executions of the program ($B_{program}$) above is now:

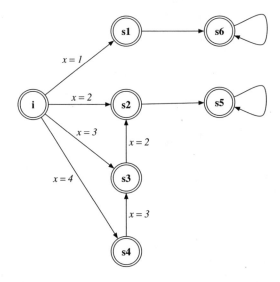

Notice how the **i, s1, s6, s6,** ... execution is possible yet does not force $x = 2$ to be true. So, taking the product of this new $B_{program}$ and $B_{\square(x \neq 2)}$ we now get:

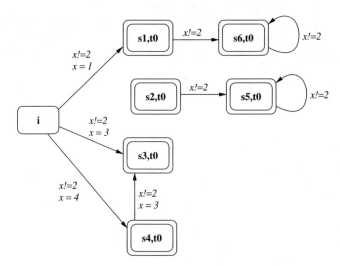

Note that there *is* now an infinite sequence, starting from an initial state, and passing through an accepting state infinitely often. Just start from the initial state above, move to '**(s1,t0)**' then on to '**(s6,t0)**', continuously looping on that.

Since the product automaton is non-empty, there is a run of the program (starting with the $x = 1$ transition) that satisfies the negation of our required property, and this represents a *counter-example*.

This is the basis of the model-checking approach, and a model-checker will typically return a sequence such as that above as a counter-example to the verified property. Notice how the counter-example provides a sequence of legitimate program steps that, at the same time, violates our required property.

This simple, but powerful, idea is the basis of many practical model-checkers, one of which we will consider next.

5.3   The Spin system

Spin [282, 283, 285] is essentially an implementation of the automata-theoretic approach described above, but with a number of enhancements and specializations, as follows.

- It uses *on-the-fly* model checking [235], which effectively involves only construct-ing the full product, $B_S \times B_{\neg\varphi}$, where necessary and keeping the B_S and $B_{\neg\varphi}$ separate as much as possible.

- It uses *partial-order reduction* [238], a method for reducing the number of paths that must be examined during automata-based verification. Specifically, Spin uses the method described in [286, 403].

- It uses a modified version of acceptance in product automata; see Section 5.3.3.

- It uses a particular high-level modelling language, called Promela, to describe both the system to be verified and the property to be checked.

- It is an *explicit-state*, rather than *symbolic*, model-checker; the latter being a popular trend (which will be discussed in Section 5.4.6).

We will explain these aspects in subsequent sections.

Spin was developed at Bell Labs during the late 1980s, and was initially aimed at verifying properties of communications protocols [282, 283, 285]. How-ever, it is now very popular and is used for checking temporal properties of a wide variety of applications involving concurrent processes. Spin is available for download from http://spinroot.com and the manual is available online at http://spinroot.com/spin/Man/Manual.html. (Indeed, some of the examples that follow are taken from this manual.) There is also a useful graphical version provided there, 'xspin'.

Recall the overall model-checking process as described below:

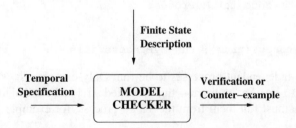

We will first examine the particular modelling language that is used to input finite-state descriptions to the Spin model-checking system. This high-level language is used to describe the finite-state system to be checked and is called Promela (from 'Process meta language').

5.3.1   The `Promela` modelling language

While `Promela` can generally be seen as a fairly standard imperative programming language, it does have some novel features. It is a *non-deterministic*, *multi-process*, language, incorporating Dijkstra's guarded commands [146] and Hoare's communication channels [275]. In particular, this communication via channels can be defined to be synchronous (i.e. *rendezvous*) or asynchronous (i.e. *buffered*).

Crucially for many contemporary applications, `Promela` also allows for the dynamic creation of concurrent processes and communication channels.

Processes

`Promela` is a language based on independent processes. Processes are the only really *active* components in the language. For example, the state of a variable or of a message channel can only be changed, or inspected, by processes. In `Promela`, processes are defined using a `proctype` declaration.

```
proctype MyProcess()
{
        /* code for MyProcess */
}
```

This process type is named `MyProcess`, while the declaration body (enclosed in '{' and '}') consists of a list of (local) declarations and/or statements. A `proctype` definition only declares process behaviour, it does *not* execute it. In order to execute a process, we must call `run`, for example:

```
run MyProcess()
```

The unary operator `run` applies to the name of a process type to be executed, but may potentially fail if execution is not possible (e.g. if too many processes are already running). The `run` statement can be used to make multiple instances of processes and pass parameters to processes, for example:

```
proctype MyProcess(bool flag)
{
        /* code for MyProcess */
}
```

```
run MyProcess(true); run MyProcess(false)
```

This generates two independent processes to be run. Only one process is *guaranteed* to be executing in a `Promela` program – this is called `init`. In order to *bootstrap* the other processes, we must run them from the `init` process, for example:

```
proctype MyProcess()
{
        /* code for MyProcess */
}
```

```
proctype YourProcess()
{
        /* code for YourProcess */
}

init
{
        run MyProcess(); run YourProcess()
}
```

Such 'run' statements can be used in any process to spawn new processes at any time, not just in the initial process. Once running, a process can only terminate when all its subprocesses have terminated. In summary, processes in Promela are encapsulated entities–each has its own private data and address space and other processes can not access this private data. It is important to note that executing 'run MyProcess()' does not necessarily immediately run the process MyProcess(). It is just a signal to the scheduler that this process is *available* to be run; it does not guarantee exactly when the process *will* be run. The choice of which process to run at which time is made by a run-time scheduler. Such a scheduler continually selects a process to be run, and then allows it to execute up until either it suspends (waiting for an external event) or until the scheduler suspends it (because the process has been given enough run time); in either case, another process to be run next is then selected. As in standard multi-processing, if the scheduler gets the balance between switching processes very often and giving the processes enough time to achieve something, then the illusion that the processes are truly executing in parallel can be maintained.

Program statements

We next give a brief overview of the syntax of selected Promela program statements; there is, of course, much more; see [457, 285].

Variables and simple types. As in conventional programming languages, *variables* are used to store information, and both globally and locally visible variable declarations are possible.

```
/* global variable declarations here */

proctype MyProcess()
{
        /* local variable declarations here */
}
```

Such declarations are all of the form

```
variable-type variable-name
```

and the basic types are fairly standard, including bool, byte, short, int, as well as arrays of all these.

Example 5.5 *Simple program statements include:*

```
int my_number;

bool my_boolean

int my_array[42];

my_number = 47;
my_boolean = true;
my_array[0] = my_array[1] + my_array[2] + 99
```

Promela expressions, declarations and assignments are similar to those in the 'C' language [312]. Thus, not only do all variables initially have the value '0', but we can also define macros

```
#define my_num 1
```

declare variables

```
bool d, e, f;
```

and assign, test and compare variables

```
d = (e == false || f == true)
```

Blocking execution.   All statements in Promela can potentially be blocked, even *tests*! Thus, the statement (status == free) has the same effect as the *busy-waiting* loop

```
while (status != free)
        skip            /* wait for status==free to become true */
```

This has several implications, specifically that

- a process can now wait for an event to happen just by waiting for a statement (such as status==free) to become executable,

- a condition can only be executed (passed) when it holds, and

- if the condition does not hold, execution blocks (and the process suspends) until it does.

Note that, if a process is suspended awaiting some external condition then, when that condition does become true, the process will again be available to run.

Simple statements.   Various types of statements will be dealt with later, specifically communication statements, but we note here that statements are arranged in a sequence using ';' or ('->'), that is

```
StatementA; StatementB

guard_test -> StatementC
```

Example 5.6 *For example, we can write both of*

```
(status == free) -> (priority = 2)

x = 3; y = 5; z = c + d
```

The basic *branching* statement within Promela is a generalized if/switch/case statement, for example

```
if
:: (status != free) -> option1
:: (status == free) -> option2
:: (status != free) -> option3
fi
```

Here, at most one option from the list will be executed. If there are several possibilities whose guards are satisfied, then a non-deterministic choice is made; if there are no choices, the process blocks. If the guard statement is 'true', then instead of writing '(true) -> option' we simplify this to 'option'. For example, below is a fragment of Promela code (taken from the Promela manual, http://spinroot.com/spin/Man) in which the counter() process will non-deterministically either increment or decrement the count variable.

```
int count;

proctype counter()
{
        if
        :: count = count + 1
        :: count = count - 1
        fi
}
```

Thus, counter() will either increment or decrement the value of the global variable count once. Which one is chosen in any particular run is, of course, up to the (probably random) choices made in the Promela execution. In the above example, we might hope that the choice is 50% each though we certainly cannot guarantee this.

Repetition is just an iterated version of the single choice given above. This uses 'do...od' rather than 'if...fi' to signify the iteration.

Example 5.7 *Here is a variation on the above 'counter' program, again taken form the Promela manual.*

```
proctype counter()
{
        do
        :: count = count + 1
        :: count = count - 1
        :: (count == 0) -> break
        od
}
```

This program randomly increments and decrements count *until the process may terminate when* count==0. *We say 'may' because, when '*count == 0*', all three options within the '*do...od*' loop are possible, but when '*count != 0*' then the third possibility cannot occur. (Note that '*break*' just exits from the '*do*' loop.)*

Improved example.   We can refine the example above to ensure that the process *really* stops when a zero count is reached:

```
proctype safer_counter()
{
        do
        :: (count != 0) -> count = count+1
        :: (count != 0) -> count = count-1
        :: (count == 0) -> goto done
        od
  done: skip
}
```

Thus, if 'count == 0' we exit from the loop. However, if 'count != 0' then we have a choice between incrementing and decrementing count, and then continue around the loop. Note that the goto statement is an alternative to 'break' if we want to exit a loop to jump to some specific statement.

Variable scope and shared variable problems.   As we have seen, global variables can be viewed (and modified) by all processes, but this can often lead to problems. For example, consider the following, taken directly from the Promela manual, http://spinroot.com/spin/Man:

```
int state = 1;

proctype MyProcess()
{
  int tmp;
  (state==1) -> tmp = state; tmp = tmp+1; state = tmp
}

proctype YourProcess()
{
  int tmp;
```

```
        (state==1) -> tmp = state; tmp = tmp-1; state = tmp
    }

    init { run MyProcess(); run YourProcess() }
```

There are several items to note here concerning the, essentially competing, processes within the example above.

- If either of the two processes completes before its competitor has started, the second process will block forever on its initial condition.

- If both pass their 'state==1' test, for example because the scheduler suspended a process just after its test, both processes will complete their execution, but the resulting value of the global variable state will be unpredictable.
  In particular, the value of state can be any of 0, 1, or 2, depending on when the processes run their internal code and when they are suspended.

Such problems (involving *race conditions*) are typical of multi-process systems with shared resources. Not surprisingly, many mechanisms have been developed for avoiding these situations. Indeed, one of the key uses of Promela/Spin is to describe and verify such algorithms.

To summarize, processes in Promela:

- are autonomous components – each has its own private data and address space and other processes can not access this private data;

- run independently and (potentially) concurrently; and

- are run by a scheduler that decides how much *run time* to allocate to each process in turn.

Larger example

Now, we consider a slightly larger example incorporating counter processes, similar to above, together with a shared, global variable. Variables declared within a process are local to that process. However, variables declared in a particular context are visible (and modifiable) by all processes defined within the context. Thus, in the example below, the variable count can be modified by both counter_up() and counter_down() processes.

```
    int count;

    proctype counter_up()
    {
        do
        :: (count < 10)   -> printf("increment\n");
                             count = count + 1
        :: (count >= 10)  -> printf("counter is 10\n");
                             break
        od
    }
```

```
proctype counter_down()
{
      do
      :: (count > 0)  -> printf("decrement\n");
                         count = count - 1
      :: (count <= 0) -> printf("counter is 0\n");
                         break
      od
}
```

```
init { count = 5; run counter_up(); run counter_down() }
```

Notice how, in this example, we set the initial value of the shared variable to be 5 and then start the two (competing) counter processes. The behaviour of this program is dependent on the particular scheduling policy of the Promela implementation. If we run this program using the Promela implementation within the Spin model-checker (usually by running 'spin *filename*'), then there are a number of possible execution paths. In principle the program *could* run forever; in practice, one process eventually 'wins out' over the other and completes.

Example 5.8 *Running the code above one potential execution sequence of outputs we get is*

```
            decrement
  increment
            decrement
  increment
  increment
  increment
  increment
  increment
            decrement
  increment
            decrement
            decrement
  increment
  increment
            decrement
  increment
            decrement
            decrement
  increment
  increment
  increment
            decrement
  increment
            decrement
            decrement
  increment
```

```
            decrement
    increment
    increment
            decrement
    increment
            decrement
    increment
            decrement
    counter is 10
            decrement
            decrement
            decrement
            decrement
            decrement
            decrement
            decrement
            decrement
            decrement
            counter is 0
3 processes created
```

Note that this is just one *execution. Running the code again can potentially give a different result if the processes are scheduled differently. In the above, outputs from the 'counter_up' and indented, and outputs from the 'counter_down' process are indented twice. Note that, at the end, Spin reports that three processes were generated: counter_up; counter_down; and init.*

Exercise 5.7 *Run the above Promela program yourself. The code is available as the file counter-tst.pml. So just run 'spin counter-tst.pml'.*

Interleaving and Atomic statements.   As we saw above, if a process is suspended mid-way between a test and an action, problems can occur. In particular, if the execution of this process is *interleaved* with that of another process, the second process might change shared data in such a way that the test is no longer true. But it is too late for the first process; it has carried out its test and proceeds with its action as soon as it is scheduled to run. Clearly in shared variables situations, this can be problematic. One way to solve this is to define atomic sequences of statements, for example

```
    atomic { x = 1; y = 2; z = 3 }
```

This operator indicates that the sequence of statements is to be executed as one indivisible unit, non-interleaved with any other processes.

Thus, we can use atomic sequences to protect concurrent access to shared variables. Atomic sequences can also be an important tool in reducing the complexity of verification models. This is because such atomic sequences restrict the amount of interleaving that is allowed in a distributed system – this reduction in complexity can be dramatic (see later).

Example 5.9 *Consider the following version of an earlier program.*

```
int state = 1;

proctype MyProcess()
{
    atomic { (state==1) -> (state = state+1) }
}

proctype YourProcess()
{
    atomic { (state==1) -> (state = state-1) }
}

init  { run MyProcess(); run YourProcess() }
```

See how the use of `atomic` *blocks removes any potential process scheduling problems. So, now, one of* `MyProcess` *or* `YourProcess` *will reach its atomic block first, resulting in the other being suspended forever (waiting for the '*`state==1`*' test to succeed).*

Channels and communication

As we saw in Chapter 3, once we have concurrent components it is vital to be able to communicate in some way. So far we have seen how `Promela` processes can communicate information through (shared) global variables; we have also seen how this is fraught with danger! Fortunately, `Promela` also provides a more sophisticated and useful form of communication, specifically the communication by *channels* that we have seen earlier. As described in Chapter 3,

> *Channels are unidirectional structures that exist between two processes (they were characterized in various algebraic approaches such as CCS [376] and CSP [275]).... For example if a channel exists between* C_1 *and* C_2, *then* C_1 *can send a message to* C_2 *down this channel:*

Although perhaps not as flexible as general message-passing, channels turn out to be very useful in many `Spin` applications. They typically describe the communication structures within applications comprising concurrent processes, and such message channels are used to model the transfer of data from one process to another. They can be declared either locally or globally, and can have various *capacities*, as we will explain below.

The `Promela` syntax used for channel declaration is

```
chan "channel_name" = ["size"] of { "message_type" }
```

For instance:

```
chan my_chan = [23] of { short }
```

which declares a channel, called 'my_chan', that can store up to 23 messages of type short.

Sending and receiving messages.   The statement

```
channel_name!expr
```

sends the value of expression expr to the channel channel_name, that is it appends the value to the tail of the channel. Only a single value is passed through the channel in the case above. The statement

```
channel_name?msg
```

receives the message within the process that executed this statement. It retrieves it from the head of the channel and stores it in a variable msg. The channels pass messages in *first-in-first-out* order. So, a typical example of this form of activity is as follows:

```
proctype MyProcess(chan my_end_of_channel)
{
    int my_value = 79;
    my_end_of_channel ! my_value;      /* send '79' into
                                          channel */
}

proctype YourProcess(chan your_end_of_channel)
{
    int your_value;
    your_end_of_channel ? your_value;      /* read from
                                              channel */
    printf("Received %d\n", your_value);

}

init
{
    chan c = [1] of { int };
    run MyProcess(c);
    run YourProcess(c);
}
```

There are a number of interesting aspects here.

- Since the channel is to be shared between MyProcess and YourProcess, it should not really be declared local to either process. It could be declared as a global variable but a more typical way is to declare it within the init process (which is guaranteed to be active throughout the whole execution) and then explicitly pass it to both the processes that wish to use it. Hence the code within the init process.

Each of `MyProcess` and `YourProcess` is thus declared with one parameter, namely the channel that the `init` process will tell it to use.

- Now `MyProcess` simply declares a local variable with value `79` and then sends this value, using '`!`', down the channel. Since the channel is originally declared with capacity of one, then the channel is now full – later we will see what happens if a process tries to write to a full channel.

- `YourProcess` will attempt to read a value, using '`?`', from the channel and store it in its local variable `your_value`. Now, several things can happen.

    If the channel contains a value, then reading the value will remove it from the channel and then `YourProcess` will continue on to print out the result it has received.

    However, depending on how the processes have been scheduled, it may be that the `YourProcess` process gets to its 'channel reading' operation *before* `MyProcess` has written anything into the channel. In general, if a process tries to read from a channel that is empty, that process suspends awaiting another process writing to the channel. So, in the above case, `YourProcess` would suspend, `MyProcess` would execute and write to the channel, and then `YourProcess` would be woken up and could then proceed with its read operation.

This might seem complex, but if you understand this, you understand much of the communication approach within `Promela`. As you can see, channels are useful for ensuring that processes wait until a certain event occurs before proceeding.

Exercise 5.8 *Try running the above* `Promela` *program yourself. The code is available as the file* `simple_comms.pml`.

Modify the code, for example by changing the integer value, to convince yourself that this is working.

Try changing the declaration of channel c *to have zero capacity, that is*

```
chan c = [0] of { int };
```

and see if the program still works (we will see later what is happening here).

As we have seen, the read/receive operation is executable only when a value is available from the channel addressed. Similarly, the send/write operation is only executable when the channel is not full. If a statement happens to be non-executable, the process trying to execute it will be suspended until progress is again possible. Thus, it is possible for a process to suspend indefinitely, for example waiting for some activity on a channel that never happens.

Processes might not want to suspend, so a useful function within `Promela` is '`len`'. Here, `len(channel_name)` returns the number of messages currently stored in channel `channel_name`. Similarly, `empty(channel_name)` returns true only when the channel `channel_name` is empty and `full(channel_name)` returns true only when the channel `channel_name` is full.

More complex channels.   So far we have just given examples that involve transferring basic values, such as integers, via a channel. However, `Promela` channels allow two further interesting aspects:

- transferal of more complex data structures; and

- transferal of channel names.

If the messages to be passed via the channel are more complex, the declaration may look as follows:

```
chan channel_name = [12] of {byte, int, chan}
```

This time the channel stores up to twelve messages, each consisting of one 8-bit value, one 32-bit value and a channel name (see below). If more than one value is to be transferred per message, they are specified in a comma separated list, for example

```
channel_name!expr1,expr2,expr3
```

If more parameters are sent per message than the message channel can store, the redundant parameters are lost without warning. Meanwhile, if fewer parameters are sent than the message channel can store, the value of the remaining parameters is undefined. As with sending, multiple parameters can be received. Similarly, if the 'receive' operation tries to retrieve more parameters than are available, the value of the extra parameters is undefined; if it receives fewer than the number of parameters that was sent, the extra information is lost.

Channels can also contain channel names. We have already seen that channel names can be passed between processes, for example as parameters in process instantiations. We can also define channels to themselves contain channel names, for example

```
chan cc = [3] of { chan };
```

When a process reads a channel name from `cc` the process can then use this new channel as if it were a local channel or a parameter. This allows us to develop quite sophisticated programs, particularly involving dynamic communication structures. The following example provides some intuition.

Example 5.10 *The idea here is as follows.* `Producer` *has data it wants to send to* `Consumer` *– in this case just an integer – but does not know which channel to send it on. However,* `Producer` *has a channel connection to* `Navigator`, *a process that does indeed know what channel* `Consumer` *will be listening on. So* `Navigator` *will pass the name of* `Consumer`'s *channel on to* `Producer` *who will then utilize this.*

```
proctype Producer(chan directions)
{
    chan new_destination;          /* new channel to write to  */
```

```
        directions ? new_destination; /* read the new channel    */
        new_destination ! 1066;         /* send 1066 to new channel */
        printf("Processor sent 1066 down new channel\n");
}

proctype Navigator(chan location, datachannel)
{
        location ! datachannel;    /* send datachannel to Producer */
        printf("Navigator sent out new channel\n");
}

proctype Consumer(chan datachannel)
{
        int x;
        datachannel ? x;                 /* read x from datachannel */
        printf("Reader received x = %d\n", x);

}

init
{
        chan datachannel = [1] of { int };
        chan location    = [1] of { chan };
        run Navigator(location, datachannel);
        run Producer(location);
        run Consumer(datachannel);
}
```

So, in the program above Consumer simply waits to read a single integer from its data (channel). Meanwhile, Producer reads the name of this data channel as an item sent by Navigator on their shared channel. Once Producer has the name of this data channel, it can send its integer down this.

The key element here is that Producer dynamically finds out where it should send data; it does not know this at initialization time.

Exercise 5.9 *Again, try out the example above it can be found as the file* forwarding .pml. *Try experimenting with variations to convince yourself it works.*

Synchronous communication. So far we have talked about asynchronous communication between processes via message channels, declared using statements such as

```
    chan channel_name = [size] of { int }
```

where `size` is a positive constant that defines the buffer size. A logical extension is to allow for the declaration

```
chan rv_port = [0] of { int }
```

to define a *rendezvous* [8] port that can pass single integer messages. Since the channel size is zero, this channel can pass, but not store, messages. Message interactions via such rendezvous ports are, by definition, synchronous in that a message can only be passed if both sender and receiver are waiting to act. The communication occurs and the information is passed from sender to receiver without being stored in the channel.

Example 5.11 *The following example exhibits several things: rendezvous over channels using a* global *rendezvous port; the use of more complex data in channels, namely a byte/integer pair; and, as we will see when it is run, timeout because of perpetually suspended processes.*

```
chan port = [0] of { byte, int };

proctype MyProcess()
{
        port ! 5,909;
        port ! 5,693;
}

proctype YourProcess()
{
        int value;
        port ? 5,value;
        printf("Received value %d\n", value);

}

init {   atomic { run MyProcess(); run YourProcess() }  }
```

If we run the above, we get 'Received value 909' output, but then a `timeout` *because MyProcess is suspended forever waiting to write 5,693 to YourProcess.*

Exercise 5.10 *Try running the above example yourself; see* `rendezvous.pml`.
See what happens if you comment out the second message send in `MyProcess`?

Clarification.   It is useful to examine a variation of the example above but now varying the size of the 'port' buffer. So, consider

```
chan port = [N] of { int };          /* no byte field this time */

proctype MyProcess()    { port!909; port!693; }
```

```
proctype YourProcess() { int value; port?value; }

init  {  run MyProcess(); run YourProcess()  }
```

What happens if N=0? What happens if N>0?

N=0: (a) The two processes will synchronously execute their first statement: a hand-shake on the first message and a transfer of the value '909' to the local variable in YourProcess.

(b) The second statement in MyProcess will be blocked, because there is no matching receive operation in YourProcess, and so MyProcess will suspend forever (or until a timeout occurs).

N>0: Here MyProcess can run to completion, writing both its two values into the channel. Then, YourProcess will remove the first of these and both processes will terminate leaving the '693' value in the channel.

N=1: Here, a typical sequence of events is as follows.

(a) MyProcess can complete its first send operation, but suspends on the second one, since the channel is now filled to capacity.

(b) YourProcess can then retrieve the first message and complete.

(c) At this point MyProcess becomes executable again (since the channel has some capacity left) and completes, again leaving its last (i.e. '693') message as a residual element in the channel.

Exercise 5.11 *Try running the example above yourself with the three different values for N, namely 0, 1, and 2. The code is in* rendezvous2.pml. *Check that the actual behaviour matches the explanation above.*

5.3.2 Running Spin

Given a Promela program, describing a system comprising several processes, we can use Spin either:

- to execute the program (via 'spin program_name')

  that is performing one run, carrying out random choices where necessary; or

- to generate a structure representing *all* the possible runs of the system (via 'spin -a program_name')

that is effectively generating a Büchi Automaton describing all possible behaviours of the program.

Aside: From Promela to C. We will examine the second of these aspects in more detail in Section 5.3.3, but just note that the automaton structure, together with the process of checking this against some logical property, is actually coded in C. Thus, if we invoke

```
spin -a program-name
```

then Spin generates a set of C files, named pan.c, pan.h etc. These can then be compiled, for example via[2]

```
gcc -o executable_name pan.c
```

and then run, typically by

```
executable_name -a
```

As we will see later, executing this binary will actually carry out exhaustive exploration of the state space, as well as the checking of relevant properties.

Moving back to the basic execution mechanism within Spin, if we simply invoke

```
spin program-name
```

then Spin simply performs a straightforward execution (i.e. random simulation) of the Promela program program-name. Thus, the process scheduler uses random choices, and non-deterministic branches within the code are resolved randomly.

Although we stated above that verification of Promela code is carried out by generating a description of the full state space through the '-a' flag, there is actually another, much simpler, way to carry out a basic form of verification. This involves straightforward local checks implemented though *assertions* within the program.

Beginning verification: assertions

There is an additional language construct in Promela called the 'assert' statement. Statements of the form

```
assert(a_boolean_condition)
```

are always executable. If the Boolean condition holds, then the statement has no effect. If, however, the condition does *not* hold, the statement will produce an error report during execution with Spin.

Recall that running Spin without options gives us a random simulation that will only provide output when execution terminates or if a print statement is encountered. In this case, if we have an assertion within the program-name program that fails then we will get the following error once we run 'spin program-name':

```
spin: line XY "program-name", Error: assertion violated
```

where XY will be some line number. Typically, the exact assertion that has been violated will also be printed out.

---

[2] In your system the name of the C compiler, as well as the way you invoke it, may well be different.

Example 5.12 *A simple assertion example is given as follows.*

```
proctype Tester()
{
    int x = 0;

    do
       :: (x < 5) -> x=x+1;
       :: (x >= 5) -> break
    od;

    assert(x == 5);
}

init { run Tester() }
```

Here, the assertion has no effect, but, if we change the assertion to `assert(x==4)`, *then an assertion violation will be given. (These two programs can be found in* `sim-ple_assertion1.pml` *and* `simple_assertion2.pml`.)

Exercise 5.12 *Consider the following* `Promela` *program.*

```
proctype Producer (chan input, output, dump)
{
       int p_count = 0;
 Send: output!(p_count+3);                  /* send to Consumer */
       input?p_count;                       /* read from Consumer */
       if
       :: (p_count < 8) -> goto Send;       /* loop around */
       :: (p_count >= 8) -> dump!p_count    /* send then exit */
       fi
}

proctype Consumer (chan input, output)
{
       int c_count;                         /* read from Producer */
 Read: input?c_count;
       assert(c_count <= 10);
       output!c_count;                      /* back to Producer */
       goto Read
}

proctype Sink (chan input)
{
       int dumped;
       input?dumped;                        /* read dumped value */
       assert(dumped == 9)
}
```

```
init
{
    chan p2c = [0] of { int };
    chan c2p = [0] of { int };
    chan d   = [0] of { int };
    atomic { run A(c2p, p2c, d); run B(p2c, c2p); run Sink(d) }
}
```

Now,

(a) *what is the communication structure between the three processes* Producer, Consumer, *and* Sink, *that is which channels link the processes and which direction does information flow between them?*

(b) *will the assertion in the* Consumer *process succeed?*

(c) *will the assertion in the* Sink *process succeed?*

5.3.3   Automata-theoretic model checking via Spin

In this section we consider more sophisticated verification of Promela code within Spin. This will essentially implement (but with some variation; see below) the automata-theoretic model-checking approach we saw earlier in the chapter. However, before we move on to this aspect, it is important to note that assertions, as described above, still work when we move on to full state-space exploration.

Back to assertions.   If we have assertions in our Promela code, but now invoke 'spin -a' generating C code, then when we compile and run the C code the assertions are still checked.

Example 5.13 *If we perform*

```
spin -a simple_assertion2.pml
```

then compile and execute the resulting C code, the output we get (among others) is

```
pan: assertion violated (x==4) (at depth 6)
```

Exhaustive verification.   Now let us turn to the full verification of more complex temporal properties, as we saw in the automata-theoretic approach earlier. Recall that, if we want to check whether $S \models \varphi$, where S is some system of processes and φ is a temporal property, then we require:

- B_S – a Büchi Automaton representing all possible executions through the system, S;
- $B_{\neg\varphi}$ – a Büchi Automaton representing all behaviours conforming to '$\neg\varphi$'; and

- a mechanism for calculating $B_s \times B_{\neg\varphi}$ and checking whether *accepted_by*$(B_S \times B_{\neg\varphi})$ is empty.

Spin does indeed generate B_S, though it is hard to see, being hidden within the internals of the C code generated. In order to construct $B_{\neg\varphi}$, Spin requires the property to be checked. Rather than inputting temporal formulae and converting these to automata, Spin expects the user to input $B_{\neg\varphi}$ as an additional Promela process, called a *never claim*, within the input code. Finally, Spin does not directly generate the product, $B_s \times B_{\neg\varphi}$, but uses a technique called 'on-the-fly' model checking [235, 283] in order to check *accepted_by*$(B_S \times B_{\neg\varphi}) = \emptyset$ while keeping B_s and $B_{\neg\varphi}$ separate.

We will look at these three aspects in the subsequent sections, but just note that the way Spin works is, diagrammatically:

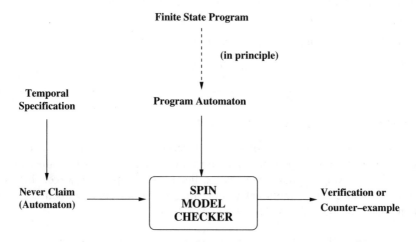

Note: Spin automata. Before we continue, however, there is an important aspect of Spin that we must consider. Recall that the product of Büchi Automata we described earlier involved a model of the system in which all states were accepting. Then, accepting states in the product automaton had to correspond to accepting states in *both* the system and property automata. In Spin it is *not* the case that all states in the system automaton are accepting; typically, most are not accepting states. Then, any accepting state in the product automaton must correspond to an accepting state in *at least one* of the system automaton or property automaton.

This has several advantages, one of which is that more complex systems can be described in which all runs are *not* equally likely. By allowing the user to specify accepting states in the system automaton, Spin ensures that quite sophisticated execution, for example involving fairness, can potentially be described. A further advantage is that the procedures that analyse the product automaton can just as well be used to analyse the system automaton, *without* any property automaton being present. This hints at Spin's ancestry in the area of reachability checking.

However, since we are only carrying out simple verifications here, and not considering more sophisticated protocols, our system automata do not need any particular accepting states added. In this case, where a system automaton consists of non-accepting states, then the effect is very similar to the basic Büchi Automata approach where all states are accepting. Thus, in our examples and exercises we will not dwell on this distinction but the interested reader is encouraged to explore the Spin documentation to see how more sophisticated protocols can be represented.

'On-the-fly' product construction. Constructing automata products such as

$$B_S \times B_{\neg\varphi}$$

can be *very* expensive. For example, the number of states in the product automaton may be huge. So, to carry out exhaustive verification, Spin generates a C program effectively containing both the automata B_S and $B_{\neg\varphi}$. Rather than combining these automata explicitly, the C program attempts to explore the possible executions *simultaneously*. Recall that we wanted $B_S \times B_{\neg\varphi}$ in order to capture all executions that are legal executions of the system, S, while at the same time being legal models for $\neg\varphi$. The basic idea with the on-the-fly approach is to explore *all* the paths through B_S and, as we do so, simultaneously check whether any path satisfies $B_{\neg\varphi}$. So, essentially:

1. we start exploring B_S, also starting at an initial state in $B_{\neg\varphi}$;

2. at every transition taken in B_S we must find a matching transition in $B_{\neg\varphi}$;

3. every time we have a non-deterministic choice in B_S (or in $B_{\neg\varphi}$), we take one branch (remembering the others) and proceed;

4. if we find no simultaneous legal move, we backtrack to a previous choice point;

5. if we reach a repeated combination of states from B_S and $B_{\neg\varphi}$, then we have found an infinite path: if this path is *accepting*, then we have found a legal path for the product automaton; if the path is *not* accepting, then we must return to an alternative previous choice.

And so on. If we have explored all possible paths through B_S but have found none that are accepting with $B_{\neg\varphi}$, then we finish, noting that the system S satisfies the property φ. However, if we *do* find such a path, as in (5) above, then this represents our counter-example.

So, essentially B_S and $B_{\neg\varphi}$ are kept separate and the C code is used to implement this simultaneous checking process. Avoiding construction of the full product is useful since this is the main complexity in model checking.

Aside: Space problems. Although model checking is a very appealing technique, for a very large class of computational systems it just will not be practical! As we have seen, checking a temporal property essentially means checking all possible executions of the system. Think of all the possible executions that we must explore, even for simple systems. Even if we have a relatively small number of possible executions, the size of each state in the sequence may be *very* big. In particular, the worst case we have is when the space needed to represent all the potential executions is *infinite*!

However, for reasonably-sized finite systems model checking is still viable. So let us look at further aspects concerning the use of Spin for exhaustive verification of Promela programs. We begin by examining where the possible choices in B_S come from.

Constructing B_S.   To give an indication of how an automaton, such as B_S, is generated from Promela program code (describing the system S), we will look at several examples. First of all, consider simple execution sequences, such as:

```
bool a, b, c;
c = true;
b = true;
a = true
```

This can be represented as a Büchi Automaton by

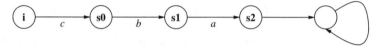

This construction seems straightforward and, indeed, for simple code fragments it is. However, an important aspect of automata is that they are essentially non-deterministic, yet have a finite number of states. This leads to (at least) two questions:

- What about the potentially *infinite* nature of numbers within programs?

- Where does *non-determinism* in the automaton come from?

Considering the first, we note that in most programs only a finite set of different numbers is used. We can, of course, generate examples that go on forever increasing the value of a variable – however, in reality these will hit a bound defined by the data space in the architecture. In both these cases, the set of numbers considered is finite and so considering only such finite sets appears to be a reasonable approach. Specifically, within Promela, only a finite number of processes are allowed, the basic data types (e.g. int) have finite ranges, and channels have fixed capacities.

Now for the second question above. How is non-determinism generated in an automaton given a Promela program? Well, there are several sources for this non-determinism. The first source of non-determinism (and, hence, branching in the automaton structure) comes from the code within each process. Consider

```
if
   ::   x = 3;
   ::   x = 8;
fi
```

Remember that the execution can non-deterministically choose between either of these assignments (since the guards for each are empty, i.e. 'true'). So, this will give us a branch in the automaton structure:

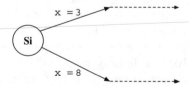

Similar branches are generated from other control constructs.

Another form of non-determinism comes from the possibility of suspension of execution awaiting external input. Consider the statement

```
c?input
```

Here, the process is attempting to read a message from channel c and store it in the variable input. However, if the channel is currently empty, then the process must suspend and wait until there is a message to extract. This might give an automaton structure of the form

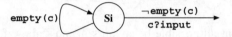

The major source of non-determinism, however, comes from the process scheduling policy within the execution system. To see this, consider two very simple processes defined as follows:

```
int x, y;

proctype MyProcess()
{
    x = 1;
    x = 3;
    x = 5;
}

proctype YourProcess()
{
    y = 2;
    y = 4;
    y = 6;
}

init { run MyProcess(); run YourProcess(); }
```

Distinct executions for each of MyProcess and YourProcess separately might appear as follows:

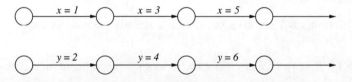

Now, since these processes will be executed depending on the policy of the scheduler, all of the following (combined) execution sequences are possible (and there are many more).

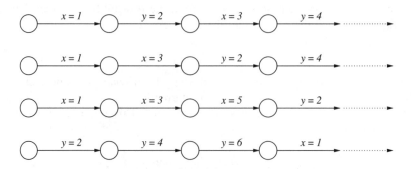

Since B_S must represent all possible behaviours, it must allow for all of these possibilities. This increases the number of branches (and, hence, paths) greatly. We will see in Section 5.4.4 that a technique called *partial-order reduction* can be used to limit these possibilities. Indeed, Spin uses this technique within its exhaustive verification.

Constructing $B_{\neg\varphi}$. In Spin, an automaton such as $B_{\neg\varphi}$ is termed a *never claim*. Thus, never claims are effectively Büchi Automata, given in Promela. In fact, such an automaton is represented as a special process, called a never process, within Promela. When the C program is generated, the never process is executed in parallel with the other processes.

But, if we are to include a never process within our Promela code, how do we generate such processes? Happily, Spin provides a mechanism for producing the code for such never processes from PTL input. Specifically, if we call

```
spin -f "..."
```

where a simple temporal formula is inserted instead of "...", then the code for the never process representing the Büchi Automaton for this formula is produced. The temporal formula input is a string comprising standard logical expression together with simple temporal operators:

'[]' for 'always in the future'
'<>' for 'sometime in the future'
'X' for 'next'
'U' for 'until'

The output from this is Promela code, with a slightly peculiar syntax.

Example 5.14 *If we run* Spin -f "<>q" *we generate a Büchi Automaton for '◇q'.*
Specifically:

```
never                  /* automaton for <>q */
{
T0_init:
          if
```

```
                  :: ((q)) -> goto accept_all
                  ::  (1)  -> goto T0_init
           fi;
    accept_all:
           skip
}
```

The labels T0_init *and* accept_all *represent names of states in the automaton, in particular accepting states have names beginning with '*accept'.

Now, ((q)) -> goto accept_all *means 'if q, then go to the state labelled by* accept_all'.

We can think of '1' as meaning 'anything', so (1) -> goto T0_init *means 'with any input, go to the state labelled by* T0_init'.

Such never claims are Büchi Automata because we can see by recalling the Büchi Automaton for ◇q:

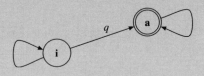

Executing never claims.   The Promela statement

```
    never { statements }
```

generates a special type of process that is instantiated once. The processes created when the program starts are actually the init and never processes, if present. An execution step of the system is a *combined*, *synchronized* step: executing an (enabled) statement from the never claim together with a step from the rest of the system. Thus, a never claim blocks execution of the system in those situations in which the never claim has no enabled statements (in the current state). This forces the overall execution of the Spin system to backtrack to a previous choice.

A never claim is intended to monitor every execution step in the rest of the system for illegal behavior and for this reason it executes synchronously. Thus, such a Promela execution can be exhaustively (but efficiently) analysed for correctness violations. In particular, Spin can search for executions satisfying general (linear time) temporal properties (or claims) defined through never claims. Since these never claims are essentially Büchi Automata, then quite complex infinite (hence cyclic) behavior can be detected.

Example 5.15 *Consider the following example showing a system and a property in both* Promela *and automaton forms.*

Code:

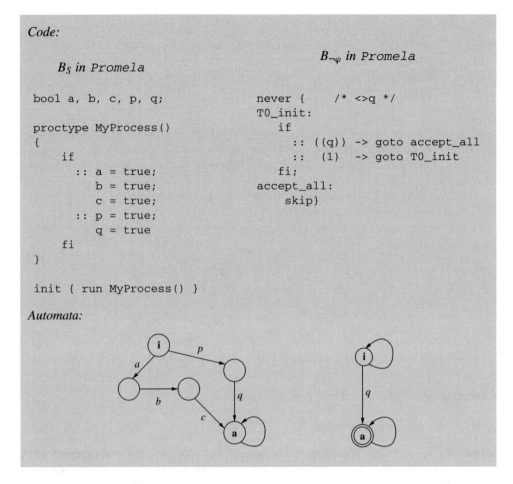

Bs in Promela

```
bool a, b, c, p, q;

proctype MyProcess()
{
    if
        :: a = true;
           b = true;
           c = true;
        :: p = true;
           q = true
    fi
}

init { run MyProcess() }
```

B₋φ in Promela

```
never {     /* <>q */
TO_init:
    if
        :: ((q)) -> goto accept_all
        ::  (1)  -> goto TO_init
    fi;
accept_all:
    skip}
```

Automata:

Let us consider verification, by hand, of the above example.

1. Start at state i in both automata.

2. choose the 'a' step in B_S together with the loop on the 'i' state in $B_{\neg\varphi}$.

3. Take the 'b' step in B_S together with the loop on the 'i' state in $B_{\neg\varphi}$.

4. Take the 'c' step in B_S together with the loop on the 'i' state in $B_{\neg\varphi}$.

5. Now loop on the 'a' state in B_S and loop on the 'i' state in $B_{\neg\varphi}$.
   This is *not* an accepting joint path, so we backtrack to a previous choice point, namely the choice made when both automata were in their 'i' states.

6. Now, take the 'p' step in B_S together with the loop on the 'i' state in $B_{\neg\varphi}$.

7. Take the 'q' step in B_S together with the 'q' step in $B_{\neg\varphi}$.

8. Now loop on the 'a' states in both B_S and $B_{\neg\varphi}$.

Since we have found an acceptable joint path, then '$\Diamond q$' is satisfied of S. Consequently, $\Box \neg q$ is not true in all executions of S. Specifically, if our system follows the p, q path.

5.3.4 Larger verification example

Recall the example we studied in Section 5.2.3 based on the simple program:

```
int x = random(1,4);      /* x is a random Integer 1, 2, 3 or 4 */

while (x != 2)            /* 'while...do' loop */
do
    if                   /* switch, or multi-way if, statement */
       :: (x < 2)  ->   x:=x+1;
       :: (x > 2)  ->   x:=x-1;
    fi
od
```

We will now follow this example again, but will code it in Promela and will carry out verification using Spin. A Promela version of the program above (which can be found in random_example.pml) is

```
int x;                  /* global so all can see it */

init
{
    if                  /* random choice x = 1, 2, 3, or 4 */
       :: x = 1;
       :: x = 2;
       :: x = 3;
       :: x = 4
    fi;

    do                      /* 'while...do' loop */
       :: (x == 2) -> break;
       :: (x < 2)  -> x=x+1;
       :: (x > 2)  -> x=x-1;
    od
}
```

Now we also need a property to check. Recall that our negated property is '$\Box(x \neq 2)$' and so, generating a never claim from this we get

```
    never                   /* automaton for [](x!=2) */
    {
    accept_init:
        T0_init:
                if
                 :: ((x!=2)) -> goto T0_init
                fi;
    }
```

If we carry out exhaustive verification on the program above with the 'never' process added, then we find no errors.

> **Exercise 5.13** *The example above including the never claim, is in* `random_` `example.pml`*. Try exhaustive verification (via* '`spin -a....`'*,* '`gcc -o` `myname....`' *and* '`myname -a`'*).*

Now, we move on to consider the erroneous version that we studied earlier:

```
int x;                       /* global so all can see it */

init
{
   if                        /* a variety of random choice */
     :: x = 1;
     :: x = 2;
     :: x = 3;
     :: x = 4
   fi;

   do                        /* 'while...do' loop */
     :: (x == 2) -> break;
     :: (x > 2) -> x=x-1;
   od
     /* will x be 2 now? */
}
```

Clearly this is problematic because a choice of `x = 1` means that none of the cases in the `do...od` loop will be used.

> **Example 5.16** *Carrying out exhaustive verification on the revised program now gives us an error:*
>
> ```
> pan: acceptance cycle (at depth 8)
> pan: wrote random_example2.pml.trail
> ```
>
> *Note that a counter-example is provided in the '*`.trail`*' file, though we will not discuss these files further here.*

> **Exercise 5.14** *The erroneous version is in* `random_example2.pml`*. Again try exhaustive verification yourself.*

> **Example 5.17** *Finally, we consider a typical system of processes coded in* `Promela` *together with a relevant* `never` *process.*
>
> ```
> int dumped = 0;
> ```

```
proctype XX (chan out)
{
    int total = 0;
    do
    :: (total >= 8) -> break;
    :: (total < 8) -> total = total + 3;
    od;
    out!total;
}

proctype YY (chan in) { in?dumped }

init {  chan c = [0] of { int }; run XX(c); run YY(c); }

never                   /* <>(dumped == 9) */
{
    T0:
            if
            :: (dumped == 9) -> goto accept
            :: (dumped != 9) -> goto T0
            fi;
    accept:
            skip
}
```

If we now carry out exhaustive verification on this, we should get `claim vio-
lated!`.

Exercise 5.15 *Run the example above yourself; it is in* `dumped_vern.pml`.
  Now change the `never` *process to represent* `<>(dumped==10)` *rather than*
`<>(dumped==9)`.
  What happens when you carry out exhaustive verification on this revised version?

5.4   Advanced topics

Algorithmic verification, particularly model checking, has become both popular and useful. Consequently, there are a huge number of advanced topics in this area. Necessarily,
therefore, we only explore a selection of topics. For *much* more detail, see [122, 117,
34, 247].

5.4.1   The story so far

We begin with a subjective view of the development of model checking. This is, of course,
very biased but at least provides a view of the evolution of this important technology.

1. **Beginnings**

   Model checking, as described earlier, was essentially developed both by Clarke and Emerson [169, 115] and Queille and Sifakis [424] in the early 1980s. They devised quite specific algorithms for (specific) logics and any implementations that existed tended to be relatively slow. An overview of the work at that time is given in [119]. At the same time, work in the *reachability analysis* area was developing similar techniques. Although not couched in terms of model checking, these techniques used automata and tools developed at that time were the direct ancestors of Spin [281]. A difference at this stage was that the 'model-checking' tools were typically targetted at hardware verification (via branching-time logics) while the 'reachability' tools were typically targetted at protocol analysis (via approaches akin to linear-time logics).

2. **Automata**

   As model checking became more popular and was developed for an increasing variety of logics, so the need for a general framework became apparent. This was provided in the mid-1980s once the link between model checking and finite automata over infinite strings was developed [454, 486, 482]. The link to various forms of automata, particularly Büchi Automata not only provided an all-encompassing framework for model checking but pointed towards new implementation techniques.

3. **Implementations**

   In the late 1980s more mature model-checking implementations appeared. These often incorporated an automata-theoretic view, together with efficiency techniques, such as *optimized search*, *on-the-fly* checking, *partial-order reduction* etc. [129, 235]. Tools such as Spin [282] appeared at this time, but many implementations still suffered from some *state explosion* problems.

4. **Encoding**

   A big step forward occurred when, in [98, 372], the *symbolic model-checking* approach was introduced. This provided a way to encode the model-checking problem as a (large) Boolean formula and then invoke efficient checkers for Boolean satisfiability. In particular, Bryant's Boolean-Decision-Diagrams (BDDs) were the initial target [93]. This approach proved to be extremely successful and was taken still further when model-checking problems were encoded as SAT problems [64, 417] and then solved by invoking efficient SAT solvers.

5. **Coverage**

   Symbolic systems, for example NuSMV [113], were indeed fairly quick but developments in this area gradually began to be concerned more with efficiency (that paths could be checked quickly) and less with completeness (that all failing paths would be found). This led on to a range of techniques aimed at improving efficiency that would not necessarily guarantee correctness. These worked well in specific areas. In particular *abstraction* [120], a variety of *abstract interpretation* [131], turned out to be very useful for more complex classes of problem involving infinite-state systems [37].

6. **Tools**

The current view is of several influential classes of tool based on algorithmic verification, each of which has spawned large, ongoing research areas.

- **Infinite systems**

  Since most 'real' programs are essentially infinite, then abstraction was found to be very useful for practical model checking. It is, of course, not guaranteed to be complete, and usually not automatic. So the predominant approach at present is to use the analysis of an abstracted system to refine the abstraction itself. This is exactly the *Counter-Example Guided Abstraction Refinement (CEGAR)* approach [121, 124] where abstractions are gradually refined to be more accurate.

- **Bounded search**

  In the quest for fast checking, the development of *bounded model checking* [123] was very important. While bounded search had already appeared in earlier explicit state model-checkers, such as Spin, the combination of bounded search with SAT techniques was a significant innovation. Thus, rather than guaranteeing to explore all possible executions, then only executions of length less than N were explored. This could be quick and, if necessary, N could be subsequently increased.

- **Dynamic checking**

  The traditional approach to algorithmic verification in general, and model checking in particular, is to have a description of the system and a description of the property and use both to carry out verification. This is typically carried out *before* any deployment or execution of the system. The idea with *run-time verification* [262] is to use lightweight verification technology to check executions *as they are being created*. In this way, errors can be spotted at run-time too. (See Section 7.3.)

Thus, current directions involve applying verification technology quickly, but without necessarily guaranteeing that all errors will *definitely* be found.

5.4.2   Model checking is ubiquitous!

Model checking has been *very* successful. It is now applied in a great many different areas and has been extended to a great many different logical frameworks.

The initial work on model checking targetted branching-time temporal logics, particularly CTL. This was understandable because the complexity of the model-checking process in CTL is low and indeed this direction has led on to very efficient tools such as NuSMV. However, it soon became apparent that different, and more powerful, logics would be required. This led to the model checking of linear temporal logics, of which Spin is an exemplar. In the *branching* field, this led to model checking of CTL* and then to both the μ-calculus and alternating temporal logics, such as ATL. ATL is interesting in that it captures a form of cooperative reasoning. Given a set (a *coalition*) of agents, A, ATL allows operators such as $\langle\langle A \rangle\rangle\varphi$, meaning that the set of agents have a collective strategy that will achieve φ. This approach has been very influential, not only on the specification and verification of open, distributed systems, but also on the modelling of the behaviour of groups of intelligent agents [480]. A particular model-checker, called Mocha [377], has been developed for such logics.

Further developments of model checking have included checking *combinations* of temporal and modal logics, for example [174], and tackling more complex models. (We will say more about the latter in Section 5.4.9.) In addition, model checking is now also applied directly to program code [287, 288, 492]. For example, there are model-checkers for high-level languages such as C [284, 36, 472] and Java [493, 302]. All this points to successful, and increasing, use of this algorithmic verification technology.

5.4.3   Model checking has problems!

Model checking is a very popular approach to the verification of certain classes of system. Many people see this technique as the 'Holy Grail' of comprehensive program verification [182, 276, 277, 344]. Indeed, as we have seen, temporal properties of realistic programs can be checked reasonably efficiently. However, there are still at least two problems with this approach.

- *Restriction to finite-state models.*
  Many real-world systems are *not* finite-state and, even though techniques such as *abstraction* (see below) have been applied to reduce infinite-state systems to finite-state ones, these techniques are not automatic. So there is still a problem in applying model checking to potentially infinite-state systems.

- *State explosion problem.*
  Even with some of the refined reduction and representation techniques we will see later, the formulae/structures required to represent the state spaces of realistic systems are huge. In addition, once we wish to represent concurrent systems with many asynchronous processes, the structures required to describe all the behaviours in non-trivial systems tend to be even bigger!

Essentially these are both problems of *space*. Consequently, a lot of work has been carried out trying to develop techniques to:

- reduce infinite-state systems down to finite-state systems; and

- reduce large finite-state systems down to finite-state systems of a more manageable size.

In the next few sections we will examine some of these techniques, specifically: *partial-order reduction*; *abstraction*; and *symbolic representations*. However, this is such an important area that there are many other techniques that have been devised [247].

5.4.4   Partial-order reduction

Recall that, in executing a program, a major source of non-determinism comes from the scheduling policy within the execution system. Previously, we considered these two processes:

```
int x, y;
```

```
proctype MyProcess()    { x = 1; x = 3; x = 5; }

proctype YourProcess() { y = 2; y = 4; y = 6; }

init { run MyProcess(); run YourProcess(); }
```

As we saw earlier, the executions for each of these separately might appear as follows:

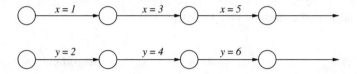

If *synchronous* execution of the two processes was considered, then we would get only one possible execution:

$$\bigcirc \xrightarrow{x = 1\ \&\ y = 2} \bigcirc \xrightarrow{x = 3\ \&\ y = 4} \bigcirc \xrightarrow{x = 5\ \&\ y = 6} \bigcirc \longrightarrow$$

But, if we consider *asynchronous* or *interleaved* concurrency, then an enormous number of possible executions are possible, for example:

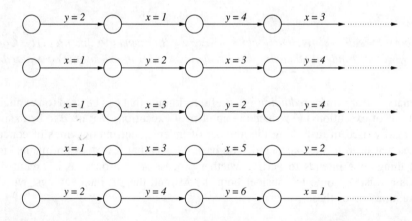

The key thing to note about the vast number of possible interleaving/executions is that many of them may be *equivalent* with respect to the formula we are checking. For example, if we consider the possible executions above, then many of these are equivalent with respect to formulae such as

$$\Diamond(x = 5) \wedge \Diamond(y = 6).$$

Here, we do not really care in which order the variables are assigned, as long as they both eventually reach their desired values. Thus, in this case we can use *partial-order reduction* and, during the course of a state space search, only explore a *subset* of the possible transitions from each state.

The basic idea is that for any two program instructions, for example $x:=1$ and $y:=2$, then, if the order of these is *not* relevant/important to the property we wish to check, then we do not need to represent, in B_S, all the possible interleavings of these two instructions. This is because `x:=1; y:=2` is equivalent to `y:=2; x:=1` with respect to our property. So, in the case above we do not need to represent all possible interleavings of `MyProcess` and `YourProcess` in B_S, simply a 'representative' selection of possibilities.

Partial-order reduction is a very useful technique, potentially reducing the search space significantly. However, implementing this might take a considerable amount of pre-processing. In particular, the decision of whether to ignore a certain execution may depend crucially on the formula being verified.

Example 5.18 *During verification, we might decide that the following executions are equivalent and so only use one of them:*

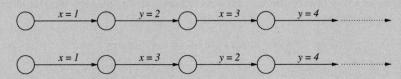

But, what if we are checking

$$\Box((y \ge x) \lor \bigcirc(y \ge x)).$$

The above executions are clearly not *equivalent with respect to this property. Consequently, we must indeed take care when reducing the set of executions considered.*

In summary, *partial-order reduction* is a useful technique that can be used (on explicit representations of executions) to reduce the number of executions that need to be examined. It is primarily used in restricting the number of independent interleavings of concurrent processes that need to be considered. The basic idea is that if it does not matter, for the kind of things one intends to verify, whether process A or process B is executed first, then it is a waste of time to consider both the AB and the BA interleavings; we should only examine one of these.

5.4.5   Abstraction

Infinite-state systems are typically too big to verify. Thus the technique of *abstraction* is commonly used to reduce an infinite-state system to a finite-state one. In this case, abstraction essentially involves replacing the infinite system with a finite system that has the same (relevant) behaviours as the infinite one. Again, this may depend on the property being checked.

There are a wide number of abstraction techniques used within model-checkers, some being automatic, the majority being semi-automatic at best. At least to give a flavour of the variety of abstraction techniques, we will just consider two approaches:

- *under approximation*; and

- *predicate abstraction*, which is a form of *over approximation*.

It is important to note that, while we have described abstraction as being a way to handle infinite-state systems, it is often also used to reduce a large finite-state system to a smaller, or at least more manageable, one.

Under approximation.   The basic idea here is to 'throw away' parts of the program, typically variables, that are irrelevant to the property being checked. Here are three (example) varieties.

Variety 1:   A program contains variables x and y, but the property being checked only concerns y and we are *sure* that y definitely does *not* depend on x.

→ remove the x variable, and all references to it!

Variety 2:   A program contains an integer variable x, but the property being checked does not depend on large (i.e. greater than 100) values of x.

→ redefine x to have the bounded type $0 \ldots 100$

Variety 3:   A program contains infinite structures, for example lists, but again the property being considered does not depend on exploring the full list.

→ replace lists with appropriate-sized fixed arrays.

There are many more examples. Such under-approximation techniques may be automatic, but often are not. They are clearly brutal and so are sometimes termed 'meat-axe abstractions'! They are also related to *program slicing* [475, 506]!!

Predicate abstraction.   The essential idea here is to replace variables that have infinite values with variables that have finite values, and ensure that each instruction that modified the infinite variable now modifies the finite variable accordingly. Since we are trying to reduce an infinite-state system to a finite-state one, any such abstraction system will generally be semi-automatic at best. The technique can be subtle, so we will provide the intuition behind this approach with a simple (and much used) example.

Example 5.19 *Consider a process that reads an integer, stores it as a local variable and applies some arithmetic operations to the variable:*

```
proctype MyProcess(chan in)
{
    int counter;
    in?counter;
    counter = counter + 77;
    counter = counter - 33;
    counter = counter * counter * 5;
}
```

In principle (though clearly not in practice), the Integer variable can take any value. So, if we were to apply our model-checking approaches so far we would have to explicitly represent all possible Integer values. However, let us imagine that we are concerned with verifying the property 'counter>0'. *We can introduce a predicate* 'positive' *aimed at capturing* counter>0, *and transform the process to:*

```
proctype MyProcess(chan in)
{
    bool positive;
    in?positive;    /* was the value received positive or not? */

    if                      /* choices from "counter = counter + 77" */
    :: (positive == true)   -> (positive = true)
    :: (positive == false) -> (positive = true)
    :: (positive == false) -> (positive = false)
    fi

    if                      /* choices from "counter = counter - 33" */
    :: (positive == false) -> (positive = false)
    :: (positive == true)   -> (positive = true)
    :: (positive == true)   -> (positive = false)
    fi

    positive = true; /* from "counter = counter * counter * 5" */
}
```

Let us consider the first choice in the process above. The original process added 77 to the variable. If the variable's value was greater than zero (i.e. positive *was* true) *then it will still be greater than zero; if the variable was* not *greater than zero (i.e.* positive *was* false) *then, once 77 is added, the variable could be either greater than zero, or not. Hence the non-deterministic choices in this case. Similarly for the second,* 'counter - 77' *instruction. Then the final instruction in the original process guarantees that* counter>0.

So, while we wanted to check ◇(counter>0) *in the original process, we can now check* ◇positive *in the abstracted process. By using a predicate, such as* positive, *we have abstracted a process dealing with Integer values to one dealing only with Boolean values.*

5.4.6   Symbolic representation

As we have seen, explicitly representing all the executions in a system (for the purposes of model checking) may take up a very large amount of space. However, the idea behind *symbolic model checking* [98, 372] is to use appropriate logical formulae to *symbolically* capture the states of a system and the transition relation between these states. This may seem like a strange thing to do, but the symbolic representation avoids explicitly building (and exploring) a very large graph/automaton.

While, in general, handling logical formulae can be expensive, an important class of logical system has been developed that can often be used to support checking of symbolic representations efficiently. So, *Boolean formulae* are used to represent the model; specifically, Binary Decision Diagrams (BDDs) [93] (or *ordered BDDs*) are used to represent and solve (often efficiently) checking problems over Boolean formulae. BDDs are a notation in which Boolean formulae can be represented as a graph structure on which certain logical operations can be very quick. This is dependent on finding a good ordering for the Boolean predicates within the graph structure.

This symbolic approach has been *very* influential, leading to a significant increase in the size of system that can be verified using model checking. Such symbolic formulations were particularly successful in the SMV [372, 99] and NuSMV [113, 393] model-checkers, which check branching temporal formulae (in CTL) over finite automata. A similar approach has been used to take advantage of a new generation of fast SAT solvers, with SAT-based model checking encoding the verification problem as a SAT problem [64, 417, 30].

5.4.7   Bounded model checking

Checking infinite paths via algorithmic verification requires exploring a graph/automaton structure until either an error is detected or a repeating state is reached. Since product automata can, in principle, be huge, then it might be a long time before a looping state is found. In addition, since we use a *depth-first* search approach, we might be exploring an unproductive path for a long time while there is an alternative path on which a much simpler error could be found quickly.

So, an obvious refinement would be to use a *breadth-first* rather than depth-first search for errors within the product automaton. This is essentially what *bounded model checking* does. Given a bound, N, then a bounded model-checker performs a (quick) analysis of all paths of length $< N$. In contemporary bounded model-checkers, this search is typically encoded as a SAT problem and a fast SAT solver is invoked.

If no error is found, the process can be repeated for larger and larger values of N until all possible violations have been ruled out. Essentially, the search for an error proceeds through *depth-first iterative deepening*, where each local search is encoded as a SAT problem. This approach is fast (SAT solvers are, indeed, very fast!), widely applicable [307] and successful (simple errors typically occur quite early in the product automaton) [30].

5.4.8   Counter-example guided abstraction refinement (CEGAR)

As we have seen above, abstraction helps us establish properties of a system by first simplifying it. The abstracted system might not satisfy *exactly* the same properties as the original, but at least we require our abstraction to be sound (properties verified on the abstracted version should be verifiable on the original). However, not all properties we could verify of the original system remain verifiable within the abstracted version. Thus, in Example 5.19 above, we clearly cannot establish properties about exact Integer values within our abstracted process.

So, it is often important to check whether an error found in an abstracted system is truly an error of the original system or whether it is the fault of using an inappropriate abstraction. If the latter, then we would ideally like to refine the abstraction to make it more appropriate and to eliminate this 'spurious' error. This is exactly what the *Counter-Example Guided Abstraction Refinement (CEGAR)* approach [121, 124] does. We begin with a coarse abstraction, then carry out verification on this. The results of verification are analysed and, if necessary, used to refine the abstraction. Then verification is carried out again. And so on, until the abstraction is sufficiently precise.

5.4.9   Real-time, probabilistic and hybrid models

The temporal models, and therefore automata, that we have so far considered have just had basic transitions labelled by simple actions or assignments. However, there are *many* more possibilities. One such is to use a *real-time* model. In our models so far, the only relationship between states is that each subsequent one is considered as the *next* moment in time. In order to develop a *real-time* version of this approach, we can consider such sequences, but with timing statements referring to particular clocks (in the following diagram [196] the clock is t) added between each consecutive moment. The diagram shows an example of a timed model (here $t < 1$ is a constraint stating that the time, t, is less than 1 on this transition, while the time t is at least 8 on the $t \geq 8$ transition).

| sunny   X > Y | $t < 1$ | rainy   ... | $t \geq 8$ | windy   ... | $\rightarrow$ ... |
| process_stopped | $\rightarrow$ | ... | $\rightarrow$ | ... | |

Where only a finite number of different states exist, Büchi Automata can also be extended to recognize these *timed sequences* [19, 20]. In practical applications of such models (see Section 5.4.10) various automata-theoretic operations, such as emptiness checking, are used. These tend to be complex [23], but vary greatly depending on the type of clocks and constraints used. This type of structure leads to the development of *timed automata* [19–21, 23, 60], which are both influential and popular.

In formalizing real-time aspects, a number of logics can be developed [21]. For instance, the standard temporal logic we have seen so far can be extended with annotations expressing real-time constraints [322]. Thus,

"I will finish writing this section within 9 time units"

might be represented by: '$\Diamond_{\leq 9} finish$'. There are a great many different real-time temporal logics (and axiomatizations [447]) and excellent surveys of this area are available by Alur and Henzinger [21, 22], Ostroff [400] and Henzinger [266]. Timed varieties of temporal logic are also appearing in industry standards, such as languages for hardware verification. These include PSL and ForSpec, which can be broadly seen as temporal logics, extended with both dynamic logic modalities and explicit clocks [28].

Yet, there is more to modelling *quantitative* aspects of systems than just real-time. Many (hard) practical problems, for example in complex control systems, require even more expressive power, and so models of *hybrid systems* were developed. Hybrid systems combine the standard discrete steps from the automata-theoretic approach with more complex mathematical techniques related to continuous systems (e.g. differential equations). Unsurprisingly, a key development here is *hybrid automata* [25].

A final, but very important variety of model involves *probabilities* between states. Thus, rather than taking a non-deterministic choice of which state to move on to next, we can have a choice of different transitions with a corresponding selection of probabilities.

The automata developed range through many varieties of probabilistic model, including *discrete-time Markov chains*, *continuous time Markov chains* and *Markov decision processes*. As with all of the approaches above, a temporal logic corresponding to the particular automaton structure has been developed. In fact there are many, but the simplest in this area is the probabilistic temporal logic, PCTL [259].

5.4.10   Real-time, probabilistic and hybrid model checking

Once we have models and logics for *quantitative* aspects, such as real-time, continuous dynamics or probabilities, then the basic model-checking idea can be extended to these. Most extensions in this way follow the automata-theoretic view we have seen already, but just adapt this to the particular form of logic/automata required.

Research and implementation in these areas has led to a number of popular and powerful model-checkers. Thus, for real-time systems there are a number of model-checkers [24, 63, 476, 405, 328, 425, 490, 308], the most successful being the UPPAAL system. This has been used to model and verify networks of timed automata, and uses model checking as a key component [342, 59].

Within the probabilistic area, the PRISM model-checker [331, 421] is the most widely used. In PRISM, we can compute the minimum or maximum probability over a range of possible configurations or parameters of a model, producing a different kind of best/worst-case analysis. As with many of the model-checkers mentioned in this section, PRISM is a symbolic model-checker, which allows it to tackle the verification of quite large systems [164].

Finally, the area of model checking for hybrid systems is an increasingly popular and important area. Model-checkers that have been successfully used here include the HyTech system [267, 300], and the RED system [431].

5.5   Final exercises

Exercise 5.16 *Try matching the automaton produced by* Spin *against the one you would expect.*

- spin -f "[]p"   *i.e.* $\Box p$
- spin -f "X p"   *i.e.* $\bigcirc p$
- spin -f "p U q"   *i.e.* $p \, U \, q$
- spin -f " XXq"   *i.e.* $\bigcirc \bigcirc q$
- spin -f "[]<>r"   *i.e.* $\Box \Diamond r$

Exercise 5.17 *Consider the following example:*

```
proctype A (chan in, out) {....... }
```

```
proctype B (chan in, out) {....... }
proctype C (chan in, out) {....... }

init {   chan a2b = [1] of { int };
         chan b2c = [1] of { int };
         chan c2a = [1] of { int };
         atomic {   run A(c2a, a2b);
                    run B(a2b, b2c);
                    run C(b2c, c2a)   }
}
```

(a) What does the communication channel structure look like for this example?

(b) Process A is

```
proctype A (chan in, out)
{
         int total;
         total = 0;
S1:      total = (total+1)%8;
         out!total;
         printf("A sent %d\n", total);
         in?total;
         assert(total != 1);
         printf("A received %d\n", total);
         if
         :: (total != 0) -> goto S1;
         :: (total == 0) -> out!total
         fi
}
```

(Recall that % is the modulo arithmetic operator – so, for example, (17%8)=1 and (15%8)=7.) Assuming we run the process above with appropriate channels given as arguments, what is happening here?

(c) Process B is

```
proctype B (chan in, out)
{
      int total;
S1:   in?total;
      printf("B received %d\n", total);
      if
      :: (total != 0) -> total = (total+1)%8; out!total;
                         printf("B sent %d\n",total);
                         goto S1;
      :: (total == 0) -> out!total
      fi
}
```

Again, assuming we run such a process with appropriate channels given as arguments, what is happening here?

(d) *Assuming that process* C *is essentially the same code as process* B, *then what will happen when the whole* Promela *system is executed? In particular what sequence of outputs can we expect? Try changing this example (in* promela3ex.pml) *to have different numbers, and see what happens.*

(e) *Will the assertion in process* A *succeed?*

(f) *If we change the assertion in process* A *to be* assert(total !=3), *what will happen?*

(g) *If we wanted to verify that it is* not *the case that the* total *variable has a non-zero value infinitely often, what temporal formula would we wish to check?*

Exercise 5.18 *Look again at our erroneous program in* random_example2.pml. *Try to add an assertion and verify by both single execution (maybe you have to try this several times) and by exhaustive verification.*

6

Execution

The future will be better tomorrow.
— Dan Quayle

In this Chapter, we will:

- [INTRODUCTION] look at several ways of generating a program from a PTL specification (Section 6.1);

- [TECHNIQUE] describe the METATEM approach to the direct execution of PTL specifications (Section 6.2);

- [SYSTEM] examine the `Concurrent MetateM` system which implements the METATEM approach, bringing in deliberation, concurrency and multi-agent systems (Section 6.3); and

- [ADVANCED] highlight a selection of more advanced work including both extensions of METATEM and alternative approaches to implementing temporal specifications (Section 6.4).

Relatively few exercises will appear in this chapter, but the solutions to those that do can, as usual, be found in Appendix B.

6.1 From specifications to programs

We have already seen, earlier in this book, how temporal formulae can be used to describe individual computational components. In addition, we have seen that temporal formulae can usefully describe properties of existing programs (and that these properties can

---

An Introduction to Practical Formal Methods Using Temporal Logic, First Edition. Michael Fisher.
© 2011 John Wiley & Sons, Ltd. Published 2011 by John Wiley & Sons, Ltd.

sometimes be checked via model-checking techniques). However, there is a gap here between the specification of a component and its actual implementation.

As in many formal methods, there are several ways to move from a formal specification of a component to an implementation of that component (that satisfies the specification). The most common techniques for achieving this are *refinement*, *synthesis* and *direct execution*.

6.1.1   Refinement of temporal specifications

As we saw earlier, temporal specifications can be gradually refined, introducing more and more detail while decreasing their inherent non-determinism, until the specification is a very close match with a 'real' program. Consequently, as part of this process, we would expect the temporal semantics of the 'real' program to be (logically) equivalent to the refined temporal specification. At this point, the program is guaranteed to implement the original specification. Thus, the art of formal software development here is to (manually) direct the refinement process, and corresponding refinement proofs, in such a way that the required program is produced. There are many works on refinement, both in formal methods in general [2, 106, 135, 301, 306, 375] and in temporal specification in particular [313, 363].

6.1.2   Synthesis from temporal specifications

An alternative, and in many ways more appealing, approach is to *directly synthesize* a program from a temporal specification. This technique, which can sometimes be automatic, leads us towards the use of temporal specifications as our programming notation. Given a specification, synthesis attempts to generate a program which will be guaranteed to implement the specified behaviour. As you can imagine, this is quite difficult in general. However, ground-breaking work has been carried out on synthesizing finite automata (which can then easily be used to generate a 'real' program) from temporal specifications [415, 416]. This led to an increasingly popular subfield concerning the synthesis of classes of automata from various temporal formalisms (see Section 6.4.8).

6.1.3   Executing temporal formulae

Thus, on the one hand, we can carry out manual refinement, which is laborious but leads to programs that implement the specification; on the other we can attempt to directly extract a program from the specification, though this is typically a very complex (and rarely automatic) process. In this chapter, however, we will look at a 'third way', namely the *direct execution* of temporal specifications. This provides a relatively lightweight mechanism for animating temporal specifications that is particularly useful when *prototyping* specifications to see how they behave.

The execution procedure is automatic and, in certain cases, complete (i.e. guaranteed to generate an execution if any is possible) though, unsurprisingly, retaining this completeness becomes more difficult as we move beyond single components described just using PTL.

Aside: Executing logical formulae. The direct execution of (non-temporal) logical formulae has been quite influential within Computer Science and Artificial Intelligence,

particularly through the *Logic Programming* approach popularized by languages such as
PROLOG [320, 321, 459]. As we will see later, the PROLOG style of execution (based on a
depth-first, backward chaining, search through a set of Horn Clauses), has been adapted
to PTL (Section 6.4.3). Although appealing, this Logic Programming approach does not
fit so easily within the temporal framework, having some technical problems and being
slightly counter-intuitive in the way that executions are constructed. So, rather than, for
example, backward chaining from some eventuality we wish to see true, the approach we
describe in this chapter instead *forward chains* from the known facts while attempting
to satisfy multiple eventualities. In this sense, the approach we use here is closer in style
to *production-rule languages* [251] than to Logic Programming.

To begin our explanation, let us first recall the relationship between temporal formulae
and temporal models. As we have seen, a formula such as

$$\bigcirc do(a) \wedge \bigcirc \bigcirc do(b) \wedge \bigcirc \bigcirc \bigcirc do(c)$$

might have models that look something like

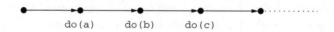

Yet, such models also correspond to executions of a program such as

```
do(a); do(b); do(c)
```

Thus, if we can find a way to extract models from a temporal specification, we are
effectively managing to *execute* that specification directly (rather than refining it to a
program and then executing the program). We can modify our earlier view of the relationship between programs, specifications and sequences, to incorporate this idea of direct
execution as follows:

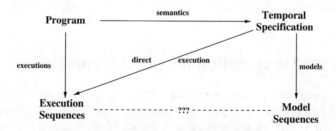

Thus, execution of temporal specifications can be seen as building sequences that are
models for the specifications [191]. It is important to note, however, that the infinite
nature of such models mean that, in some cases, execution corresponds to *attempting*
to build these models. In addition, a particular feature of the way we build our models,
called the '*Imperative Future*' approach [50], is that they are built *from the beginning*; the
model is constructed step by step, starting from the initial state. This is important because
it corresponds closely to the way programs are executed, where execution sequences are
built in a similar way.

Aside: Execution as interpretation.   Just as there are several ways, in traditional Computer Science, to take a program and implement it so we can see the possibilities for implementing temporal specifications as being comparable (without being too exact in this analogy):

- *specification refinement* $\simeq$ *program transformation*
  Transformation, including techniques such as *partial evaluation*, replaces the current program with another that still achieves what we require, but in a more efficient way.
  Refinement involves replacing the current specification by another that at least does what the current one does, but does it in a more specific way.

- *specification synthesis* $\simeq$ *program compilation*
  Compilation involves producing a description in a target language that ensures the new description generates all required runs of the program.
  Synthesis involves producing a description in another language, typical a finite automaton, that satisfies the original specification.
  (Although this is, perhaps, a little tenuous, we also note that compilation techniques involve significant pre-processing, analysis and optimization. Similarly, synthesis techniques are typically the most complex and difficult aspects we discuss here.)

- *specification execution* $\simeq$ *program interpretation*
  Interpretation attempts to build a run of the program as the syntactic structure of the program is being explored.
  Execution involves using the specification as it is, and applying a lightweight process to attempt to build an execution as the specification is explored.

Thus, rather than carrying out a great deal of pre-processing to ensure that the resulting code will work in all/many situations (as compilation and synthesis both do), interpretation and execution attempt to build a run/execution of the system as the program/specification is being explored. Interpretation is lightweight but might not always generate an appropriate execution (especially if the interpreter needs to make an arbitrary choice); similarly, the direction execution of temporal specifications that we will see below has similar properties.

6.2   METATEM: executing temporal formulae

Given a PTL specification of an individual component, how do we execute this specification in order to provide an appropriate 'run'? It is clear that, for example, formulae such as '$a \wedge \bigcirc b \wedge \square c$' should result in a sequence of the form

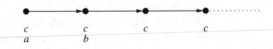

But what if we have disjunctions? Or temporal choices, as are inherent in '$\Diamond$' and 'U'? And what if the PTL formula is itself unsatisfiable?

    To cater for these aspects, we require a general algorithm for constructing a model from a PTL formula, if one exists. To do this we will use the basic 'METATEM' algorithm, which provides a simple mechanism for executing PTL formulae in order to build such executions [49]. The basic METATEM approach involves:

1. transforming the PTL specification into SNF (as in Section 2.6);

2. from the initial rules, which define an initial state of the execution, *forward chaining* through the set of step rules that describe the *next* state of the execution; and

3. constraining this execution by attempting to satisfy eventualities (sometimes termed *goals*), such as $\Diamond g$ (i.e. '*g* eventually becomes true') introduced by the *sometime* rules. (This, in turn, may involve some strategy for choosing between such eventualities.)

Recall the three forms of rule in SNF:

$$\textbf{start} \quad \Rightarrow \quad \bigvee_{b=1}^{r} l_b \qquad \text{(an \textit{initial} rule)}$$

$$\bigwedge_{a=1}^{g} k_a \quad \Rightarrow \quad \bigcirc \bigvee_{b=1}^{r} l_b \qquad \text{(a \textit{step} rule)}$$

$$\bigwedge_{a=1}^{g} k_a \quad \Rightarrow \quad \Diamond l \qquad \text{(a \textit{sometime} rule)}$$

As stated above, we use initial rules to provide choices for the first state in the execution, then use an algorithm for forward chaining through the step rules which we will describe below.

Aside: Constructing temporal models.   The execution of the PTL specification φ, is an iterative process of labelling a sequence of states with a propositional valuation in each state, that eventually yields a model for φ (if the formula is satisfiable). The model structure produced is a sequence of states, with an identified start point. Our approach begins at state 0, and steps through each state in the structure in turn. In principle, because the sequences are infinite, then this process *could* go on forever. However, a property of PTL is that any execution will eventually 'loop' (i.e. reach the same state/formula configuration at some point in the future) and so we only need to explore a finite number of states. If we succeed in building a structure, σ, then we know that it will satisfy $\langle \sigma, 0 \rangle \models \varphi$.

6.2.1 METATEM execution: initial state

We begin by constructing the *first* state, at $t = 0$. For this, we only need to consider initial rules, that is rules of the form

$$\textbf{start} \quad \Rightarrow \quad \bigvee_{b} l_b .$$

Together, all such rules provide us with a choice of labellings for the initial state. The execution mechanism simply chooses one of these labellings and moves on to construct the next state ($t = 1$).

Example 6.1 *Assume we have only the initial rules*

$$\textbf{start} \;\Rightarrow\; a \vee b$$
$$\textbf{start} \;\Rightarrow\; \neg a$$

If we conjoin all the right-hand sides of the initial rules, we get

$$(a \vee b) \wedge \neg a$$

which simplifies to

$$b \wedge \neg a .$$

Thus, in this case, there is no choice and the first state must have 'a' false and 'b' true.

If it turns out that this initial choice was *bad*, in that it later leads to a contradictory situation, then the execution will eventually *backtrack* to this choice point (see Section 6.2.5) and another option will be selected.

Example 6.2 *A more interesting set of initial rules is*

$$\textbf{start} \;\Rightarrow\; c \vee d$$
$$\textbf{start} \;\Rightarrow\; \neg e \vee \neg f$$

Again we conjoin all the right-hand sides of the initial rules, giving

$$(c \vee d) \wedge (\neg e \vee \neg f) .$$

Simplifying and rewriting to DNF gives us

$$(c \wedge \neg e) \vee (c \wedge \neg f) \vee (d \wedge \neg e) \vee (d \wedge \neg f)$$

which represents the four basic choices we have. We will see later how we might prefer certain choices but, whichever one we now choose, we can construct the first state based on this (while making assignments to other propositions) and move on to $t = 1$. If this execution subsequently fails, then we come back to choose a different first state. If we have explored all the above four possibilities and each one leads to failure, we terminate the procedure reporting that the formula is unsatisfiable.

6.2.2   METATEM execution: forward chaining

Once we have selected a first state we then proceed to build the rest of the sequence. To give some idea of this part of the execution process, we will consider forward chaining via a step rule of the form $(L \Rightarrow \bigcirc R)$ executed at an arbitrary, $t > 0$, point in the execution.

Now, since we wish to construct state t, and given that state $t - 1$ has already been constructed, the antecedent of the step rule, L, is evaluated in the model at the previous state, that is we check $\langle \sigma, t - 1 \rangle \models L$. Since, in SNF step rules, L is just a conjunction of literals, then this check is relatively simple. There are, of course, two possible outcomes:

- L evaluates to **false**, that is $\langle \sigma, t - 1 \rangle \not\models L$
  Here there is nothing more to be done with this rule, by virtue of the meaning of '$\Rightarrow$', because **false** $\Rightarrow \bigcirc R$ is true regardless of R. So, we now have no constraints on the new state we are constructing and can make an arbitrary choice. (However, see later for preferences on such choices.)

- L evaluates to **true**, that is $\langle \sigma, t - 1 \rangle \models L$
  Now, since we also know that '$L \Rightarrow \bigcirc R$' is true everywhere, then L being true at $t - 1$ requires us to ensure

$$\langle \sigma, t - 1 \rangle \models \bigcirc R$$

which, in turn, requires

$$\langle \sigma, t \rangle \models R .$$

Recall that as R, in SNF, is of the form

$$\bigvee_{b=1}^{r} l_b$$

then we clearly might have a choice of assignments to make to propositions within state t.

Aside: Eventualities.   For the moment we will ignore the possible presence of eventualities, such as $\diamondsuit z$. However, we will return to these in Section 6.2.3.

When the specification comprises several step rules, then these are all evaluated simultaneously. Of course, if the antecedents of more than one rule evaluate to true, the consequents of the successful rules will be conjoined to form a single formula which must be made true (as above).

Example 6.3 *A straightforward example of basic execution concerns the following set of SNF rules.*

$$
\begin{aligned}
\textbf{start} &\Rightarrow p \\
\textbf{start} &\Rightarrow \neg q \\
p &\Rightarrow \bigcirc q \\
q &\Rightarrow \bigcirc r
\end{aligned}
$$

Here, the first state $t = 0$ is labelled with $\{p, \neg q\}$. Then we assess all the step rules in this state and find that the only constraint we have on $t = 1$ is that 'q' must be true. (The choice of whether p or r are true remains unspecified.) Moving forward to construct $t = 2$, we check the step rules in $t = 1$. We know for certain that since q is true in $t = 1$ then r is now forced to be true in $t = 2$. And so on.

Example 6.4 *In the above example we had few choices, but now let us consider the following rules.*

$$\text{start} \ \Rightarrow \ s$$
$$s \ \Rightarrow \ \bigcirc(x \lor u)$$
$$s \ \Rightarrow \ \bigcirc(\neg v \lor \neg w)$$

Now, at $t = 0$, s is certainly true. In constructing $t = 1$ we find that both step rules fire and the execution mechanism is left with the choice of disjuncts from

$$(x \land \neg v) \lor (x \land \neg w) \lor (u \land \neg v) \lor (u \land \neg w) .$$

One of these is selected (with an appropriate choice for the remaining propositions) and we move on to $t = 2$.

Note: This approach contrasts with Logic Programming where we follow only one rule at a time.

6.2.3   METATEM execution: eventualities

Now, what happens if we have a *sometime rule* such as $c \Rightarrow \Diamond d$? Well, recall that a property of the '$\Diamond$' operator is

$$\Diamond \varphi \ \Leftrightarrow \ (\varphi \lor \bigcirc \Diamond \varphi) .$$

If we apply this to the above sometime rule, then we get

$$c \ \Rightarrow \ (d \lor \bigcirc \Diamond d)$$

which simplifies to

$$(c \land \neg d) \ \Rightarrow \ \bigcirc \Diamond d .$$

So, to some extent, such rules just look like standard step rules. We check the left-hand side in the last state constructed and, if this is satisfied, make the right-hand side true in the current state. Of course the difference here is that we have to decide what to do about $\Diamond d$ in the current state. If d is already constrained to be false, then we must postpone satisfaction of $\Diamond d$, effectively making $\bigcirc \Diamond d$ true. However, if d is currently unconstrained, then the execution mechanism has a choice between making d true now and postponing this. We will see later that this is among the many strategic choices that the METATEM execution mechanism has.

Example 6.5 *Consider the simple set of rules below and let us assume that we will never postpone satisfaction of an eventuality when we have a choice.*

$$\text{start} \ \Rightarrow \ k$$
$$\text{start} \ \Rightarrow \ \neg m$$
$$k \ \Rightarrow \ \bigcirc \neg m$$

$$k \Rightarrow \bigcirc l$$
$$l \Rightarrow \bigcirc \neg m$$
$$k \Rightarrow \Diamond m$$

At $t = 0$ we have k true and m false. At $t = 1$ we certainly know that m is false and l is true. So, we still cannot satisfy $\Diamond m$ and must postpone again. Now, in $t = 2$, m is still false (because l was true at $t = 1$) so we again postpone $\Diamond m$. Finally, assuming no other restrictions are added, then $\Diamond m$ can be satisfied at $t = 3$ by making m true.

Now that we have the execution of eventualities such as '$\Diamond m$', then we also have the possibility of postponing satisfaction of an eventuality forever! This is obviously bad!

Example 6.6 *Consider the following simple example where we get the same configuration forever.*

$$\textbf{start} \Rightarrow \neg a$$
$$\textbf{true} \Rightarrow \Diamond a$$
$$\textbf{true} \Rightarrow \bigcirc \neg a$$

So, every state is labelled by '$\neg a$' and the satisfaction of '$\Diamond a$' is postponed forever.

Clearly, an important aspect is to be able to detect such situations and decide what to do.

6.2.4   METATEM execution: loop checking

A '*loop*' in METATEM is where the same state recurs along a sequence together with the same unsatisfied eventualities. For example, in Example 6.6 above, a state labelled by '$\neg a$', but with '$\Diamond a$' outstanding, repeats forever. However, it is important to note that loops can, in some cases, be perfectly acceptable. Consider

$$\textbf{start} \Rightarrow t$$
$$\textbf{start} \Rightarrow v$$
$$t \Rightarrow \bigcirc t$$
$$v \Rightarrow \bigcirc v$$

This gives a recurring state labelled with $\{t, v\}$. This is a perfectly acceptable model as the specification essentially captures $\Box (t \wedge v)$.

So, it is useful to be able to detect 'loops' but then we must decide whether they are 'bad' or 'good'. Essentially the difference between the two is that in 'bad' loops, one or more eventualities are continually outstanding (i.e. never satisfied and so always postponed), while in 'good' loops, any eventualities that occur are satisfied at some point within the loop.

Example 6.7 *Here is a distinguishing example.*

$$\textbf{start} \Rightarrow f$$
$$f \Rightarrow \Diamond \neg f$$
$$\neg f \Rightarrow \bigcirc f$$

Here, assuming we always satisfy '$\Diamond\neg f$' as soon as we can, then the execution loops, perpetually producing

$$\{f\} \longrightarrow \{\neg f\} \longrightarrow \{f\} \longrightarrow \{\neg f\} \longrightarrow \dots$$

But with '$\Diamond\neg f$' required throughout the loop.

The above represents a 'good' loop since exactly the same eventuality does not remain postponed throughout the whole of the loop. In fact the $\Diamond\neg f$ eventuality is satisfied and another, identical one, is then generated.

So, some form of 'loop checking' is essential and, when such loops are found, the execution must make suitable decisions about what to do (see below). It is also interesting to note that similar loop checks occur in many other techniques for satisfiability checking in temporal logics.

Aside: Loop size. As PTL has the finite-model property, then typically the METATEM execution mechanism will loop eventually. How long we have to wait before this loop occurs and, indeed, how big this loop might be, is dependent on the PTL formula we are executing. We can use a coarse form of *abstract interpretation* [130] to generate an upper bound on this size which is, in the worse case, $2^{5|\varphi|}$, where $|\varphi|$ is the size of our specification, φ; see below and [47] for more details.

6.2.5   METATEM execution: backtracking

Backtracking is a standard procedure that is used when exploring logical properties. Whenever we have a choice, for example between disjuncts, we make a selection but record the other possibilities. If we later find that our choice led to some inconsistency we stop, undo the current construction, and revert back to the last point where a choice was made and select a different option. If, in a particular scenario, all the choices have been explored and each leads to an inconsistency, then we go back to the last choice before this one. And so on. If we eventually get back to our initial starting point, then we know that the formula is inconsistent because we have explored all potential models. So, when we are attempting to construct our temporal sequence, we make a choice between disjuncts but record the other possibilities in case we backtrack to this choice point.

Example 6.8 *Recall the set of rules from Example 6.4. Let us extend this with two additional step rules, giving*

$$
\begin{array}{rcl}
\textbf{start} & \Rightarrow & s \\
s & \Rightarrow & \bigcirc(x \vee u) \\
s & \Rightarrow & \bigcirc(\neg v \vee \neg w) \\
x & \Rightarrow & \bigcirc\textbf{false} \\
\neg v & \Rightarrow & \bigcirc\textbf{false}
\end{array}
$$

> *Now, at* $t = 1$ *we have the choice of disjuncts from*
>
> $$(x \wedge \neg v) \vee (x \wedge \neg w) \vee (u \wedge \neg v) \vee (u \wedge \neg w).$$
>
> *Let us assume we select the first one, labelling* $t = 1$ *with* $\{x, \neg v\}$. *Now, as we move forward to* $t = 2$ *we find* **false** *is required. This is clearly inconsistent and so we must backtrack to consider one of the other choices. Back in* $t = 1$ *we might choose* $\{x, \neg w\}$ *but this again leads us to* **false** *in* $t = 2$. *Similarly, a choice of* $\{u, \neg v\}$ *also leads us to a contradiction. Thus, we are left with the choice of* $\{u, \neg w\}$ *at* $t = 1$ *which allows us to move on to* $t = 2$ *without immediate inconsistency.*

As we can see from the above example, there can be quite a large amount of backtracking. In current METATEM implementations such backtracking is quite simple and naive. In particular, the more sophisticated forms of backtracking developed over many years by researchers tackling Logic Programming implementations [329, 458] are *not* yet used within the METATEM implementation. There are two reasons for this: first, the implementation technology of executable temporal logics is still at a relatively early stage; and second, as we will see later, the predominant use of METATEM is in a multi-agent environment which itself heavily limits the amount of backtracking that can be undertaken.

6.2.6   Detailed execution example

We now present a more detailed execution example, using choices, backtracking and one eventuality. This is similar, in style, to the 'car agent' example in [201] and gives a low-level explanation of the basic METATEM execution. In this example, the propositions we can control are *begin*, *typing*, *almost_there*, *tired* and *thinking*. A simple specification, already in our normal form, might be:

1. **start** $\Rightarrow \neg typing$

2. $\neg typing \Rightarrow \bigcirc begin$

3. $begin \Rightarrow \bigcirc (tired \vee thinking)$

4. **true** $\Rightarrow \bigcirc (\neg tired \vee \neg typing)$

5. $begin \Rightarrow \Diamond typing$

6. $tired \Rightarrow \bigcirc tired$

7. $(thinking \wedge typing) \Rightarrow \bigcirc almost\_there$

8. $almost\_there \Rightarrow \bigcirc (tired \vee typing)$

Informally, the meaning of these rules is as follows:

1. *typing* is false at the beginning of execution.

2. If *typing* is false, *begin* will be true in the next moment.

3. Whenever *begin* is true, then either *tired* or *thinking* will be true in the next moment in time.

4. At any future moment in time, we cannot have both *tired* and *typing* being true.

5. Whenever *begin* is true, a commitment to eventually make *typing* true is given.

6. If *tired* is true, then it will also be true in the next state (note that this effectively means that, once *tired* is true, then it will *always* be true).

7. If both *thinking* and *typing* are true then, in the next moment, *almost_there* is made true.

8. Finally, once *almost_there* is made true, then either *tired* or *typing* will again be made true in the next moment in time.

We will briefly show how execution is attempted so how model construction proceeds. In the following diagrams

  ○ represents a previously constructed state,

  ⊙ represents the current state being developed, and

  ⊗ represents a state we have backtracked from.

Step 1: from rule 1, build an initial state in which *typing* is false. (There are, of course, other propositions but, for clarity, we omit these from the diagrams below. Note, however, that the omitted propositions are typically false.)

¬ *typing*

Step 2: from rule 2, build a next state in which *begin* is true.

¬ *typing*   *begin*

Step 3: rule 3 gives us a choice of making either *tired* or *thinking* true in the next state. We begin by exploring the '*tired*' possibility, and note that rule 5 means that we must also make *typing* true somewhere down this branch, while rule 4 says that *typing* must be false here.

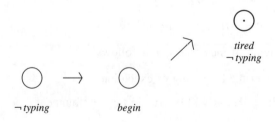

Step 4: after some further execution we recognize a loop where *tired* is true forever along this branch, forcing *typing* to be false everywhere along this branch (from rule 4). Since we therefore cannot satisfy ◇*typing*, we fail in exploring/executing this branch. We backtrack and return to the other option in Step 3, namely that *thinking* becomes true in the next state. We can now also satisfy ◇*typing* from rule 5, by making *typing* true here.

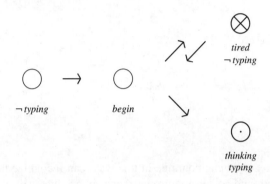

Step 5: since *thinking* and *typing* are both true, then rule 7 leads us to make *almost_there* true.

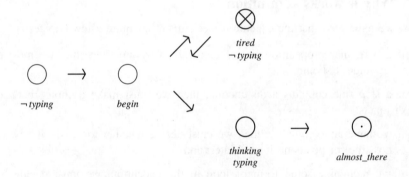

Step 6: now, rule 8 provides us with a choice and, again, if we take the *tired* option, this leads us to problems (because of rules 2, 3, 5, and 6).

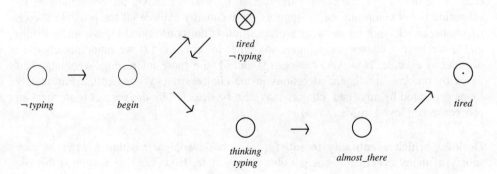

Step 7: however, if we take the *typing* choice, then execution (and model construction) can continue.

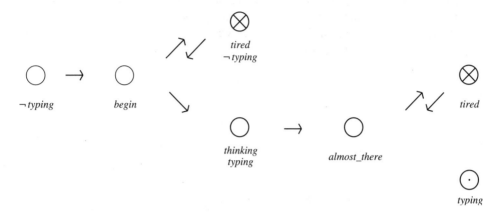

And so on. Execution can continue in this way, can recognize that a previous state has re-occurred (and so can loop round), or can terminate (if all branches explored lead to contradictions).

6.2.7   Why it works – intuition

By now we have seen that the execution mechanism has quite a few choices:

1. if a particular proposition in *not* constrained by any step rule, we must decide what value it should take;

2. if a step rule contains a disjunction, then we must make a choice between the disjuncts;

3. if we have an eventuality, then we must decide whether to satisfy it now (if that is possible) or postpone it until later; and

4. if we recognize some form of loop in the execution, we must decide how to handle this.

Most of these choices do not affect the correctness of the execution, but often affect either its efficiency or clarity. In particular, in (1) and (2) above, the correctness of the execution is not compromised, as long as we eventually explore all the possible choices (if we backtrack). So, typically, if a proposition is not constrained (1) we make it false and if we have a disjunction without any other information (2) we randomly choose a disjunct to execute. However, choices (3) and (4) are more interesting. Avoiding 'bad' loops by making 'intelligent' decisions in the choice strategy is essential. Fortunately, loops generated by any 'bad' choices may also be detected by the general loop-checking procedure employed in METATEM.

Deciding which eventuality to satisfy.   The basic approach within METATEM is to satisfy as many eventualities as possible in any state. However, we also need to make sure that we do not continually postpone satisfaction of any particular eventuality. Indeed, the correctness of the execution mechanism depends on the essential *fairness* of this selection [197]. In the basic METATEM approach this is handled by attempting to satisfy eventualities in a list ordered by the *'age'* of the outstanding eventualities. Thus, if a particular eventuality has been outstanding for quite a number of states, it is likely to be the first one attempted as execution proceeds. Certainly it should be attempted *before* any newly generated eventualities.

Example 6.9 *Consider the set of rules*

$$
\begin{array}{rcl}
\textbf{start} & \Rightarrow & \neg f \\
\textbf{start} & \Rightarrow & \neg g \\
\textbf{true} & \Rightarrow & \Diamond f \\
\textbf{start} & \Rightarrow & a \\
a & \Rightarrow & \Diamond g \\
\textbf{true} & \Rightarrow & \bigcirc(\neg f \vee \neg g)
\end{array}
$$

Here, we have '$\Diamond f$' generated at every state – thus, even when f is satisfied, there will be another '$\Diamond f$' to satisfy in the next state. However, 'g' only needs to occur once to satisfy the single '$\Diamond g$'.

Now, when we come to $t = 1$ we must choose to satisfy one eventuality and postpone the other (since, by the last rule, f and g can never be true at the same time). Imagine that we satisfy f and postpone $\Diamond g$ to the next state. When we get to $t = 2$, we have a new $\Diamond f$ together with the $\Diamond g$ postponed from the last state. Again we need to choose between them.

If the choice is unfair, it is possible that f will be chosen at every one of these states, and so $\Diamond g$ will be postponed forever. Although our loop checking might detect such a situation, it is clearly desirable to avoid getting to that stage. By using a 'fair' choice we ensure that $\Diamond g$ regularly gets a chance to be satisfied. In particular, the (fair) METATEM approach of attempting the oldest *outstanding eventualities first would ensure that g is satisfied by $t = 2$.*

Exercise 6.1 *Consider the following set of rules, similar to Example 6.9 but with $\Diamond g$ also occurring infinitely often:*

$$
\begin{array}{rcl}
\textbf{start} & \Rightarrow & \neg f \\
\textbf{start} & \Rightarrow & \neg g \\
\textbf{true} & \Rightarrow & \Diamond f \\
\textbf{true} & \Rightarrow & \Diamond g \\
\textbf{true} & \Rightarrow & \bigcirc(\neg f \vee \neg g)
\end{array}
$$

Assuming we invoke the 'oldest eventuality first' strategy, what two sequences of states represent possible METATEM executions for the above?

Checking loops.   Earlier, in Section 6.2.4, we described a mechanism for checking for loops in the execution. This involved keeping a detailed history and comparing the current state (including outstanding eventualities) against all previous states. This is quite difficult to implement efficiently! Now, if we assume that the execution is indeed using the above 'oldest eventuality first' strategy, then we can do something a little simpler.

Recall that the main reason we need to carry out loop checking is to catch the 'bad' situations where an eventuality is continually postponed. Since we now know that, in this case, the eventuality will keep on being the first one attempted (because it will gradually become the oldest) then all we need to check is whether any given eventuality has been continually unsatisfied for N states. In the implementation, each eventuality is 'tagged' with the number of the state in which it was invoked and then the tag is removed once it is satisfied. So we can easily see if any eventuality has been outstanding (i.e. invoked, but not yet satisfied) for an unnecessarily large number of steps.

In such a case, we force backtracking. Either the eventuality cannot possibly be satisfied, and so this backtracking will lead us back towards declaring a contradiction, or there *is* a way in which it can be satisfied. By the properties of the logic we know that such a *satisfying execution* could have been reached by an earlier choice, and so backtracking here will force us to find it.

6.2.8   Why it works – formal description and properties

Execution of a PTL formula, φ, means constructing a sequence, σ, that satisfies $\langle \sigma, 0 \rangle \models \varphi$. In our case, we are attempting to construct a model for the formula corresponding to the set of SNF rules. Now, a formalization of the METATEM execution mechanism is given below, where we attempt to construct a model for a PTL formula described by sets of Initial, Step and Sometime rules. In this, two elements characterize the execution: the valuation of propositions in each state (S_i, for state i); and lists of currently outstanding eventualities (E_i, for state i).

Note.   For simplicity, we will assume that all the *sometime* rules within Sometime are of the form $A \Rightarrow \Diamond^+ B$, where '$\Diamond^+$' means 'at some point from the next moment in time onwards'. As we saw earlier, $\Diamond p \Leftrightarrow (p \vee \Diamond^+ p)$. Now, the execution mechanism proceeds as follows:

1. Make a (consistent) choice of Boolean assignments for propositions as described by the Initial set of rules; label this as S_0 and let $E_0 = \langle \rangle$.

2. Given S_i and E_i, proceed to construct S_{i+1} and E_{i+1} as follows:

    (a) Let $C = \{F \mid (P \Rightarrow \bigcirc F) \in$ Step and $S_i \models P\}$
    that is, C represents the constraints on S_{i+1} derived from the Step clauses.

    (b) Let $E_{i+1} = E_i \widehat{\,} \langle G \mid (Q \Rightarrow \Diamond^+ G) \in$ Eventuality and $S_i \models Q \rangle$
    that is, E_{i+1} is the previous list of outstanding eventualities, with all the newly generated eventualities appended (we should, of course, carry out some simplification here).

    (c) For each $V \in E_{i+1}$, starting at the head of the list,
    if $(V \wedge C)$ is consistent,
    then update C to $(C \wedge V)$ and remove V from E_{i+1}.
    Otherwise, leave V within E_{i+1}.

    (d) Choose an assignment consistent with C and label this S_{i+1}. If there are no consistent assignments, then backtrack to a previous choice point.

   (e) [loop check] **if** V has occurred through all of $E_i, E_{i-1}, \ldots, E_{i-N}$
                          then fail and backtrack to a previous choice point
                          else go to (2)

The core MᴇᴛᴀᴛᴇM execution mechanism, as defined above, with the strategy combining

S1: attempting the oldest eventualities first at the choice in 2(b), and

S2: forcing backtracking at 2(e) if the same outstanding eventualities occur for N states continually (where N is related to the bound on the size of a finite model for the logic),

is *complete*, that is an execution (a model) will be produced if, and only if, the original formula is satisfiable. The proof of this theorem for basic MᴇᴛᴀᴛᴇM can be found in [49, 50], but the intuition behind this can be given as follows. Essentially, the proof proceeds by showing that the above execution mechanism will (eventually) explore all the tableau structure for PTL [500, 501]. The tableau structure is similar to the Büchi Automaton representation we saw earlier. Thus, the condition (S1) above ensures that we potentially explore all possible paths through the tableau/automaton, while (S2) ensures that we do not continue on paths that are not (quickly) leading to an acceptable sequence. So, if we have seen an eventuality outstanding continually for a number of states, which is greater than the number of different tableau/automaton states, then we know that either there is *no* acceptable path, or there is a shorter one that we have missed. In either case, we backtrack.

Aside: Is completeness important?   A central result concerning the execution of SNF rules (and so, PTL formulae) is that, if the above execution procedure is followed, then a model (an execution sequence) will be constructed if, and only if, the formula (specification) is satisfiable [49]. In this respect, propositional MᴇᴛᴀᴛᴇM is a complete theorem-prover for PTL: we can attempt to execute a formula – if we fail we know that the formula is *unsatisfiable*. However, even though MᴇᴛᴀᴛᴇM has this property, it is actually rarely used in this way. In particular, as soon as we allow multiple executions and communication, or we extend the basis to first-order temporal logic, then the execution mechanism becomes incomplete anyway. We will see more of this in Section 6.3.

Exercise 6.2 *Choose an example from Chapter 4 where a set of SNF rules are shown to be unsatisfiable. Now, use* MᴇᴛᴀᴛᴇM *to directly execute this set of rules. What is the expected behaviour? And does this occur? (It is advisable to choose a* small *example!)*

6.2.9 Executing MᴇᴛᴀᴛᴇM rules – reactive behaviour

Now that we have the basic MᴇᴛᴀᴛᴇM execution mechanism, what can we do with it? We will next present a few different ways in which the flexibility of the approach allows us to represent some interesting behaviours. We will begin with simple *reactive* behaviour.

    If we wish to describe a scenario where, when one element becomes true, another must surely follow, then we can simply code this as

$$stimulus \quad \Rightarrow \quad \bigcirc response .$$

However, such rules can also be more complex in terms of both temporal delay,

$$stimulus \quad \Rightarrow \quad \bigcirc \bigcirc \square \bigcirc response$$

and outcomes,

$$stimulus \quad \Rightarrow \quad \bigcirc(\bigcirc response1 \ \lor \ \bigcirc \bigcirc \square response2).$$

More importantly, the forward-chaining aspect of METATEM execution allows us to enforce several of these at once, for example with situations such as

$$\begin{aligned} stimulus1 \quad &\Rightarrow \quad \bigcirc response1 \\ stimulus2 \quad &\Rightarrow \quad \bigcirc response2 \\ stimulus3 \quad &\Rightarrow \quad \bigcirc response3 \end{aligned}$$

If, for example, *stimulus1* and *stimulus3* both occur at the same state, then *response1* and *response3* will both be true at the next state. As well as being potentially useful for applications such as process control [184], where we must keep track of several different activities, this aspect forms the basic, underlying reactive behaviour within autonomous *agents*. (We will examine these further in Section 6.2.12.)

Example 6.10 *Consider the following set of reactive rules*

$$\begin{aligned} \mathbf{start} \quad &\Rightarrow \quad open\_the\_door \\ open\_the\_door \quad &\Rightarrow \quad \bigcirc shock1 \\ open\_the\_door \quad &\Rightarrow \quad \bigcirc shock2 \\ shock1 \quad &\Rightarrow \quad \bigcirc jump \\ shock2 \quad &\Rightarrow \quad \bigcirc shout \\ (jump \land shout) \quad &\Rightarrow \quad \bigcirc scream \end{aligned}$$

Executing these should give us the sequence

$$\{open\_the\_door\} \longrightarrow \{shock1, shock2\} \longrightarrow \{jump, shout\} \longrightarrow \{scream\} \longrightarrow \ldots$$

6.2.10   Executing METATEM rules – planning

Logic Programming mechanisms, such as PROLOG, with its depth-first search for a refutation, are typically good for planning but less suited to describing multiple reactive behaviours as in Section 6.2.9 above. Consequently, we might expect that a forward-chaining approach, as exhibited by METATEM, would struggle to handle planning activities. However, while they must be represented in a slightly unorthodox way, METATEM can indeed tackle (at least simple) planning.

Consider, for example, some *goal* situation we wish to achieve once a trigger condition occurs. Now, if we want to represent the fact that we aim to reach this situation, then we can simply have

$$trigger \quad \Rightarrow \quad \Diamond goal.$$

Here, there is nothing to stop METATEM just making '*goal*' true. However, we must describe exactly what is involved in actually achieving this goal. So, let us imagine that we have three constituents that must be true in order for '*goal*' to be achieved. We might formalize this as (recalling that '●' means 'at the last moment in time')

$$goal \;\Rightarrow\; ●(requirement1 \wedge requirement2 \wedge requirement3).$$

Rewriting this formula, we get, successively

$$goal \;\Rightarrow\; ●(requirement1 \wedge requirement2 \wedge requirement3)$$
$$\downarrow$$
$$\bigcirc goal \;\Rightarrow\; (requirement1 \wedge requirement2 \wedge requirement3)$$
$$\downarrow$$
$$\neg(requirement1 \wedge requirement2 \wedge requirement3) \;\Rightarrow\; \neg\bigcirc goal$$
$$\downarrow$$
$$(\neg requirement1 \vee \neg requirement2 \vee \neg requirement3) \;\Rightarrow\; \bigcirc\neg goal$$

And, transformed to SNF rules,

$$\neg requirement1 \;\Rightarrow\; \bigcirc\neg goal$$
$$\neg requirement2 \;\Rightarrow\; \bigcirc\neg goal$$
$$\neg requirement3 \;\Rightarrow\; \bigcirc\neg goal$$

So, since we are trying to satisfy '$\Diamond goal$', then if any of the three requirements remain unsatisfied then we will not be able to succeed. So, if we consider

$$\bigcirc goal \;\Rightarrow\; (requirement1 \wedge requirement2 \wedge requirement3)$$

above then '$\Diamond goal$' essentially requires us to establish

$$\Diamond(requirement1 \wedge requirement2 \wedge requirement3).$$

Now, depending on the particular scenario we have, it might be (though it is not always) possible to transform this to

$$\Diamond requirement1 \;\wedge\; \Diamond requirement2 \;\wedge\; \Diamond requirement3$$

effectively generating three new *subgoals*. Unless these subgoals can be achieved then the overall '*goal*' cannot itself be achieved. (We will see in Section 6.2.12 how we might deal with multiple, potentially conflicting, goals.)

Example 6.11 *Successfully making a cup_of_tea requires that we have tea, hot_water, and a cup. We might formalize this as follows.*

$$thirsty \;\Rightarrow\; \Diamond cup\_of\_tea$$
$$\neg tea \;\Rightarrow\; \bigcirc\neg cup\_of\_tea$$
$$\neg hot\_water \;\Rightarrow\; \bigcirc\neg cup\_of\_tea$$
$$\neg cup \;\Rightarrow\; \bigcirc\neg cup\_of\_tea$$

In addition, we can legitimately add

$$thirsty \implies \Diamond tea$$
$$thirsty \implies \Diamond hot\_water$$
$$thirsty \implies \Diamond cup$$

to generate relevant subgoals. (Note that there may be further constraints on hot_water etc.)

So, if a *goal* has three requirements which must be met, then these lead to three subgoals. An alternative is that there may be several different ways to achieve our *goal*:

$$goal \implies \bullet(requirement1 \lor requirement2 \lor requirement3).$$

Thus, in order to achieve our *goal*, we must have achieved one of these three requirements. So, '$\Diamond goal$' effectively requires us to establish

$$\Diamond requirement1 \lor \Diamond requirement2 \lor \Diamond requirement3.$$

In this case, we must decide which option to choose in order to satisfy our '*goal*'. These choices correspond to different ways to achieve the goal, and so we might try to satisfy one of these requirements. This might succeed, but might also fail, forcing us to backtrack to consider another option. And so on.

Thus, we can utilize the deductive and backtracking aspects of METATEM in order to achieve the construction of a suitable sequence (i.e. a plan) to achieve our goal. However, simply using the METATEM execution mechanism in order to search for a plan is likely to be quite inefficient. A more plausible route would be to add meta-level constraints to direct this execution (see, for example [48]).

Example 6.12 *There may be several different ways to achieve our goal of getting to Manchester:*

$$at\_Manchester \implies$$
$$\bullet(Manchester\_by\_train \lor Manchester\_by\_plane \lor Manchester\_by\_car).$$

Now our specification is

$$visiting \implies \Diamond at\_Manchester$$
$$\left(\begin{array}{l} \neg Manchester\_by\_train \land \\ \neg Manchester\_by\_plane \land \\ \neg Manchester\_by\_car \end{array} \right) \implies \bigcirc \neg at\_Manchester$$

We need not make $\Diamond Manchester\_by\_train$, $\Diamond Manchester\_by\_plane$ *and* $\Diamond Manchester\_by\_car$ *all true, but we should at least choose* one *of these to pursue as a* subgoal.

6.2.11   Agents and METATEM

As we saw above, we are able (at least to some extent) to describe both reactive and planning aspects within METATEM. These capabilities have led to METATEM being a potentially useful language in which to describe and implement *rational agents* [190, 206]. But what is an *'agent'*? And what is a *rational* agent? And why might METATEM indeed be appropriate?

What is an agent?   This is a question that has taken up far too much time during the brief history of this area [219]. Agents began to be developed around 25 years ago as a natural convergence of AI techniques, particularly planning, communication and distributed computing [79]. To keep things simple let us consider an agent to be simply an *autonomous computational entity*. Thus an agent not only encapsulates data, as an 'object' does, but also encapsulates *behaviour* so that the agent makes its own decision about what to do in any given situation.

What is a *rational* agent?   A rational agent is an agent that acts in a 'reasoned' and explainable way. Recall that the basic concept of an agent is quite general, for example encompassing computational components that have complex neural networks, genetic algorithms or stochastic automata inside. These might indeed be autonomous, but it is often very difficult to explain exactly why such an agent chooses the particular course of action it does. For example, it might be just a random effect of the internal probabilistic computation.

On the other hand, rational agents must have explicit *reasons* for acting the way that they do [89]. Typically, these might be described as 'goals', 'intentions' or 'desires', all of which are categorized as *motivations*. The agent must then decide what to do based on its current motivations. A common view of a rational agent is that it is a computational entity that is *reactive* (i.e. it can respond to stimuli from its environment or other agents) *pro-active* (i.e. it can decide for itself how to proceed) and *social* (i.e. it can communicate with, and work together with, other agents in order to cooperate, compete, or coordinate [503].)

Why METATEM?   We might consider using METATEM to describe rational agents because:

- as we saw above, *reactive* behaviour can simply be represented within step rules;

- similarly, *pro-active* behaviour can be represented by the need to satisfy various eventualities; and

- as we will see later, METATEM components can communicate with others and so can develop sophisticated *social* interactions.

Specifically, we can use eventualities (or variations on them) as *motivations* that the agent has for doing something. Indeed, basic eventualities, such as

$$\Diamond at\_home$$

can be seen as very coarse motivations. Thus, in the presence of such eventualities/goals then, whenever the execution mechanism has a choice, it tries to satisfy its eventualities/goals.

Aside: Goals in METATEM.   Unsurprisingly, the use of '$\Diamond g$' to mean that 'g is our goal' is simple, but sometimes a little *too* coarse. For example, once we commit to making $\Diamond g$ true, then we really *must* make g true at some time in the future. In reality, agents sometimes decide to relinquish their goals, whereas METATEM components cannot choose never to satisfy g if they have committed to $\Diamond g$. It turns out that '$B\Diamond g$' is quite a useful, but initially strange, way to represent the fact that 'g is our goal' [200]. Here, 'B' represents the agent's *belief* (see Section 6.4.2) and so '$B\Diamond g$' might be seen as representing 'the agent believes that eventually g will be achieved'. This is used as a motivation for the agent itself but can also double as a motivation for a group of agents working together since the above does not prescribe exactly *which* agent should achieve the goal. Indeed, our agent might ask another agent to ensure 'g', thus also satisfying $B\Diamond g$.

6.2.12   Deliberation

In practice, what makes METATEM useful for representing and implementing rational agent computation is *deliberation*. Essentially, in rational agency, deliberation is the process of reasoning about, and manipulating, both goals and potential ways to solve them. It is this ability to control when, and how, goals are attempted that captures the form of deliberation we are interested in. To motivate further the need for deliberation, let us describe a simple example.

Example 6.13 *Consider an agent controlling a survey vehicle (for example, a robot rover) that is attempting to explore an area of land for mineral deposits. The agent receives information about the (general) terrain of the area and about initial target areas to explore. It has several motivations:*

- *to explore potential mineral deposits as soon as possible;*

- *to avoid dangerous terrain (for example, a ravine);*

- *to find new potential mineral sites; and*

- *to ensure enough fuel is available to return to base.*

These can be together seen as the goals that the agent has. Thus, the agent must dynamically

- *assess its information for veracity and, if necessary, revise the information held (after all, sensors are never perfect and the terrain maps might well be inaccurate),*

- *deliberate over its (possibly conflicting) goals in order to decide what actions (for example, movement) to take, and*

- *based on its current state, generate new goals (for example to explore a new target site) or revise its current goals.*

As we can see from the above example, goals must be manipulated in quite flexible ways. But how does this abstract view of agent *deliberation* relate to METATEM execution?

Well, as we have already seen, there are several choices that the MᴇᴛᴀᴛᴇM execution has concerning eventualities, the two most important of which are

- if we have an eventuality that must be satisfied, then we must decide whether to satisfy it now (if that is possible) or postpone it until later, and

- if we have several eventualities that must be satisfied, yet not all of them can be satisfied at once, which ones to satisfy now and which to postpone?

In general, MᴇᴛᴀᴛᴇM execution strategies for the first try to follow the usual strategy of satisfying eventualities as soon as possible. For the second, however, there are a range of different approaches. Below we will describe one, based on general eventuality ordering functions [194], while in Section 6.3.4 we consider the simpler approach currently used in MᴇᴛᴀᴛᴇM implementations.

To explain 'eventuality ordering functions', let us view the outstanding eventualities that MᴇᴛᴀᴛᴇM has, at any moment in time, i, as a list (specifically, the E_i from the execution mechanism in Section 6.2.8). In general, this list of eventualities will be attempted at each state 'in list-order'. Thus, in the basic MᴇᴛᴀᴛᴇM strategy we would order the list based on the age of the eventuality. When an eventuality is satisfied, it is removed from the list; when a new eventuality is generated, it is added to the end of the list. In this way, we ensure that the oldest outstanding eventuality is attempted first at each step.

However, now that we have a *list* of eventualities, what is to stop us re-ordering this list before we start trying to execute them? In particular, if we are representing an agent with multiple goals, then what is to stop us re-ordering our list so that the most 'important' goals appear earlier in this list? Thus, if the agent can *re-order* this list between states then it can have a quite sophisticated strategy for *deliberation*, that is for dynamically choosing what goals to tackle next. We will describe a simple example below. However, it is important to note that, unless we put some constraints on the re-ordering that we might apply, then there is a strong danger that the completeness of the execution mechanism will be compromised [197].

Example 6.14 *Consider the following scenario, from [194, 197]. Let us first begin with a list of eventualities, which can be considered as high-level goals or, in BDI [428, 429] terminology,* desires:

$$[\Diamond\text{be\_famous}, \Diamond\text{sleep}, \Diamond\text{eat\_lunch}, \Diamond\text{make\_lunch}]$$

The basic approach would be to execute these oldest-first. However, before proceeding to (attempt to) build the next state in the execution, we might re-order these eventualities so that the most 'important' appears first. Again, in BDI terminology, we might view these as intentions:

$$[\Diamond\text{be\_famous}, \Diamond\text{eat\_lunch}, \Diamond\text{sleep}, \Diamond\text{make\_lunch}]$$

Thus, the deliberation here captures the move from 'desires' to 'intentions', and partially characterizes the agent's view of what goals are currently most important.

Imagine now that we can assess which goals are the 'easiest' to achieve, for example because we have a set of plans to consult. We might next re-order the eventualities/goals based on this aspect – for example we might not know how to satisfy 'be_famous' and might recognize that 'make_lunch' must be satisfied before 'eat_lunch'. So, we might then generate our final sequence of goals to attempt:

$$[\Diamond \text{make\_lunch}, \Diamond \text{eat\_lunch}, \Diamond \text{sleep}, \Diamond \text{be\_famous}]$$

This then provides the list of eventualities that we attempt in order. In this way, the list of outstanding eventualities can be manipulated via ordering functions corresponding to some relevant aspect of the agent, and this re-ordering can potentially occur at every step of the execution.

6.2.13   METATEM and multi-agent systems

The basic METATEM approach is useful for animating temporal specifications of individual computational components. However, most applications that are either interesting or realistic tend to have *multiple* distinct computational components. Thus, Concurrent METATEM overlays METATEM with an operational framework providing both concurrency and communication [189]. Within this framework, our METATEM processes execute *independently* and *asynchronously*.

Independent, asynchronous execution is the simplest approach to implement as well as being the most realistic. Taking the theme of simplicity further, then communication is achieved through *broadcast* message-passing. Here, a message is sent from one process/agent but it is not directed to any particular destination. Instead it is sent to *all* processes/agents[1]. This mechanism not only provides a simple method of communication for use in open systems (because we do not necessarily need to know the names/addresses of all other processes or agents) but also maintains the close link between execution and distributed deduction [193].

We might visualize such a system as follows:

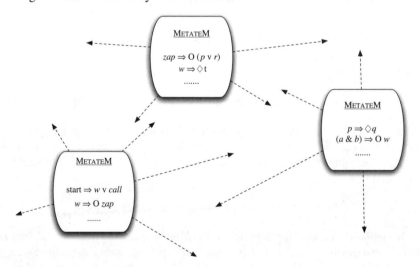

---

[1] As we will see later, there are limits to this broadcast provided by the agent's *context* and so we are really using a form of *multicast* message-passing.

Here, each METATEM agent is executing its own specification independent of the other agents. At certain points an agent might broadcast a message to the others. Similarly, it might receive broadcast messages from these agents. When a message is broadcast, then the message is guaranteed to *eventually* arrive at each other agent, though not necessarily at the same time. Think of the specification as being

$$broadcast(msg) \Rightarrow \forall A \in Agents. \; \Diamond \, received\_by(msg, A).$$

In earlier work on Concurrent METATEM, the actual mechanism for invoking such broadcast was different to that we will see later in the current implementation. Nevertheless, we will explain the original approach since it clarifies how concurrent computation in METATEM corresponds to distributed deduction. Here, the definition of which messages an agent recognizes and which messages an agent may itself produce is given through an *interface definition*.

Example 6.15 *A sample* METATEM *interface definition is*

$$car$$
$$in: \quad go, stop, turn$$
$$out: \quad empty, overheat$$

Here, {*go, stop, turn*} *is the set of messages the* '*car*' *agent recognizes, while the agent itself is able to produce the messages* {*empty, overheat*}.

These 'in' (sometimes termed 'environment') and 'out' (sometimes termed 'component') sets effectively partition the set of propositional symbols used in the agent's rules. Thus the alphabet is partitioned via

$$\text{Propositions} = \text{Input} \cup \text{Output} \cup \text{Internal}$$

where the Input, Output and Internal sets are all disjoint. Now, when we write our SNF rules,

- Input propositions can *only* appear on the left-hand side of a rule,
- Output propositions can *only* appear on the right-hand side of a rule, while
- Internal propositions can appear on either side.

Example 6.16 *The rules for the above* '*car*' *example might be*

$$start \Rightarrow \neg moving$$
$$(moving \wedge go) \Rightarrow \bigcirc (overheat \vee empty)$$
$$go \Rightarrow \Diamond moving$$

The side-effects of execution are now clear:

- *go*, *stop*, and *turn* are all atoms whose values are supplied by the agent's environment, that is other agents – these propositions are made true once a message of the same name is received;

- once either *empty* or *overheat* is made true, a message of the same name is broadcast to the agent's environment.

In this way, communication is incorporated seamlessly into the logical interpretation of the system.

Example concurrent program

The following example has been used, in various guises, in many earlier papers on Concurrent METATEM. It is a version of the resource controller examples we saw earlier and concerns three 'professors' (i.e. agents) bidding to a 'funding body' (again represented by an agent) for resources. Each professor's behaviour is summarized by its name: eager; mimic; and jealous. The actual definitions of each agent are given below (note that we use predicates over a finite domain, rather than propositions, in order to shorten this description and that X and Y represent universally quantified variables over this domain).

```
funder
    in :   apply
   out :   grant
 rules :   apply(X) ⇒ ◇grant(X)
           grant(X) ∧ grant(Y) ⇒ (X = Y)
```

```
eager
    in :
   out :   apply
 rules :   start ⇒ apply(eager)
           true ⇒ ○apply(eager)
```

```
mimic
    in :   apply
   out :   apply
 rules :   apply(eager) ⇒ ○apply(mimic)
```

```
jealous
    in :   grant
   out :   apply
 rules :   grant(eager) ⇒ ○apply(jealous)
```

The basic idea here is that each of the three professor agents may broadcast an *apply* message in an attempt to secure a grant from the funder. These agents send their own name as the only argument to this message. The funder agent must then decide which applications to grant, given that it is only allowed to grant to at most one professor at any moment in time (to do this, we use '=' over our finite domain).

Again, this example exhibits both safety and liveness rules in funder, ensuring that at most one *grant* will be made at every moment (*safety*) and that each *apply* will eventually be fulfilled (*liveness*). This example also shows the power of broadcast communication because not only can several agents send and receive the same message, but the behaviour of one agent can be defined to be dependent on messages sent to/from others. For example, mimic and jealous both base their behaviour on eager's communications. Specifically, mimic only applies when it sees eager applying, while jealous only applies when it sees eager receive a grant.

Backtracking in multi-agent systems.   Now that each agent, which is executing using a METATEM engine, is itself part of a multi-agent system, we must take care with *backtracking*. Specifically, once an agent has broadcast a message, it *cannot* backtrack and undo this. (Note that this is a general, and obvious, feature of communication in open, distributed and concurrent systems since, once a message has been sent, there is no way to undo all its potential effects.) Consequently, within Concurrent METATEM, we do not allow any agent to backtrack past the broadcast of a message. So, once an agent has broadcast a message, it has effectively *committed* its execution to that path.

This can be seen as a limitation, because the completeness of individual METATEM execution is lost, since we cannot always backtrack to find a satisfying path. However, a typical approach is for the user to specify the agent in such a way that it only commits to an execution path (i.e. via broadcast communication) once the local search for a solution to some problem has terminated. Thus, using METATEM, agent execution is often characterized as periods of internal (backtracking) computation, interleaved with communication events. If we think of 'real' agents, for example humans, then longer-term 'thinking' is often interleaved with 'acting'.

Now that we have explored the basic aspects of METATEM execution, we will move on to consider one particular implementation.

6.3   The Concurrent MetateM system

Although the basic METATEM execution algorithm has been implemented in a number of ways, its predominant use at present is as part of the Concurrent MetateM system [188, 268]. This system utilizes the METATEM execution centrally and so implements the algorithms described in Section 6.2, but has quite a few extensions and enhancements, as follows:

- It executes (a limited form of) *first-order* temporal logic, rather than just PTL; see Section 6.3.2.

- It takes a *multi-agent* view. Specifically, a Concurrent MetateM system comprises autonomous, asynchronously executing METATEM agents which can communicate with each other; see Section 6.3.3.

- It provides a range of *meta-level* directives for guiding both *deliberation* and execution in general; see Section 6.3.4.

- It provides higher-level organizational/structuring mechanisms that allow not just individual components and straightforward collections to be implemented, but also highly structured *teams* and *organizations*; see Section 6.3.5.

- Being implemented in `Java`, it also provides mechanisms to utilize the underlying `Java` code and, to some extent, coordinate `Java` execution; see Section 6.3.6.

`Concurrent MetateM` was originally developed in the 1990s, being a natural extension of the basic METATEM approach [188, 189]. However, at that stage, it was essentially a simple prototyping language for PTL specifications. As work on *agent-based systems* grew to prominence [503, 504], it became clear that the `Concurrent MetateM` approach might be useful in this area [190, 503]. Thus, developments over the last decade have tended to emphasize the agent aspects, such as *deliberation*, *belief*, *resource-bounded reasoning*, *reasoning under uncertainty*, *groups*, *teams*, and *organizations*.

6.3.1   The `Concurrent MetateM` distribution

The current version of `Concurrent MetateM` was developed by Anthony Hepple and is available from

$$\texttt{http://www.csc.liv.ac.uk/~anthony/metatem.html}$$

as well as at

$$\texttt{http://www.csc.liv.ac.uk/~michael/TLBook/MetateM-System}$$

This distribution includes examples and both command-line (the default) and graphical versions of `Concurrent MetateM`. The graphical visualization tool provides a dynamic visualization of the relationships between agents during execution as well as a monitoring facility that provides the ability to isolate individual agents and monitor their logical state.

The system

The `Concurrent MetateM` system is a `Java` application that has no dependencies other than a `Java` Run-time Environment supporting `Java 6.0` code. Defining a multi-agent system involves creating the following (plain text) source files: a *system* file; at least one *agent* file; and, potentially, several *rule* files.

System file.   This file, with suffix '`.sys`', declares the agents involved together with any *content* or *context* relationships (see Section 6.3.6). For example [203]:

```
agent fred   : "example/robot.agent";
agent barney : "example/robot.agent";
agent wilma  : "example/boss.agent";
```

Thus, this system comprises three agents, with `fred` and `barney` being instances of the `robot` agent, and `wilma` being an instance of the `boss` agent.

Example 6.17 *Clearly there can be many instances, for example:*

```
agent environment: "Surveillance/environment.agent";
agent sensor1: "Surveillance/sensor.agent";
agent sensor2: "Surveillance/sensor.agent";
agent sensor3: "Surveillance/sensor.agent";
agent sensor4: "Surveillance/sensor.agent";
agent sensor5: "Surveillance/sensor.agent";
agent sensor6: "Surveillance/sensor.agent";
agent sensor7: "Surveillance/sensor.agent";
agent sensor8: "Surveillance/sensor.agent";
agent target1: "Surveillance/fusion.agent";
agent target2: "Surveillance/fusion.agent";
```

Agent file. An agent file, with suffix '.agent', is provided for each agent type. It describes a series of directives and METATEM rules that define the program. The directives include the definition of any run-time options, agent abilities, meta-predicates and 'include' statements. A fragment of an example agent file (the *traffic light controller* example from the distribution) is shown below. (Note that all these elements will be described in more detail in the next few sections.)

```
type agent;                             // file describes an agent

ability timer: metatem.agent.ability.Timer; // define a Timer

at_least 1 red;                         // various directives
at_most  1 green;
at_most  1 amber;

startblock:                             // some initial rules
{
   start => red(east_west);
   start => red(north_south);
}

ruleblock:                              // step/sometime rules
{
   amber(X) => ~green(X);               // illegal combinations
     red(X) => ~green(X);               // illegal combinations

   // red must eventually turn to amber...
     red(X) => SOMETIME amber(X);

   //... but not before the other light is red only
     red(X)  & amber(Y)  & (X!=Y) => NEXT  amber(X);
     red(X)  & green(Y)  & (X!=Y) => NEXT  amber(X);
     .............
}
```

Note that the use of 'rule blocks' is a programming convenience allowing multiple agent types to re-use METATEM code.

Running the system

To see how Concurrent MetateM executes we will next present a range of simple examples and show what output is generated. To begin with look at the file 'start_choice.sys', containing just

```
agent chooser: "start_choice.agent";
```

This declares the agent named chooser to be of the type defined in the file 'start_choice.agent'. This file, in turn contains the definition of this agent type:

```
type agent;                             // file describes an agent

ability print: metatem.agent.ability.Print;     // agent can print

//  logging FINE;                               // logging commented out

ruleblock:
{
  start => s ;
      s => NEXT print("Chose: x")  | print("Chose: u");
      s => NEXT print("Chose:  v") | print("Chose:  w");
}
```

Note that this gives us four choices at $t=1$. Now, executing the above system simply involves invoking[2]

```
run_metatem.sh start_choice.sys
```

This execution gives

```
[chooser] Chose: x
[chooser] Chose: ~v
```

showing that the $\{x, \neg v\}$ combination was chosen. Note that, if we uncomment the 'logging' line, we get *much* more detail.

Example 6.18 *Once we have choices, we of course have the possibility of taking the* wrong *choice. If this has occurred, then we would like to fail and backtrack to a previous choice. So, now consider the basic* 'start_choice.agent' *from above, and extend it with a single additional rule:*

```
print(X) => NEXT false;
```

---

[2] run_metatem.sh is a script which calls Java with the appropriate flags and paths. An alternative script, metatem_gui.sh starts up the graphical interface.

(This version can be found in `start_choice_failing.sys`.*) Now, if we execute this, we find that the agent makes a choice, prints its message, then fails in the next state, and so backtracks. It does this for all the four initial choices until it has exhausted its execution possibilities and fails:*

```
[chooser] Chose: x
[chooser] Chose: ~v
[chooser] Chose: x
[chooser] Chose: ~w
[chooser] Chose: u
[chooser] Chose: ~v
[chooser] Chose: u
[chooser] Chose: ~w
                    Feb 12, 2010 18:30:00
                    metatem.agent.BasicAgent run
SEVERE: [chooser] Inconsistent state - backtracking options
                    exhausted.
                    An end state will now be generated.....
```

So far we have not used eventualities very much. Let us now begin with a simple test of the 'fairness' of the core METATEM execution mechanism. Recall the METATEM rules from Example 6.9 where there are two eventualities that can never be satisfied at the same moment: '$\Diamond f$' is invoked infinitely often, while '$\Diamond g$' is just invoked once. To provide a check of the basic fairness of the deliberation mechanism, we can execute this example (code in `FfFg.sys` and `FfFg.agent`), giving

```
[my_agent] Executed 'f'
[my_agent] Executed 'g'
[my_agent] Executed 'f'
[my_agent] Executed 'f'
[my_agent] Executed 'f'
    .......      ...      ...
```

Here, even though f is chosen first, g is then chosen in the next state. Now, if we execute the modified program from Exercise 6.1, where *both* $\Diamond f$ and $\Diamond g$ occur infinitely often (see `FfFg_fair.sys` and `FfFg_fair.agent`), then we get

```
[my_agent] Executed 'f'
[my_agent] Executed 'g'
[my_agent] Executed 'f'
[my_agent] Executed 'g'
[my_agent] Executed 'f'
[my_agent] Executed 'g'
[my_agent] Executed 'f'
[my_agent] Executed 'g'
[my_agent] Executed 'f'
    .......      ...      ...
```

Finally, in this initial selection of examples, we consider the more detailed example given in Section 6.2.6. Recall that the example essentially (we have modified this slightly to ensure `typing` is false to begin with) comprises the rules

```
        start          =>   begin;
        start          =>   ~typing;
        begin          =>   NEXT tired | thinking;
        true           =>   NEXT ~tired | ~typing;
        begin          =>   SOMETIME typing;
        tired          =>   NEXT tired;
 thinking & typing =>   NEXT almost_there;
  almost_there       =>   NEXT tired | typing;
```

The output from this is an execution as expected:

```
state 0: [begin]
state 1: [thinking, typing]
state 2: [almost_there]
state 2: [typing]
....
```

We will see more examples later, but we now move on to consider how `Concurrent MetateM` differs from, and extends, the basic METATEM approach.

6.3.2   Rule syntax

Rather that just executing PTL, `Concurrent MetateM` allows execution of a limited form of *first-order* temporal logic. To highlight this, we give below each of the main rule types executed within the system, correspond them to general SNF, and show what sample code looks like. You will also see that there is an additional *present-time* rule form allowed within `Concurrent MetateM`. This form is allowed in many varieties of SNF, often so that '$\forall x.p(x) \Rightarrow q(x)$' does not have to be encoded as the pair of rules

$$\textbf{start} \quad \Rightarrow \quad \forall x. \neg p(x) \vee q(x)$$
$$\textbf{true} \quad \Rightarrow \quad \bigcirc(\forall x. \neg p(x) \vee q(x))$$

(Of course, the two forms are equivalent.) So, below, we now look at *four* basic rule forms.

Initial rules: Initial rules only allow *grounded* literals. As with all of our four forms, we will first describe the *general* form of the (extended) rule, show an *example* of such a rule, and then give the actual `Concurrent MetateM` *code* for this example.

General: $\textbf{start} \Rightarrow \bigvee_{i=1}^{a} p_i(c_1, \ldots, c_{n_i})$

Example: $\textbf{start} \Rightarrow p(c) \vee q(d)$

Code: `start => p(c) | q(d);`

Step rules:

General: $\forall \overline{x}.\left[\left[\overset{a}{\underset{i=1}{\bigwedge}} p_i(\overline{x}) \wedge \exists \overline{y}. \overset{b}{\underset{j=0}{\bigwedge}} q_j(\overline{y}, \overline{x})\right] \Rightarrow \bigcirc \exists \overline{z}. \overset{c}{\underset{k=1}{\bigvee}} r_k(\overline{z}, \overline{x})\right]$

Example: $\forall x.\left[\left[p(x) \wedge \exists y.q(y, x)\right] \Rightarrow \bigcirc \exists z.[r(z, x) \vee s(z, x)]\right]$

Code: `p(X) & q(Y,X) => NEXT r(Z,X)|s(Z,X);`

So, there are two new aspects here. Existential quantification is now allowed on the right-hand side of rules, and universal quantification is allowed across rules. Let us begin with the first of these. Once we have existentially quantified predicates, we must decide how to satisfy these. The basic approach is as follows:

1. if we have $\exists x.p(x)$ but some other rules specify at least one actual instance of 'p' that has to be true, for example '$p(a)$', then the existential rule has no effect;

2. however, if no other constraints on 'p' exist then skolemization is invoked and a new constant, for example '$c1$', is generated so that '$p(c1)$' can now be made true.

A second important aspect of quantification in step rules is that universal quantification is allowed *across* rules, effectively allowing one rule to be invoked for each relevant instance.

Example 6.19 *When we run* `Concurrent MetateM` *on*

```
        start => q(c,a)
        start => p(a)
        start => q(d,b)
 p(X)  & q(Y,X) => NEXT   r(Z,X) | s(Z,X)
```

we do indeed get either '`r(v1,a)`*' or '*`s(v1,a)`*' in the next state (where '*`v1`*' is a new constant). If, however, we run*

```
        start => q(c,a)
        start => p(a)
        start => p(b)
        start => q(d,b)
 p(X)  & q(Y,X) => NEXT r(Z,X) | s(Z,X)
```

then we generate both `r(v1,a)` *or* `s(v1,a)`*, and* `r(v2,b)` *or* `s(v2,b)`*. (Note that* `v1` *could legitimately be the same as* `v2`*!) The code for this can be found in* `exist_step.sys`*.*

Exercise 6.3 *Try running* `Concurrent MetateM` *on both*

$$\textbf{true} \implies \bigcirc(\exists x.\ p(x) \vee q(x))$$

and on

$$\textbf{true} \implies \bigcirc(\exists x.\ p(x) \vee q(x))$$
$$\textbf{true} \implies \bigcirc p(c1)$$

Do both execute as expected?

Sometime rules:

General: $\forall \overline{x}. \left[\left[\bigwedge_{i=1}^{a} p_i(\overline{x}) \wedge \exists \overline{y}. \bigwedge_{j=0}^{b} q_j(\overline{y}, \overline{x}) \right] \Rightarrow \Diamond r(\overline{x}) \right]$

Example: $\forall x. \left[\left[p(x) \wedge \exists y. q(y, x) \right] \Rightarrow \Diamond r(x) \right]$

Code: `p(X) & q(Y,X) => SOMETIME r(X);`

Example 6.20 *If we run*

```
        start => q(f,a);
        start => p(a);
        start => p(b);
        start => q(g,a);
        start => q(g,b);
 p(X)   & q(Y,X) => SOMETIME r(X);
```

then we generate both `SOMETIME r(a)` *and* `SOMETIME r(b)`*, both of which are satisfied as soon as possible. (The code for Example 6.20 can be found in* `simple_sometime.sys`*.)*

Non-temporal (present-time) rules:

General: $\forall \overline{x}. \left[\left[\bigwedge_{i=1}^{a} p_i(\overline{x}) \wedge \exists \overline{y}. \bigwedge_{j=0}^{b} q_j(\overline{y}, \overline{x}) \right] \Rightarrow \bigvee_{k=1}^{c} r_k(\overline{x}) \right]$

Example: $\forall x. \left[\left[p(x) \wedge \exists y. q(y) \right] \Rightarrow \left[r(x) \vee s(x) \right] \right]$

Code: `p(X) & q(Y) => r(X) | s(X);`

It is important to note that existentially quantified variables are not allowed in the consequent of a present-time rule, preventing circumstances in which present-time rules fire repeatedly (possibly infinitely) as a result of new terms generated by repeated grounding by skolemization.

Implicit quantification.   As we have seen in the examples above, we are able to omit explicit quantification symbols from the program code. We achieve this by making the following assumptions:

- Any variable appearing positively in the antecedents is universally quantified.

- Any variables that remain after substitution of matching, and removal of non-matching, universal variables are existentially quantified.

- Of the existentially quantified variables, we ignore those that appear only negatively in the antecedents.

- Existential variables in the consequent of an otherwise grounded rule, which cannot be matched, are grounded by skolemization.

Example 6.21 *If you are* very *keen on either functional or logic programming you might want to try* append.sys *containing the following set of rules.*

```
start => append(a, list(b, nil));
append(X, Y)  & append_result(X, Y, Z)
    => NEXT print("appended ", X, " to ", Y, " giving ", Z);
append(X, Y) => append_request(X, Y);
append_request(X, list(H, T)) => append_request(X, T);
append_request(X, nil) => append_result(X, nil, list(X,nil));
append_request(X, list(H, T))  & append_result(X, T, XT)
    => append_result(X, list(H, T), list(H, XT));
```

Running this gives:

```
[append] appended a to list(b,nil) giving list(b,list(a,nil))
```

Aside: Built-in predicates. Within the basic rules, Concurrent MetateM has a number of 'built-in" predicates. These are described in detail in the documentation but essentially comprise standard arithmetic predicates such as[3] is/2 and lessThan/2, as well as predicates for manipulating *sets*. We will see later that sets of agents are particularly important and so there are a variety of set expressions involving operations such as *union*, *intersection* and *difference*. Finally, along with the manipulation of strings, Concurrent MetateM also provides built-in predicates that allow a temporal formula to be represented as a string term if required. This aspect is particularly useful when we need to pass temporal formulae between agents.

Next, we turn to the multi-agent aspects of Concurrent MetateM.

6.3.3   Concurrency and communication

In Concurrent MetateM, a multi-agent approach is implemented, with each separate METATEM agent being executed essentially as described in Section 6.2.13. However, there is an important difference in the way that communication is implemented. First, there is no explicit 'in :' and 'out :' declarations and so no partition of the propositions/predicates within the system. Instead there are explicit send and receive primitives that can be used in agent rules.

It is unsurprising that agent specifications in the language can contain literals such as

```
receive(Sender, message(X,Y,Z))
```

that become true when a message matching 'message(X,Y,Z)' is received. Here, 'Sender' is a variable that matches to the sender of the message. Thus, if agent A contains the rule

```
receive(Sender, message(X,Y,Z)) => NEXT do(X,Z,Y)
```

---

[3] As with PROLOG, predicates such as p/2 are called 'p' and take two arguments.

and if it receives a message 'message(aa,bbb,cccc)' from agent B, then A will attempt to execute do(aa,cccc,bbb) in the next state. Perhaps more surprising, given that we have previously used *broadcast* message-passing, is the form of the message-sending used in Concurrent MetateM. Here, when a literal such as

```
send(context, broadcast(message(self,"yin","yang")))
```

is made true, the message

```
broadcast(message(self, "yin", "yang"))
```

is *sent* to the agent called 'context'. Although this looks like point-to-point message passing (i.e. the message is sent to a specific destination), it will effectively get the context agent to broadcast the message(self,"yin","yang") message to all agents (within that context).

While we will consider these aspects in more detail later, we just note there that in order for such a mechanism to work, the agent needs

1. a suitable agent to send its message to – we will see in Section 6.3.6 how agents are organized within *contexts* and how such contexts can distribute a 'broadcast(...)' message; and

2. the ability to 'send' a message – this will involve declaring the agent to have such an ability with the following declaration within the agent's definition file (see Section 6.3.5).

```
ability send: metatem.agent.ability.Send;
```

Aside: 'self'.   The meta-variable 'self' expands during execution to be the name of the agent currently executing the predicate in which 'self' appears.

Multi-agent example

We will now look at one particular multi-agent example we are familiar with and show how it is executed using Concurrent MetateM. Consider the multi-agent example of three professors and a funder given earlier in Section 6.2.13. We can implement this, with the overall structure (grants.sys) being

```
agent funder:   "Grants/funder.agent";
agent eager:    "Grants/eager.agent";
agent mimic:    "Grants/mimic.agent";
agent jealous: "Grants/jealous.agent";
```

Each of the agents is defined in an appropriate way, as follows:

funder agent:

```
    award(X)   & award(Y)  => (X=Y);
award(X)   & in(Y,content) => NEXT send(Y, grant(X));
                   award(X) => NEXT print("Awarding to ", X);
             requested(Z) => SOMETIME award(Z);
```

Note that the in(Y,content) aspect allows the agent to broadcast its message, in this case the 'grant' message. The rest of the rules essentially ensure that no two distinct agents can be awarded a grant at the same moment, and that any agent that makes a request will eventually be awarded.

eager agent:

```
                    start => send(funder, broadcast(apply(self)));
                     true => NEXT send(funder,
                                        broadcast(apply(self)));
receive(S, grant(self)) => NEXT print("Eager receives grant");
receive(S, grant(X))
              & X\=self => NEXT print("Eager unhappy");
```

As before, eager applies for a grant at every step.

mimic agent:

```
receive(S, apply(eager)) => NEXT send(funder,
                                        broadcast(apply(self)));
receive(S, apply(eager)) => NEXT print("Mimic unhappy");
 receive(S, grant(self)) => NEXT print("Mimic receives grant");
```

Here, mimic applies for a grant only when it sees eager applying.

jealous agent:

```
  receive(S, grant(eager)) => NEXT send(S, broadcast(apply(self)));
  receive(S, grant(eager)) => NEXT print("Jealous unhappy");
  receive(S, grant(self)) =>  NEXT print("Jealous receives grant");
```

Here, jealous applies for a grant only once it sees eager being awarded.

Now, when we run the full system we get the expected outputs (remember the asynchronous nature of agent execution here):

```
    [mimic] Mimic unhappy
    [funder] Awarding to eager
    [eager] Eager receives grant
    [mimic] Mimic unhappy
    [jealous] Jealous unhappy
    [funder] Awarding to eager
    [jealous] Jealous unhappy
    [eager] Eager receives grant
    [mimic] Mimic unhappy
    [funder] Awarding to eager
    [eager] Eager receives grant
    [jealous] Jealous unhappy
    [funder] Awarding to mimic
    [mimic] Mimic receives grant
    [eager] Eager unhappy
    [funder] Awarding to eager
    [funder] Awarding to jealous
```

```
[eager] Eager unhappy
[eager] Eager receives grant
[jealous] Jealous unhappy
. . . . . . . . . . . . . . . . . . . . . . .
```

Note that the interleaving of threads in `Java` can sometimes make agent outputs appear in strange orders.

Next we will move on to consider various meta-level directives that help the user control execution in `Concurrent MetateM`.

6.3.4   Deliberation and control

As described earlier, the basic METATEM execution mechanism leaves a great many aspects undefined. In particular, when choosing between disjuncts or between eventualities there are many possibilities. In this section we outline some of the features of the `Concurrent MetateM` system that allow us to, if we wish, control the choices relating to deliberation within the agent. Specifically, we will look at two such *meta-level* directives:

- 'prefer', which allows us to say which possibilities we would *prefer* to be satisfied (if we have a free choice); and

- 'atLeast' and 'atMost' which provide the system with guidance about how many instances of a predicate to allow.

It is also useful to know that, currently, `Concurrent MetateM` allows the system to be extended via the implementation of further *deliberation meta-predicates* as 'plug-ins'. (See the system documentaiton for more details.)

Deliberation using 'prefer'

Recall that, if there are multiple outstanding eventualities, but only a strict subset of these can be satisfied at the same time, then the basic METATEM approach is to choose to satisfy the eventuality that has been outstanding longest. In addition, the `Concurrent MetateM` implementation provides the meta-level predicate 'prefer' which is used to direct (or at least influence) the system's choice between conflicting eventualities. The use of 'prefer' was developed in earlier papers on METATEM as a lightweight and flexible way to provide some control over deliberation [200].

During METATEM execution the consequents of all the rules that fire are conjoined (and transformed into disjunctive normal form) to represent the agent's choices. One of these will provide the interpretation of the next temporal state, from which the next set of choices are derived. In addition to the set of disjuncts, we also have a list of outstanding eventualities. Each agent will always satisfy an eventuality if it is consistent to do so, and will avoid introducing new eventualities (via 'sometime' rules) if able to. In addition, the `prefer` construct allows the developer to modify the outcome of METATEM's choice procedure by re-ordering the list of choices according to the order of a pair of predicates, for example `prefer(fame,fortune)`.

Since just using `prefer` on its own can be quite 'brutal', we are also able to qualify such preference directives with additional information concerning (a) what *conditions* should be true for the preferences to apply and (b) how *strong* this preference should be.

Specifically, we can state

prefer X over Y with weight W when Condition is true.

Here, 'Condition' is a formula that can be evaluated in Concurrent MetateM and 'W' is an Integer in the range 1 . . . 99 (weighting values of 0 and 100 are reserved for *system* preferences). The exact syntax of the above, within Concurrent MetateM rules, would be

```
prefer("X","Y","Condition",W)
```

Actually there are *two* ways to define preferences: to specify preferences within our temporal rules; or to provide these preferences within the agent definition file. The first form occurs via rules such as

```
start => prefer("g","f","true",75);
```

The second form is a general directive such as

```
prefer g to f when true weight 75;
```

inserted in the agent's definition, and separate to the actual rules.

Let us consider these two alternative applications of the prefer construct, as might be used in the 'lift' example provided within the Concurrent MetateM distribution. Here, the aim of these preferences is to encourage the lift to continue moving in the same direction when appropriate. Thus, we might have

```
prefer downTo to upTo    when moving(down) weight 55;
prefer upTo    to downTo when moving(up)    weight 55;
```

in the agent definition file, or we might have

```
trigger => NEXT prefer("downTo","upTo","moving(down)",55)
trigger => NEXT prefer("upTo","downTo","moving(up)",55)
```

in the actual set of rules for the agent. In the first case, the preferences apply from $t=0$ onwards; in the second, the preferences only apply from the state after that when 'trigger' occurs. Thus we can see that the first form is equivalent to using

```
start => prefer("downTo","upTo","moving(down)",55)
```

Example 6.22 *This simple example should explain how preferences make an impact – it will also show how basic choice is handled. If we have the code (see prefer_basic.sys)*

```
start => f | g;
start => ~f | ~g;
true => NEXT ~f | ~g;
true => NEXT f | g;
```

then we have a straight choice between f and g at each state. Actually, the implementation takes a naive view and just chooses f every time here. However, if we now add a 'prefer' directive, that is:

```
prefer g to f when true weight 99;
```

then g will be chosen at every step. Note that, if we also add

```
true => NEXT ~g;
```

then no contradiction is generated, but f will now be chosen at every next state.

Once a preference is applied it persists for all future states. While there is no mechanism for explicitly deleting them, such preferences can be *overridden* by an otherwise identical preference which declares a higher priority value or *counteracted* by an opposing preference.

Example 6.23 *Consider again the example above, but now modified with a contradictory preference*

```
prefer g to f when true weight 55;
prefer f to g when true weight 99;

start => ~f | ~g;
start => f | g;
true => NEXT ~f | ~g;
true => NEXT f | g;
```

Here, 'f' is always chosen.

If we move on to think about *eventualities*, we see that there are now several heuristics for deciding which eventualities to make true now and which to postpone. In particular, there is an interesting interaction between the '*oldest-first*' strategy and the '`prefer`' directive. The simplest way to think of this is as follows. Recall that preferences have weights of from 0 to 100. Effectively, the '*oldest-first*' ordering has a weight of '50'. So, if we add '`prefer`' directives with weights less than 50, they will not override the '*oldest-first*' approach. If, however, we add '`prefer`' directives with weights over 50, then these do indeed override this ordering.

Example 6.24 *Let us take our earlier example with two eventualities, $\Diamond f$ and $\Diamond g$ and add a stronger preference:*

```
prefer g to f when true weight 55;

start => ~f;
start => ~g;
true => NEXT ~f | ~g;
```

```
        true => SOMETIME f;
        true => SOMETIME g;
```

Without the preference directive then f *and* g *would alternate. However, with the above directive, then* g *will occur at* every *step and* f *will be postponed forever! If, however, we reduce the weight of the preference to be less than 50, then* f *and* g *again alternate.*

This is clearly *dangerous* and so care should be taken when using such strong preferences!

Aside: Context-dependent preferences. We will see later that the *context* in which each agent operates is often vital. One of the useful aspects of the ability to define a 'Condition' under which a preference is to be used is that these conditions can capture context information. As we will see in Section 6.3.6, such conditions might contain inContext or inContent predicates describing context dependencies. This use of context-dependent preferences is certainly to be encouraged because leaving a context provides the effect of deleting a preference but with the benefit that the preference will be reinstated on re-entering another appropriate context. This is both a natural and simple interpretation of context-dependent preferences.

Guiding choices using 'atMost' and 'atLeast'

Our second set of meta-level guidance predicates particularly concern the first-order aspects of the implementation. For example, if we execute '$\exists x.p(x)$' then how many instances of p should be true? We might make $p(c1)$ true for some (possibly new) constant '$c1$', but we might also make $p(c2)$ true. The aim of the atLeast and atMost directives is to constrain how many instances of a predicate we should choose to make true at any state. The atLeast predicate places a minimum constraint on the number of instances of positive predicates, while atMost places a maximum constraint on the number of instances of positive predicates in a given temporal state, in the style of the capacity constraints described by [160]. Besides providing the developer with the ability to influence an agent's reasoning, when applied judiciously, atMost and atLeast can simplify the fragment of the logic considered and hence can increase the execution performance of a Concurrent MetateM agent.

To exemplify this, consider the following, again derived from the 'lift' example in the Concurrent MetateM distribution. Here, the lift responds to calls from the floors above and below it and, when more than one call remains outstanding, it must decide which call to service first, changing direction if necessary. It is desirable that the lift visits a floor in each state of our model. This behaviour could be specified by the rule

```
        true => NEXT atFloor(X);
```

representing '$\bigcirc(\exists X.atFloor(X))$'. This states that there must exist an instance X such that atFloor(X) is satisfied in each moment in time. As it stands we note that having both atFloor(1) and atFloor(3) together is actually consistent with the above. So, to put further constraints on the number of atFloor instances allowed we can declare

```
        at_most  1 atFloor true;
        at_least 1 atFloor true;
```

This actually obviates the need for a general rule such as `true => NEXT atFloor(X)` since we know there is always at least one instance of `atFloor(X)` true at any moment. In fact, the two directives above together ensure that *exactly* one instance of `atFloor(X)` is true at every moment.

Example 6.25 *In this simple example we show how the* `atLeast` *directive works. Consider (see* `atLeast_ppp.sys`*)*

```
at_least 3 ppp;

true   => ppp(a) | ~ppp(a);
true   => ppp(b) | ~ppp(b);
true   => ppp(c) | ~ppp(c);
true   => ppp(d) | ~ppp(d);
```

Here, three or four (most likely, three) instances of 'ppp' will be made true. If we then also add

```
at_most 3 ppp;
```

then exactly *three instances will be made true.*

As mentioned above, judicious use of `atMost` and `atLeast` predicates can aid both efficiency and clarity. We will not provide further examples here, but note that these directives appear regularly in the examples within the distribution.

6.3.5 Using Java

`Concurrent MetateM` also allows interaction with the underlying Java classes. Mainly this is achieved by defining *abilities* for the agent. Abilities are special predicates that are directly linked to Java code.

Example 6.26 *As a default,* `Concurrent MetateM` *provides* `Send` *and* `Timer` *classes and so the relevant abilities can be defined in the agent file:*

```
ability send: metatem.agent.ability.Send;
ability timer: metatem.agent.ability.Timer;
```

These allow agents to send messages to other agents immediately and to themselves after a period of delay, respectively.

We can see, from the above example, how abilities link to Java classes. Defining further abilities requires the programmer to use the `Concurrent MetateM` API and to create a Java class that extends the API's `AgentAbility` class for each action.

Once declared in the header of an agent file, abilities (such as 'send' or 'timer' above) may appear in rules wherever it is valid for a conventional predicate to appear.

Generally, abilities are treated like normal atomic predicates but a special 'external' ability type exists which, when executed in a state, prevents Concurrent MetateM from backtracking over that state. Thus, internal abilities represent actions that are reversible (i.e. have no side-effects) while external abilities represent actions that cannot be reversed by the agent. The difference between internal and external abilities is characterized in [203] by the following example: *a database selection query might be implemented as an internal ability whereas an update query might be considered external*.

Aside: Coordination.   To some extent Concurrent MetateM can be seen as *coordination language* [230] in that the underlying Java code can have an agent 'wrapper' that undertakes and controls interaction between the agents. In our case, this wrapper will be Concurrent MetateM code; this aspect is explored further in [311].

6.3.6   Multi-agent organization

So far we have described a Concurrent MetateM system as comprising a set of agents that can communicate together, typically via broadcast message-passing. However, further structure is needed within such systems, not only to constrain the extent of broadcast communication so that it is restricted to a particular group of agents (and so may be more efficient), but also to provide more sophisticated ways of representing and organizing agent behaviour. Agents often work together to achieve some goal, so it makes sense to encapsulate the set of cooperating agents within some structure. Similarly, such groups of cooperating agents might compete with other groups. And so on. Dealing with such organizations becomes quite difficult when we only have a 'flat' agent space. In addition, providing some way of structuring agents might allow them to work together in unforeseen ways.

Unsurprisingly, the development of more sophisticated agent organizations have been a very important area of research, leading to very many different approaches to teamwork [126], including: *joint intentions* [125]; *teams* [423, 474]; *roles* [177, 290]; *groups* [367]; and *institutions* [419]. Our approach is to provide a very simple, yet flexible and powerful, mechanism for structuring agents that allows many of the above constructions to be represented [204, 209, 269] and also provides the programmer with significant freedom in defining new organizational behaviour.

The basic principle we use is that *everything* is an agent, including teams, roles, groups etc. In addition, all agents have behaviour. This might be quite trivial behaviour, such as the ability to forward messages, but is essentailly any behaviour we can describe using METATEM rules. The additional step we make now is that each agent can *contain* other agents. So, every agent has

- *behaviour*, described using our temporal rules,

- *contents*, being the set of agents contained within this agent, and

- *context*, being the set of agents this agent occurs within.

We have seen how the behavioural rules are defined in Concurrent MetateM, but the initial contents/context relationships are also defined in the system file.

Example 6.27 *The declarations within a '.sys' file provide initial organizational information, for example*

```
Agent_A  { Content: Agent_C, Agent_E, Agent_G; }
Agent_B  { Content: Agent_C, Agent_E, Agent_D; }
Agent_F  { Context: Agent_G; }
```

The declarations from Example 6.27 would give the structure of

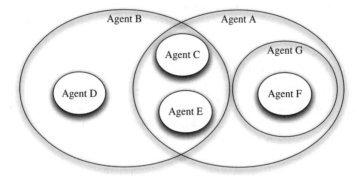

Now, once we have such a structure there are three important aspects concerning the agents' behaviours:

- agents can move in and out of contexts/contents dynamically;

- an agent can detect when another agent has joined/left its content/context and so base its behaviour on this; and

- movement between contexts can trigger communications between the agent representing the context and the agent that is moving – as we will see later, this allows us to develop quite sophisticated context-dependent behaviours.

Monitoring contents/contexts.   In order for the agent to base its behaviour on what happens in its content/context, not only does it have access to both the variables content and context, but also it can use a range of built-in predicates. The variables content and context are sets of agents, dynamically bound to the current set within the agent's content or context. The predicates that an agent might invoke include inContent(Agent), addToContent(Agent), leftContent(Agent), and enteredContent(Agent); and similarly for context.

Moving contexts.   In addition to detecting activity within contents/contexts, agents can actually take actions to move/modify these. Thus, there are a number of predicates such as addToContent(Agent) and removeFromContent(Agent) that have a side-effect of changing the content/context structure. Note that invoking such predicates actually takes one step to take effect, so 'addToContent(a)' adds agent a to the current agent's content but only at the next step, that is ○ inContent(a).

Programming context-dependent behaviour.   But what does it mean for an agent to move into another agent's content or context? The answer is that this is entirely dependent on the behaviour of the agents, as specified by their SNF rules. It may be that one agent acts as a leader and instructs agents within its content what to do. It may be that the agents within the content work together to provide general behaviour of the context agent which is then seen by other agents. Or it may be that an agent represents some role and that the agents contained within it guarantee to enact this role. And so on. All of these aspects are both possible and useful.

Communication.   We saw earlier that an agent actually sent a message to its 'context' in order for the message to be broadcast. We can now see why. Each agent has rules for handling communication requests. Then, if any agents within the content send a message explicitly to the containing agent, this agent will have rules to deal with this. In the case of broadcast, a typical rule is

```
receive(X, broadcast(M))  & in(X,content)
      & in(Y,content) & (X\=Y)
      => NEXT send(Y, M);
```

Thus, when the agent receives a 'broadcast(M)' message from an agent within its content, it then sends it to all *other* members of its content. However, although this might well be the default behaviour, we can write *any* rules here. For example, we might only forward the message to *trusted* agents:

```
receive(X, broadcast(M))  & in(X,content)
      & in(Y,content)  & (X\=Y)  & trusted(Y)
         => NEXT send(Y, M);
```

or might only forward the message if it was generated by a *trusted* agent:

```
receive(X, broadcast(M))  & in(X,content)
      & in(Y,content)  & (X\=Y)  & trusted(X)
         => NEXT send(Y, M);
```

or might even change the message in some way:

```
receive(X, broadcast(M))  & in(X,content)
      & in(Y,content)  & (X\=Y)  & modify(M,New)
         => NEXT send(Y, New);
```

It is important to note that the name 'broadcast' is *not* reserved. An agent subsystem could happily use another name, for example

```
receive(X, abracadabra(M))  & in(X,content)
      & in(Y,content)  & (X\=Y)  & trusted(Y)
         => NEXT send(Y, M);
```

And so on. The possibilities are endless. The key point is that it is the behaviour of the receiving agent, as specified via the temporal rules, that describes the semantics of the message-passing.

The 'Known' set.   So far we have seen that an agent only knows of agents that are either in its `content` or `context` sets. However, there are occasions when other agents outside these structures make contact. Thus, each agent has access to a third set, called 'known' that retains references to all agents the agent has ever interacted with. Here, 'interacted with' means has received a message from, has sent a message to, has had in its contents, has had in its context, or has received a reference to.

Aside: Manipulating sets.   It is clear that we often need to manipulate sets of agents, such as subsets of `content` or `context`. Thus, `Concurrent MetateM` allows set expressions to be used within rules. These can involve the basic `content` or `context` sets but can also be more complex expressions built up using operations such as `UNION`, `INTERSECTION` and `WITHOUT`.

> **Example 6.28** *If an agent wants to send a message m to all agents in either its* `content` *or* `context` *sets, it might invoke*
>
> ```
> in(X, content UNION context) => NEXT send(X, m);
> ```

6.3.7   Programming multi-agent organizations in `Concurrent MetateM`

The above view of computation as agents, within other organizational agents, communicating through multicast message-passing is very appealing. It is particularly useful for dynamic, open and complex systems. In order to get a view of how quite complex multi-agent behaviour can be specified and implemented the reader might consult the 'shopping' or 'surveillance' examples within the distribution, and might consult some of the papers referred to earlier.

However, just to give a glimpse of what can be done, we will outline one context-aware application. This is a version of that presented in [271], but updated slightly with more recent elements of `Concurrent MetateM`.

Scenario.   Imagine that we are modelling an 'active' museum. Here, one agent (representing a visitor) navigates through several rooms in the museum. The rooms themselves communicate with the agent (most likely the visitor's smart-phone or PDA) to suggest what exhibits to look at. Our agent will then decide what to look at next. In our `Concurrent MetateM` description, the visitor, the rooms, and possibly even the exhibits, are all agents.

Even in this simple version, the visitor agent must decide between various different possibilities. However, we can make the scenario more interesting, as follows. Imagine we have a '*sculpture critic*' role and that undertaking this role ensures that the visitor agent will only ever look at sculptures. In addition, the visitor agent prefers green exhibits! What, now, should the visitor agent choose to view?

The model.   Not only are the visitor and the rooms represented as agents, but the *sculpture critic* role and the *green preference* view are also agents! Thus, some agents represent real aspects, such as physical locations, while others represent virtual aspects, such as roles or constraints.

A sample execution.   Imagine that we start with our visitor agent in 'room1'. This will be modelled as the visitor being within the content of room1. Now the visitor moves to room2, possibly through some area that overlaps both rooms (for example, a doorway or corridor):

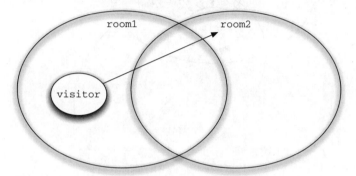

Once the visitor moves into the room2 context, the room2 agent sends messages to visitor describing the exhibits it can see there:

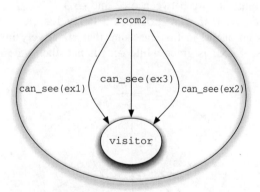

Note that room2 actually broadcasts these messages to all new members of its content.

Now, what if the visitor wants to look at the exhibits, but one at a time? An obvious way to do this is with look_at(X) where

```
atMost 1 look_at;
```

and where we constrain X to be one of the exhibits the agent can see.

So, now, the visitor agent has to decide which exhibit to look at. In the basic scenario, some random choice will be made.

But now, what if our agent takes on the role of being a *sculpture critic*? This role will also be represented as an agent and, once our visitor moves into it, the agent is sent various messages by the role. One such message might be that the visitor should *only* look at sculptures. So, let us say that ex1 is not a sculpture, while ex2 and ex3 are:

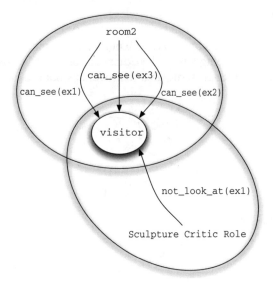

So, now, visitor must choose between ex2 and ex3.

Taking this further, what if visitor has a liking for *green* exhibits? Again, we can represent this as an agent, in this case one that effectively invokes preferences in its contents. Thus, if ex2 is green while ex3 is not, then visitor now prefers to look_at(ex2):

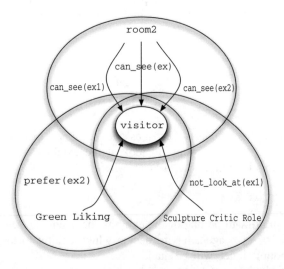

And so on. Further agents can be created dynamically so we could continue in this way. However, we will not go any further with these examples, but just comment that the core approach whereby everything is an agent and where one agent's behaviour is modified by the other agents it finds itself under the influence of, is both flexible and powerful. In particular, the agent has its own goals and choices but, movement to/from contexts can provide additional constraints, information and preferences. Crucially, the agent code is not written to cater for all possible contexts it might find itself in; the contexts just give more scope for the agent's deliberation processes. This is important in allowing us to write agent descriptions that are context-dependent but do not have specific details of the contexts 'wired-in'.

6.4 Advanced topics

6.4.1 Summary: Benefits and drawbacks of METATEM

The METATEM approach is very appealing:

- specifying behaviour using temporal formulae is expressive and intuitive;

- broadcast message-passing is very flexible, particularly in open systems;

- asynchronous, open and evolving computations are encouraged; and

- the content/context approach is a flexible and powerful way to build complex systems.

The basic execution mechanism is complete for PTL in the sense that [49, 50]:

Theorem 6.1 *If a set of SNF rules, R, is executed using the* METATEM *algorithm, with the proviso that the oldest outstanding eventualities are attempted first at each step, then a model for R will be generated if, and only if, R is satisfiable.*

Once we add deliberation, for example re-ordering of outstanding eventualities, then this theorem becomes:

Theorem 6.2 *If a set of SNF rules, R, is executed using the* METATEM *algorithm, with a fair[4] ordering strategy for outstanding eventualities, then a model for R will be generated if, and only if, R is satisfiable.*

Thus, the *fair ordering strategy* restriction imposes a form of *fairness* on the eventuality choice mechanism [197].

However, there are problems. If there is no significant backtracking within each component, then the semantics of a Concurrent METATEM system is fairly straightforward [192]. But, once complex backtracking is allowed, the detailed semantics of non-trivial systems becomes *much* more difficult. In particular, once components are not allowed to backtrack past the broadcasting of a message, then the semantics of the whole system becomes quite complex.

---

[4] By *fair* we mean a strategy that ensures that if an eventuality is outstanding for an infinite number of steps, then at some point in the execution that eventuality will continually be attempted.

It is also important to note that, once we move to the execution of full first-order temporal specifications, or we consider the execution of an agent in an unknown environment, then completeness *cannot* be guaranteed. At this stage we are only *attempting* to build a model for the temporal formula captured within the program.

If, however, we have specifications of all the participants in the multi-agent system, together with a strong specification of the communication and execution aspects, then we can, in principle, develop a full specification for the whole system. If this specification is within a decidable logical fragment, then we may be able to analyse such specifications automatically.

6.4.2   Beliefs in METATEM

The extension of METATEM with *belief* predicates and epistemic properties was originally explored in [194] and are expected to be included in future `Concurrent MetateM` implementations. Essentially, this involves extending the temporal basis of METATEM with standard modal logics, in particular a modal logic (typically KD45) of belief [254]. SNF, extended in this way, looks like

$$
\begin{aligned}
\textbf{start} &\Rightarrow \bigvee \dots \\
\bigwedge \dots &\Rightarrow \bigcirc \bigvee \dots \\
\bigwedge \dots &\Rightarrow \Diamond \dots \\
\bigwedge \dots &\Rightarrow B_1 \bigvee \dots \\
\bigwedge \dots &\Rightarrow \neg B_1 \neg \bigvee \dots \\
\bigwedge \dots &\Rightarrow B_2 \bigvee \dots \\
\bigwedge \dots &\Rightarrow \neg B_2 \neg \bigvee \dots \\
&\quad \vdots \qquad\qquad \vdots
\end{aligned}
$$

where B_i is a belief modality meaning 'agent i believes'. Again, arbitrary specifications can be transformed into this normal form, thus giving a clear basis for specifying dynamic agents that can reason about their (and other agent's) beliefs [157].

While we can, in principle, execute agent specifications given in the above temporal logics of belief, this can be quite expensive. In addition, we would like to be able to limit the amount of reasoning that a logic-based agent could carry out in certain situations. Thus, in more recent work we have investigated the extension of the agents'' descriptions to incorporate an explicit multi-context belief dimension [199, 201]. In many ways, such multi-context logics of belief are similar to standard modal logics of belief, the key difference being that we can closely control the number of belief contexts (and, so, depth of reasoning) allowed.

The syntax of this belief extension is the same as that given above and, again, arbitrary specifications can be transformed into our normal form, with the key addition of a bound on the depth of nesting of B operators allowed. This logic is investigated in [202] and part of a travel assistant example (again shortened by using predicates) taken from that paper is

$$ ask(you, X) \Rightarrow B_{me} \Diamond go(you, X). $$

This is interpreted as (where B_{me} is a multi-context belief operator)

> "if you ask for information about destination X, then I believe that you will go to X in the future".

Thus, such a logical extension allows each agent to be represented in a logic incorporating belief and provides the agent itself with a mechanism for reasoning, in a resource-bounded way, about its (and other agents'') beliefs.

As we saw above, the normal form extends that of basic temporal logic with B_i and $\neg B_i \neg$ rules characterizing belief, where 'B_i' is now a multi-contextual belief modality with standard KD45 properties. To simplify this further, we add the axiom

$$\vdash \neg B_i \neg \varphi \Rightarrow B_i \varphi$$

which ensures that each agent has beliefs about everything. This reduces the normal form to

$$
\begin{array}{rcl}
\textbf{start} & \Rightarrow & \bigvee \ldots \\
\bigwedge \ldots & \Rightarrow & \bigcirc \bigvee \ldots \\
\bigwedge \ldots & \Rightarrow & \Diamond \ldots \\
\bigwedge \ldots & \Rightarrow & B_1 \bigvee \ldots \\
\bigwedge \ldots & \Rightarrow & B_2 \bigvee \ldots \\
\vdots & & \vdots
\end{array}
$$

We can now extend execution to this variety of normal form, thus allowing resource-bounded reasoning about beliefs [202]. This involves executing the normal form for temporal logic combined with a multi-context logic of belief. Formulae are thus represented by exploring a belief model at every moment during the temporal execution.

The ability to dynamically modify the depth bound on reasoning about beliefs can potentially lead to many applications in resource-bounded systems [202]. One particularly interesting scenario involves a RoboCup agent specified using our logic in such a way that it must reason about its beliefs in order to decide whether to 'pass' or 'shoot'. Allowing a large amount of reasoning enables the agent to infer that it should pass; limiting reasoning leads the agent to shoot quickly.

Finally, we note that a more recent extension has involved the modification of the METATEM approach to incorporate *probabilistic belief* [138, 139]. Again, this is an aspect that we expect will appear in future METATEM implementations.

In the next few sections we will explore some alternative ways of executing temporal formulae, both in logics such as PTL and in interval temporal logics such as ITL [386]. While it is interesting to see how other approaches work, this is almost entirely a historical overview – very few of the approaches we describe are being actively pursued! There are also other approaches that we do not describe below, but have also been proposed for executing temporal formulae. These include the work by Brzoska on linking TEMPLOG to *Constraint Logic Programming* [94] and work by Balbiani *et al.* on extending Logic Programming with various non-classical logics [35]. The first example of this direction that we will explore is to implement TEMPLOG as a meta-level extension to PROLOG.

6.4.3   Temporal logic programming – TEMPLOG

We now consider the extension of standard Logic Programming to temporal logics. We assume the reader is familiar with the concepts of Logic Programming, as exemplified by PROLOG [459]; see Appendix A. As with classical logics, discrete temporal logics can be executed using the Logic Programming paradigm. Unfortunately, the Horn fragment of temporal logic is still incomplete [1, 5, 54, 56][5] and thus unsuitable for Logic Programming. Hence the search for a subset of temporal Horn Clauses logic that can be executed successfully using the logic programming paradigm.

The predominant approach to the extension of the Logic Programming paradigm to temporal logic is TEMPLOG [5]. Here, Temporal Horn Clauses (see Section 2.8.8), which can be categorized as either *initial* clauses (effectively if they contain **start**) or global clauses, are restricted still further using the following constraints.

- The '$\Diamond$' operator can not occur in the head of a clause.

- The '$\Box$' operator can only occur in the head of what are termed *initial definite permanent* clauses, which can be characterized as

$$\Box e \leftarrow d, \textbf{start}, c.$$

Essentially, these restrictions ensure that no positive occurrences of $\Diamond$-formulae appear in the program clauses. Thus, the idea here is to allow goals and bodies of program rules to consist of formulae such as p, $\bigcirc q$ and $\Diamond r$, while heads of program rules may consist of formulae such as p, $\bigcirc q$ and $\Box r$.

Given a particular goal, together with a set of rules, the goal can be successively reduced using reduction rules based on a variety of temporal resolution [6]. These are the temporal analogue of the standard resolution rule and are termed *Temporal SLD* (TSLD) resolution rules [5, 6]. While we will not give a full description of the TSLD rules, we provide several, in a simplified form, below.

GOAL:	$\leftarrow \bigcirc p, G$	$\leftarrow \Diamond p, G$	$\leftarrow \Diamond p, G$
REDUCTION RULE:	$p \leftarrow B$	$p \leftarrow B$	$\bigcirc p \leftarrow B$
NEW GOAL:	$\leftarrow \bigcirc B, G$	$\leftarrow \Diamond B, G$	$\leftarrow \Diamond B, G$

Thus, given a goal which is a next-time formula, and a rule which allows us to reduce the predicate to which the '$\bigcirc$' operator applies, we can replace the predicate by the new body within the context of the next-time operator. If there is a goal of '$\Diamond p$', and we have a rule that enables us to reduce p to B, then we can derive the new goal '$\Diamond B$'. Finally, if we again have '$\Diamond p$' as a goal, but instead have a rule reducing '$\bigcirc p$' to B, then we can legitimately reduce the goal to '$\Diamond B$'. This is because we must still ensure that B occurs at some point in the future in order to ensure that p is satisfied in the successor state. Other TSLD rules for TEMPLOG follow a similar pattern.

As in PROLOG, when there are a variety of rules that can be used to reduce a particular goal, an ordering of the rules is used. If the goal cannot be successfully reduced using

---

[5] The reasons why are complex but, simply, first-order temporal logic is incomplete since a form of induction (and thus arithmetic) can be coded in it – thus Gödel's incompleteness theorem shows that the logic is expressive enough to be incomplete [1]. In first-order temporal Horn Clauses, similar inductive formulae can still be represented [373].

the rules chosen, backtracking is employed as usual. In essence, TEMPLOG introduces computation over the temporal dimension into the TSLD rules, so this backtracking can sometimes be over the construction of the *temporal* sequence.

Propositional example [56]   In this example, the propositional TEMPLOG rules are simply

1.   *breakfast* ← **start**

2.   ○*coffee* ← *breakfast*

3.   □*tired* ← *breakfast*, **start**

4.   ○*lunch* ← *coffee*

Given the goal '← ◇*lunch*, ◇*tired*, **start**', the reduction sequence continues as follows.

← ◇*lunch*, ◇*tired*, **start**

← ◇*coffee*, ◇*tired*, **start**

← ◇*breakfast*, ◇*tired*, **start**

← ◇*start*, ◇*tired*, **start**

← ◇*tired*, **start**

← *breakfast*, **start**

← **start**

←

First-order example [5]   The following provide more complex first-order program rules for representing temporal information. Imagine an employee database. Some of the temporal queries that we might invoke can be represented in TEMPLOG as follows:

- Representing the salary increases for various employees:

$$increase(X, I) \leftarrow salary(X, Y), \; ○salary(X, Z), \; I = Z - Y.$$

- Founding employees are always employees:

$$□employee(X) \leftarrow employee(X), \; \textbf{start}$$

For further examples, refer to [5].

Properties of TEMPLOG   There are various technical and practical results relating to TEMPLOG that concern its expressive power, characterization and implementation. Baudinet [54, 55] provides two alternative formulations of (first-order) TEMPLOG's declarative semantics and, by relating these to TEMPLOG's operational semantics (as given by the TSLD rules of the form above), shows that these rules are complete. Further, she establishes that propositional TEMPLOG programs represent a fragment of the temporal fixpoint calculus [39]. These important results show that, not only does TEMPLOG represent a (relatively) efficient execution mechanism for a fragment of PTL, but that, as in standard Logic Programming where a variety of semantic definitions coincide, both minimal model and fixpoint semantics coincide with TEMPLOG's operational semantics.

While much of the above work was carried out many years ago, there has recently been a revival of interest is such approaches, for example [12, 102].

6.4.4   Executing interval temporal logic – TEMPURA

TEMPURA [253, 382] is an executable temporal logic based on the forward chaining execution of the interval temporal logic, ITL (see Section 2.8.10), similar to that of METATEM described above. In this sense it was the precursor of the METATEM family of languages and provides a simpler and more tractable alternative to these approaches. In addition to the standard ITL operators, TEMPURA incorporates various derived constructs, often with connotations in imperative programming, as follows:

$$
\begin{aligned}
empty &\equiv \neg \bigcirc \textbf{true} \\
fin(p) &\equiv \Box(empty \Rightarrow p) \\
if\ C\ then\ P\ else\ Q &\equiv (C \Rightarrow P) \wedge (\neg C \Rightarrow Q) \\
for\ n\ times\ do\ P &\equiv if\ n = 0\ then\ empty\ else\ [p;\ for\ n - 1\ times\ do\ p]
\end{aligned}
$$

Thus, for example, an empty interval is one where there is no *next* element in the interval, while *fin* is an operator that ensures that its argument is satisfied at the end of each interval (i.e. when the interval is empty). Many more operators can be derived from the basic ITL primitives (see [382] for details). These give a wide variety of constructs for expressing properties of computations (via properties of intervals). A particular advantage of using this language, rather than those based on discrete temporal logics is the ability to represent both the sequencing of processes (via ';') and a form of 'real-time' (via a combination of '$\bigcirc$' and ';').

TEMPURA execution. TEMPURA programs are written using combinations of the statements above. In general, the TEMPURA execution mechanism constructs intervals imperatively, rather than declaratively. The basic execution consists of a transformation of the statement to be executed into a simple normal form (see below) and an iterative process of constructing the appropriate intervals.

Note, however, that TEMPURA executes a restricted version of ITL, one that is essentially deterministic. In particular, the '$\Diamond$' and '$\vee$' operators are not defined in TEMPURA. Thus, a TEMPURA program, φ, is translated (through quantifier elimination and macro expansion) to a canonical form [253], that is

$$
\varphi \equiv \bigwedge_{i=0}^{n} \bigcirc^i p_i
$$

where p_i is a conjunction of assignments to basic propositions. This effectively provides constraints on the current moment, the next moment, the one after that etc. Each state in the interval is then constructed iteratively using these, necessarily deterministic, formulae. Since the formulae are deterministic, no backtracking is required during execution.

TEMPURA examples.   Simple examples of TEMPURA programs, together with the interval created when they are executed, are given below (taken from [382]).

Statement: *fin(p)* ∧ ○ ○ ○ *empty*
Result: an interval of length 3 at the end of which *p* is satisfied.

Statement: *while t do s ≡ if t then (s; [while t do s]) else empty*
Result: example of the definition of new operators – the *while* construct.

Statement: *while (M ≠ 0) do ([M ← N modulo M] ∧ [N ←M])*
Result: *N* is assigned the greatest common divisor of *M* and *N*
(where '←' represents assignment).

Statement: *for 12 times do [if month(2) then fin(28) else ...]*
Result: intervals representing months (for example, month number 2 is February) are sequentially composed.

Applications of TEMPURA included the simulation of hardware designs [382], the representation of multimedia [314] and parallel programming in general [253].

6.4.5   Executing interval temporal logic – TOKIO

TOKIO [220, 319] was a Logic Programming language based on the extension of PROLOG with ITL formulae. It provided a powerful system in which a range of applications could be implemented and verified. While TEMPURA executed a deterministic subset of ITL, TOKIO executed an extended subset incorporating the non-deterministic operators '◇' and '∨'.

TOKIO execution.   Since TOKIO was an extension of PROLOG, if no temporal statements are given, the execution of TOKIO reduced to that of PROLOG. When ITL statements *are* present, the execution mechanism is extended to construct intervals satisfying those temporal constraints. As described in [220], *facts* are true for all intervals, whereas rules, such as

$$a \leftarrow b, c, d$$

specify that the interval *a* must satisfy the constraints *b*, *c* and *d*. In processing constraints on intervals, the initial constraints (those satisfied 'now') are processed before the constraints relating to the next element of the interval, and so on. This iterative process continues until the construction reaches the end of an interval (i.e. *empty* is generated). This is guaranteed because only finite intervals are allowed in TOKIO.

TOKIO examples.   The following simple goals produce the output below within TOKIO (taken from [220]).

Goal: ← *write*(1)
Result: 1
Goal: ← ○ *write*(2), *write*(1)

Result: 12

Goal: $\leftarrow length(5), \Box write(1)$

Result: 11111

Goal: $\leftarrow length(8), (length(5), \Box write(0); write(1))$

Result: 00000111

The more complex goal below [319] states that either p and q must occur within the five states of the interval, or the interval is six states long and s occurs in the next interval. In either case, r is true throughout the following interval.

$$
\begin{aligned}
&\leftarrow \quad a; \Box r \\
a &\leftarrow \quad \Diamond p, \Diamond q, length(5) \\
a &\leftarrow \quad length(6); s
\end{aligned}
$$

Applications of TOKIO included the prototyping of user interfaces [303], hardware simulation and verification [319].

6.4.6 CHRONOLOG

The basic TEMPLOG form of Temporal Horn Clause, together with its operational model based on TSLD-resolution, can be characterized, and thus implemented, in a variety of different ways, as follows.

Orgun and Wadge [397, 398] describe a fragment of Temporal Horn Clauses (called CHRONOLOG) that is expressively equivalent to TEMPLOG and that can be seen as a form of *intensional* Logic Programming. This fragment includes all the restrictions of TEMPLOG but, in addition, restricts clauses so that no '$\Diamond$' operators are allowed in the body of any clause and no '$\Box$' operators are allowed in *any* heads. CHRONOLOG is equivalent in expressive power to the restricted form of TEMPLOG called TL1 [5, 54].

CHRONOLOG(L) is essentially a temporal logic programming language based on (linear) unbounded past and future [396]. This involves devising a new proof procedure called TiSLD-resolution but, as with TEMPLOG, restricts the use of more expressive temporal operators. CHRONOLOG(Z) is a further extension that has been shown to be useful in parallel contexts [351].

An alternative approach, which can also be extended to *branching-time* temporal logic programming is provided by Gergatsoulis *et al.*, who introduce another proof procedure, *CSLD-resolution* [233, 234].

6.4.7 LEADSTO

The *LEADSTO* system [86] shows many similarities with executable temporal specifications. It was developed in order to 'model and simulate dynamic processes in terms of both qualitative and quantitative concepts' [86]. Based on a classical language for states, LEADSTO also provides a language for specifying the temporal dependencies between successive states. When executed, the system attempts to build a suitable temporal execution, while also providing simulation details such as data-files concerning the states. The execution model is forward-chaining, very similar to the METATEM style, but the overall system provides additional tools such as for visualization, and has been used to provide simulation traces for a number of different applications within Artificial Intelligence.

6.4.8   Synthesis from temporal specifications

As we saw earlier, it is possible to *directly synthesize* a program from a temporal specification. Following early research by Pnueli and Rosner on generating automata from
temporal specifications [415, 416], much work has been carried out in this area. This has
led to a range of techniques and a range of different classes of problem. Typically the
basic synthesis problem for a reactive system (say, with two processes) is quite complex
(2-EXPTIME) and intricate. For certain specific communication architectures then the
synthesis problem can be easier but, in general, this remains difficult and can often
be undecidable. Just as model-checking technology has developed, so the techniques
used have been refined. Typically, multi-player games or automata-theoretic techniques
for *tree* automata are used, but often via symbolic techniques such as BDDs. Recent
work in this area has concentrated on either restrictions in the architecture or on the
specificaitons allowed, or reductions of full search to bounded search [406, 443]. Some
tools have even been released [430], and this area promises to be an important and
popular one for the future.

7

Selected applications

The future is here. It's just not widely distributed yet.
  – William Gibson

We have now seen a range of theories, techniques and tools based on our temporal logic, and have indicated some of the potential application areas for, and extensions of, these. Clearly we cannot hope to describe all the possible ways in which these can be applied in detail; indeed, we would not wish to do that here. Instead, this chapter consists of a little more detail concerning a *selection* of application areas utilizing temporal approaches. This is quite a *subjective* selection, reflecting areas that the author finds particularly interesting or has been closely involved with. Of course, there are many useful application areas that have not found their way into this chapter.

7.1 Model checking programs

In our examination of model checking we assumed that a finite state model of the program execution was available. In our practical examples, the Promela *programming language provided these models. However, most programs are not written either as finite state models or as* Promela *programs. In spite of this the basic idea behind model checking, specifically the on-the-fly automata-theoretic approach, can be extended to the direct verification of widespread programming languages such as* C *and* Java.

Let us recall the on-the-fly model-checking approach employed by Spin as described in Section 5.3.3. Here, the temporal property to be checked is negated and this is encoded as a Büchi Automaton recognizing 'bad' sequences. As the system, that is an automaton describing all possible executions, is explored the 'bad sequence' automaton monitors the

An Introduction to Practical Formal Methods Using Temporal Logic, First Edition. Michael Fisher.
© 2011 John Wiley & Sons, Ltd. Published 2011 by John Wiley & Sons, Ltd.

sequences produced. If it *ever* reaches a repeated accepting state, then it has recognized a 'bad' sequence and reports back. If it fails to reach an accepting state, it forces backtracking within the system automaton. This continues until an error is found or until all paths through the system automaton have been explored. Recall Example 5.15 from Chapter 5, where we see the 'bad sequence' automaton (called a 'never claim' in Spin) running in parallel with a system exploration process.

This is a very appealing approach and can not only be extended more generally beyond the Spin framework, but can form the basis for *program model checking*. This is where the verification is carried out *directly* on the program itself, not on some model of the program constructed separately [493]. We will highlight this approach by considering Java model checking below, but note that this style of verification has led to many model-checkers for high-level languages; not just Java [302, 492], but even languages such as C [36, 472].

7.1.1 Java programs

Let us again recap how the on-the-fly model-checking approach works and what we need in order to implement it. As we have seen, we need:

1. a mechanism for extracting all possible runs of a system;

2. some way to move the monitoring automaton forwards, as each run proceeds; and

3. a way of recognizing good/bad looping situations.

With Spin these were achieved by (1) an automaton representing all system executions, (2) a never process running synchronously with the main program execution, and (3) recognition of Büchi acceptance. Now that we wish to tackle a high-level language such as Java we need these again.

The particular approach we consider here is implemented as the Java PathFinder system (actually, Java PathFinder 2 (JPF2)), which is an explicit-state open source model-checker for Java programs [302, 493] . The key aspects that allow Java PathFinder to achieve (1)–(3) above are that it incorporates a modified virtual machine and that *listener* threads are used.

Virtual machine

Programs in Java are normally compiled to a set of *bytecodes* which are then executed, when required, by a *virtual machine*, called the Java Virtual Machine (JVM). Now, in order to allow this execution to be controlled, and indeed backtracked if necessary, Java PathFinder provides a special, modified JVM that explores all executions including all non-deterministic choices, thread interleavings etc. Importantly, this new JVM records all choices made and can backtrack to explore previous choices. Note that this modified JVM is actually implemented in Java and so runs on top of a standard JVM.

Listeners

A Java listener is a mechanism within the Java language allowing the programmer to 'watch' for events. Typically this is used for GUI events, but can also watch for state changes etc. Listeners are defined within Java's Listener hierarchy.

So, Java PathFinder uses Java's *listener* mechanism in order to provide a representation of an automaton that is attempting to build a model based on the program execution. As the program proceeds, the listener recognizes state changes in the execution and checks these against its automaton representation. At certain times the listener may be reset, forcing the JVM to backtrack. If the listener recognizes an execution sequence, then it reports this. Since we define the listeners to correspond to 'bad' sequences, then the reported sequences are counter-examples.

A general, pictorial, view of Java PathFinder is shown below. Java PathFinder is now quite well developed and is used for many Java applications [302].

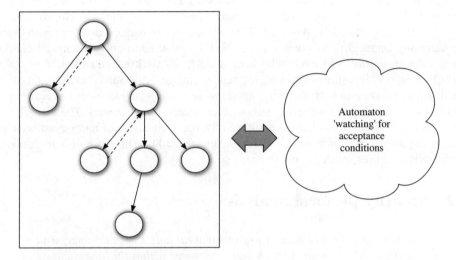

Modified Java virtual machine exploring all
possible execution branches, not only by
forward execution but by backtracking

Efficiency

While extremely useful, Java PathFinder is inherently slow. It is built on Java itself so, for example, code that is running executes on the modified JVM, which in turn runs on the standard JVM. In addition, it is an explicit-state model-checker. In order to improve efficiency, Java PathFinder employs a variety of sophisticated techniques. As well as standard partial-order reduction, which we have seen earlier, two additional aspects are interesting. The first is that, rather than just exploring the runs through a program in arbitrary order, the user can specify a '*choice generator*' that will explore branches in a specific order. The other main enhancement involves ensuring that the listener is only forced to move forward if *important* changes occur in the Java execution. Thus, 'unimportant' state changes/operations are collected together into one 'important' state. The *Model Java Interface* (MJI) is a feature of Java PathFinder that effectively allows code blocks to be treated as atomic/native methods. Consequently, since new states are *not* built by Java PathFinder for calls to atomic/native methods, these code blocks are effectively hidden from the listener.

7.1.2   Model-checking agent programs

We have seen already that *rational agents* provide an appealing and useful approach to modelling autonomous systems. Unsurprisingly, this has led to many agent programming languages being developed and applied [81, 84], with most of these being based on the BDI metaphor [428, 429] and many being inspired by Rao's AGENTSPEAK [427]. Clearly, it would be useful to employ model-checking techniques for agent programming languages. The central novelty of rational agents is *autonomy*, so we need then to verify not just temporal properties, but also logical properties relating to the agent's *beliefs* and *goals*. This is because we wish to verify not just *what* the agent does, but *why* it chooses to do it. (Without this distinction, autonomy reduces to something relatively trivial.)

Not surprisingly, model-checking techniques for agent programs is an area attracting a great deal of attention, and many tools and techniques are being developed [29, 80, 82, 85, 305, 353, 354, 371, 426]. In [145] a common semantic basis for many agent programming languages was given, and in [83] this was used to provide a model-checker for this common core. The model-checker, called AJPF, is itself an extension of Java PathFinder which allows checking of beliefs and goals of agents in addition to the usual temporal properties. In this way, agent programming languages can be verified in terms of logical properties involving temporal and motivational aspects. The flexibility of the common semantic basis has even allowed for the verification of multi-agent systems comprising agents programmed in different languages [144]. This form of *heterogeneous* verification promises much for the future.

7.2   Security protocol analysis

> *As we have seen, a very natural use of temporal logics is in the specification and analysis of situations that change over time. Within this, the analysis of how information or knowledge develops across a computational system is particularly appealing and has a natural home in the area of automated security analysis. We will here (briefly) discuss how temporal logics of knowledge (and belief) are developing into useful formalisms for this area.*

Temporal logics of knowledge [174] are useful for reasoning about situations where the *knowledge* of an agent, process or component is important, and where change in this knowledge may occur over time. A logic of knowledge on its own does not have this dynamic aspect, just allowing us to reason about knowledge at a particular moment. Adding a temporal component is one way to extend this description to the evolution of dynamic knowledge. Basically, a linear-time temporal logic of knowledge is a logic combining PTL together with a modal logic of knowledge. We already know about PTL, so let us examine a particular *modal logic of knowledge* [174, 325, 379].

7.2.1   Modal logic of knowledge

The logic we consider is an S5 multi-modal logic [325]. Modal logics are used to describe sets of 'worlds' connected through an 'accessibility' relation [68, 226]. The 'S5' aspect ensures that the relation used is actually an *equivalence relation*. This has some impact on the properties of the logic, as we will see below. We will view the worlds as describing the knowledge that an agent has at that point, and can follow the accessibility relation to reach other worlds, potentially describing different knowledge that the agent has in those worlds.

The syntax of a logic of knowledge is very simple extending propositional logic with just modal operators K_i, for every agent i. A formula such as $K_i\varphi$ is simply read as 'agent i knows φ'. The S5 nature of the logic then constrains it with axioms such as

$\vdash K_i(\varphi \Rightarrow \psi) \Rightarrow (K_i\varphi \Rightarrow K_i\psi)$

that is if agent i knows that '$\varphi \Rightarrow \psi$', then, if agent i knows 'φ', then it knows 'ψ'.

$\vdash K_i\varphi \Rightarrow \neg K_i\neg\varphi$

that is knowledge is consistent (an agent cannot know something *and* know its negation).

$\vdash K_i\varphi \Rightarrow K_i K_i\varphi$

that is if agent i knows something, then agent i knows that agent i knows it (often termed: 'positive introspection').

$\vdash \neg K_i\varphi \Rightarrow K_i\neg K_i\varphi$

that is if agent i does not know about something, then agent i knows that it does not know it (often termed: 'negative introspection').

$\vdash K_i\varphi \Rightarrow \varphi$

that is if agent i knows something, then it is true.

There is a vast amount of literature about such logics, in Computer Science, Artificial Intelligence, Philosophy and Logic, for example [174]. The utility of such a logic is in representing nested knowledge concerning agents, for example

$$K_{Tom} K_{Dick} \neg K_{Harry} football\_score$$

that is *'Tom knows that Dick knows that Harry doesn't know the football score'*.

7.2.2   Temporal logic of knowledge

Now that we have our logic of knowledge (multi-modal S5) and our temporal logic (PTL) we can simply[1] put the two logics together. So we can then write, for example

$$K_{me}\Diamond K_{you}me\_attack\_you \Rightarrow K_{me} \bigcirc me\_attack\_you$$

which might be described, in English, as:

if I know that eventually you will know that I will attack you, then I know that I should attack you next

---

[1] We say 'simply' here, but there is much more to this – a lot of work has been carried out on different ways to combine and handle various modal/temporal logics. *Temporalization*, *fusion (or independent combination)* and *product (or join)* are among the popular forms of logic combination [61, 183, 226, 330].

Aside. In the area of *rational agents*, we might re-cast our above formula in a more sophisticated way (where $G_a\psi$ is a modal formula meaning that agent a has the goal ψ) as:

$$K_{me}\Diamond K_{you}me\_attack\_you \;\Rightarrow\; K_{me}\bigcirc G_{me}me\_attack\_you$$

or even, if we are to pre-empt an opponent

$$K_{me}\Diamond G_{you}you\_attack\_me \;\Rightarrow\; \bigcirc G_{me}me\_attack\_you$$

In other words

> *if I know that eventually you will have a goal to attack me, then in the next moment I will have a goal to attack you.*

This area of logical modelling of agent behaviour via combined modal/temporal logics is a *very* popular one. Many more details are available, for example through [174, 258].

7.2.3   A little about proof, mechanisation and complexity

The decision problem for propositional multi-modal logic, such as S5 above, is not *too* complex: NP-Complete for one agent; PSPACE-Complete for more than one agent [174]. As we have seen, PTL is also PSPACE-Complete. The straightforward combination of the two logics together again is also not *too* bad because there are no axioms involving both S5 and PTL operators. However, as soon as we add axioms that provide close *interaction* between the temporal and epistemic dimensions then complexity can become *much* worse [256]. Two particular axioms have been studied in detail:

$$\begin{aligned}
\text{(synchrony+) perfect recall :} \quad &\vdash \quad K_i\bigcirc\varphi \Rightarrow \bigcirc K_i\varphi \\
\text{(synchrony+) no learning :} \quad &\vdash \quad \bigcirc K_i\varphi \Rightarrow K_i\bigcirc\varphi
\end{aligned}$$

If you think about it the names 'perfect recall' and 'no learning' make some sense. For example '$\bigcirc K_i\varphi \Rightarrow K_i\bigcirc\varphi$' implies that if the agent will know something in the next moment, then it already knows that aspect of the future now. And so it will never learn anything new. The 'synchrony' aspect means that all agents have the same concept of 'the next moment' [258].

Various different proof techniques and decision procedures exist. As with earlier chapters, we can apply clausal resolution techniques to modal logics of knowledge (and to modal logics in general) [392, 394, 389] and so to temporal logics of knowledge [155]. There are, as we would expect, tableau methods for various modal logics and combinations [133, 213, 226, 240, 244, 254, 505]. The complexity issues above mean that none of the proof methods are *very* efficient, but can be used to prove a few useful examples (see below). Another (popular) alternative is to translate the knowledge part into first-order logic (FOL) [294, 445], so that we have TL combined with FOL rather than TL combined with S5 [250]. This has the advantage that, if we are careful, we can retain decidability and use an automated prover for FOTL such as TSPASS.

7.2.4 Formalizing security protocols

There are many applications for temporal logics of knowledge (again, see [174]), but we here just mention one important one. We can use such logics to describe aspects of common communication protocols, primarily in terms of how the participants' knowledge evolves due to interactions [159]. For example [159], we might formalize part of a protocol as follows (much simplified)[2].

$$\bigcirc K_i\, information\,(xxx)$$
$$\Rightarrow$$
$$\left[\begin{array}{c} (K_i\; information\,(xxx)) \\ \vee \\ (rcv\,(i, Msg, pub\_key1\,) \wedge contains\,(Msg, xxx) \wedge K_i\; value\,(private\_key1\,)) \end{array} \right]$$

The idea here is that, for an agent i to know information about xxx in the next moment, then either agent i already knows this now, or agent i has just received a message which contains this information and, though encrypted by a public key, i has the private key counterpart of this (and can therefore decode the message).

We can then move on to describe more complex properties or requirements of the whole system, for example

$$[K_{me}\, K_{you}\, key(me) \wedge K_{me}\, send(me, you, msg)] \Rightarrow K_{me}\diamondsuit K_{you}\, contents(msg)$$

that is

> if I know that you know my public key, and I know that I have sent you a message, then I know that at some moment in the future you will know the contents of that message

We can specify all the interactions between the participants in such a way, explicitly describing how their knowledge changes. For example, the semantics of receiving a message might be that *if* a message is received, and *if* the message content can be decrypted, *then* the receiver gains the knowledge encoded within the message.

Once we carry out a specification for all such communications, together with any global properties, then we can attempt to prove requirements such as

$$K_{she}secret \Rightarrow (\diamondsuit K_{you}secret \wedge \square \neg K_{he}secret)$$

that is

> if **she** knows a secret then eventually **you** will know the secret but **he** never will.

If this proof is successful, then we know the above property is true of any interaction using this protocol. However, any failure to prove this will provide a counter-example comprising a possible attack on the protocol.

---

[2] We again use predicates here, but just for notational convenience. With finite numbers of agents, keys and messages, the problem is essentially propositional.

Thus, we can potentially use temporal logics of knowledge to reason (automatically) about security protocols. For example, in [159, 181], part of the Needham-Schroeder protocol [390], a well-known and simple communications protocol, is specified using a temporal logic of knowledge and various properties are proved by clausal resolution (using a predecessor of TSPASS).

Aside: First-order aspects.   A little first-order notation is useful in our temporal logic of knowledge. A simple and intuitive example is as follows. Imagine we have private (cryptographic) keys and that, if we are able to see the value of the key, then we can use it to unlock private data. We have a predicate '*unlock*' that is true when we can, indeed, see the value of the key. Now consider

$$\exists key. \; K_{intruder}(unlock(key) \wedge steal\_using(key)).$$

This says: 'there is a key, and the intruder knows how to unlock this and use it to steal data'. Alternatively, consider

$$K_{intruder}(\exists key. \; unlock(key)) \wedge steal\_using(key).$$

Here, the intruder knows that a key exists and that it can be unlocked. However, the scope of the '$\exists key$' is limited and so '$steal\_using(key)$' is not necessarily true because the *key* here is *not* necessarily the same one that unlocks the data. This example distinguishes between there existing some data element (a key) and the agent knowing some property about it, and the agent knowing that there exists a data element that has that property but the agent does not know which element that is.

Aside: Knowledge and belief.   Finally, a comment about belief. We typically characterize a logic of belief in a very similar way to that of knowledge [254]. The main difference is that the axiom

$$\vdash K_i\varphi \; \Rightarrow \; \varphi$$

does *not* hold for belief, that is just because we believe something does not necessarily mean it is true. So, we can have a temporal logic of belief, and can extend the earlier clausal temporal resolution calculus to this [157]. The combination of belief and knowledge (and time) is also quite useful in protocol analysis because we can specify that:

1. if we receive a message, seemingly from Alice, then we believe that Alice sent it, that is

$$B_{me} \, Alice\_sent\_msg$$

2. but we do not yet *know* that Alice sent it – it is only after we have gone through some (secure) authentication process that we can deduce

$$K_{me} \, Alice\_sent\_msg$$

Thus, the distinction between knowledge and belief is very useful for reasoning about interactions. Initially an agent might only *believe* something about other agents but, through interaction/communication, might eventually come to *know* that it is, indeed, the case.

7.3   Recognizing temporal patterns

We have seen throughout this book how temporal formulae correspond to temporal patterns on sequences. Indeed, when we examined model-checking we saw how a temporal property was assessed against a set of sequences. So this idea of checking a temporal formula to see if it is satisfied on a particular linear sequence is already very useful. However, when we considered model-checking we assumed that the 'full' sequences could be examined. But, what if the full sequence is currently being produced? There is an increasing use of temporal logics, or techniques akin to temporal logic, to recognize temporal patterns on a (continually extending) stream of data. In one respect, this can be seen as the dual of the direct execution approach we described earlier. This recognition of temporal patterns covers runtime verification, temporal data mining and temporal stream processing. We will here outline the basic idea, particularly with respect to runtime verification, and then point towards some of the many practical applications of this approach.

Verification is a complex process. We have seen that *deductive verification* typically requires a logical formula, say S, describing the behaviours of the system in question. Then we need a logical characterization of the property we would like to verify, say P. We use these together in an attempt to prove $\vdash S \Rightarrow P$ via deductive techniques. Since S can capture a whole set of systems satisfying these properties, and since each system itself is likely to be non-deterministic, then the whole deductive verification process may well be long and complex.

Algorithmic verification techniques, such as model-checking, can be simpler and faster. Here we have one system in mind, but it can have many executions through this system. As we saw earlier, model-checking essentially involves checking a temporal property against each possible execution of the system in turn. If the 'check' fails for any particular execution, then an error is flagged and that execution is provided as a counter-example. If the 'check' succeeds for all executions, then the system is verified with respect to the required property.

At this point it is instructive to recall the on-the-fly model checking approach employed by systems such as Spin, and discussed further in Section 7.1. Here, the temporal property to be checked is typically encoded as a Büchi Automaton. Indeed, the automaton used usually encodes the *negation* of the required property. This gives us an automaton that we do *not* want to recognize successfully any input (since that will be an execution with an error in). As we saw in Section 7.1, this automaton is usually executed in synchrony with the program/system. If the automaton fails to reach an accepting state, the system/automaton pair is forced to backtrack to try an alternative execution path. If the program/system is free from such errors, then the automaton will fail to find a bad program/system execution.

7.3.1   Lightweight verification

In spite of its sophisticated technology, model-checking is *still* quite expensive and requires that we have access to all possible executions of the system. Two much simpler and lightweight approaches are *testing* and *runtime verification* [236, 262, 440]. Each of these are simpler because we only consider *some* of the executions. In testing for a property, we typically carry out checks throughout at least one, but certainly not all, execution sequences.

Thus, runtime verification is *even* simpler than model checking. Not only do we only consider one execution sequence, but do not even check a *full* execution sequence. We actually only check a finite prefix of the (potentially) infinite run. The intuition is that runtime verification is used to monitor an execution sequence *as it is being constructed*! Informally, we might view the decreasing complexity involved in checking property ψ as:

- deductive verification – check that ψ is satisfied on all possible executions of all systems with a certain specification, that is those described by the *Class(S)*

$$\forall \Sigma \in Class(S). \ \Sigma \models \psi$$

- model checking – check that ψ is satisfied on all possible executions of our system S

$$\forall \sigma \in S. \ \langle \sigma, 0 \rangle \models \psi$$

- testing – check that ψ is satisfied on m executions of our system S

$$\bigwedge_{i=0}^{m} \langle \sigma_i, 0 \rangle \models \psi$$

- runtime verification – check that ψ is satisfied on the first n states of one execution, σ, of our system S

$$\langle \sigma, 0 \rangle \models_n \psi$$

Two comments about runtime verification. First, the above clearly requires us to use a modified semantics, $\models_m$, assessing the truth of temporal formulae over finite sequences. This, as we have seen in both Section 2.3 and Section 5.4.7 is not unusual. A second comment is that carrying out such checks should be *very* quick.

7.3.2   Runtime verification technologies

So, as we have seen, runtime verification aims to provide a lightweight verification technique that can be utilized as the execution proceeds. By now it should be clear how such runtime verification might work, in general:

1. choose a property that we want the system to satisfy, say φ, and negate it;

2. build a Büchi Automaton corresponding to $\neg\varphi$;

3. watch the execution being created and, every time a new state is generated, make a move in the automaton;

4. if the automaton *ever* reaches a repeated accepting state, then report an error.

The following diagram depicts an automaton monitoring ongoing execution:

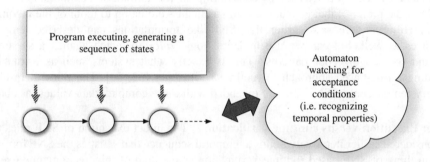

Aside.   If we recall Example 5.15 from Chapter 5, we saw a never claim running in parallel with a system exploration process. To some extent, runtime verification is similar if we assume B_S, the Büchi Automaton describing the system to be verified, is deterministic.

Thus, two expected benefits of runtime verification are:

• it typically carries out its checks very quickly (or else the system executes too fast for it); and

• it can be used to actively monitor the system as it is executing – indeed, in some cases the runtime verification system runs as a parallel process.

The runtime verification approach is very popular, providing a lightweight analysis technique for use in complex systems. It provides tools not only for Java, including Java PathExplorer [263], Java Logical Observer [70] and LARVA [128], but even for C [261]. It supports a successful *Runtime Verification* workshop, and the core techniques have been extended and applied in many different ways [345], though we only mention a few below.

Extending the patterns recognized

Within runtime verification we wish to recognize certain aspects of temporal states, and then use these aspects together to recognize temporal patterns within those states. But, so far, we have just thought of recognizing *propositional* values. Mainly, this was because of our earlier complexity concerns about handling first-order temporal logics. Yet, now, as we are purely carrying out a matching procedure, then we need worry much less about using first-order and even arithmetical aspects. Consequently, most runtime verification systems recognize temporal patterns in more complex numerical data. This is clearly important especially with applications in the lightweight verification of systems involving real-world data, such as aerospace applications [239], and can often extend to full statistical/numerical measures.

Rule-based runtime verification

We have motivated our discussion by considering the recognition of a temporal execution, or at least the prefix of that execution, using a Büchi Automaton to monitor the incoming stream. However, we saw earlier that SNF-like formulations can describe temporal sequences as well. So, can we use temporal logic *rules* to match against a stream of data and so recognize our patterns? This is exactly what systems such as EAGLE do, providing a rule-based approach to runtime verification [52, 53]. The rules themselves are very reminiscent of SNF, but are extended with more complex data values, as above.

Aside: Execution versus runtime verification.   The direct execution of SNF rules that we considered earlier iteratively builds a temporal sequence that satisfies the SNF formula. We can think of SNF-based runtime verification as the dual of this, recognizing a set of SNF rules from (the prefix of) a temporal sequence. As we know, a set of SNF rules corresponds directly to a temporal formula and so we might use the PTL $\rightarrow$ SNF translation in reverse to try to recognize specific temporal formulae from the set of rules triggered.

Three-valued temporal logics

So far we have only considered formulae with truth values of 'true' or 'false'. However, in [57] the authors suggest that a third value, 'inconclusive', should be used. The reason for this is that, since we have not yet seen all of the temporal model we cannot, in some cases, definitively say a specific formula is true or false – we might just need to wait for more states to be generated in order to be able to decide.

Before we leave this general topic, we note that the recognition of temporal sequences is actually very general and techniques for achieving this are very common across a number of fields. In the following section we will provide an overview of some of these, highlighting that runtime verification is just one instance of a very general pattern recognition problem.

7.3.3   Data mining and database querying

In the field of databases, *data mining* is primarily concerned with searching huge amounts of data to discover interesting, useful or even unexpected patterns or relationships [11]. The step from querying extremely large databases to querying incoming streams of data is not so surprising and so the field of *data stream mining* is also very well developed. In this case, data from powerful sources such as telecommunications systems or sensor networks is assessed as it is received, possibly analysed for a short time, but then discarded. This process continues indefinitely. The core element in this work is to recognize or detect patterns within the incoming stream of data [9].

In turn, the area of *temporal* data mining is again concerned with the analysis of data streams, but specifically with the aim of detecting temporal relationships and temporal inter-dependencies [27, 343]. The techniques developed in this area have proved to be particularly powerful and useful, especially since most interesting data does, indeed, incorporate temporal patterns [469]. Temporal data mining has also extended to areas such as *spatio-temporal data mining*, which even has its own specialized research event, that is the *International Workshop on Spatial and Spatiotemporal Data Mining*.

Since recognizing patterns within data is clearly important, and since highly efficient techniques are required for this, it is not surprising that much work has been carried out on developing efficient *stream query processing* languages. Much of this originated in the Database research community, and has led to very efficient systems with vast numbers of applications, from fraud detection in banking real-time decision support in healthcare. The development of this technology has led to extremely efficient data stream processing, or *event processing* systems, such as `StreamCruncher` [http://www.streamcruncher.com], `Esper` [http://esper.codehaus.org] and `StreamBase` [http://www.streambase.com]. Such systems provide very powerful and flexible query languages for data streams, seemingly providing all the power needed to recognize any of the temporal patterns we might be concerned with. From the `StreamCruncher` documentation:

> *"StreamCruncher is an Event Processor. It supports a language based on SQL which allows you to define 'Event Processing' constructs like Sliding Windows, Time Based Windows, Partitions and Aggregates. Such constructs allow for the specification of boundaries (some are time sensitive) on a Stream of Events that SQL does not provide. Queries can be written using this language, which in turn can be used to monitor streams of incoming events."*

There appear to be many such systems available, providing practical stream/event processing capabilities, but with active ongoing research initiatives and with semantics even given in terms of interval temporal logics [10, 41, 323, 402, 497].

7.4   Parameterized systems

> *Our earlier specifications allowed us to describe a finite number of components/processes possibly with some communications constraints between them. But, what if we need to specify an arbitrary number of components? While this is difficult in general, it becomes potentially tractable if we know that all the components are essentially of the same form. But, even if we can specify such infinite-state systems, how can we verify them? We might apply some form of abstraction and then turn to model-checking. An alternative is to use first-order temporal logics to specify such systems and then use our deductive techniques, such as the temporal resolution approach described earlier, to verify their properties. We will outline this approach, which has potential applications in areas such as multi-processor architectures, swarm robotics and open, distributed consensus protocols.*

We wish to verify protocol behaviour. As we have seen, model-checking is a feasible way to do this, at least for finite-state systems. However, if we are considering *infinite-state* systems, then the use of model-checking becomes more problematic, often requiring abstraction. We here explain how a particular, but common, class of infinite-state system can be captured using temporal logic and how verification can be carried out automatically [211]. The class of systems we are interested in is *parameterized systems*, typically where the systems themselves have a finite number of states.

A simple way to describe such parameterized systems is an infinite set of identical processes, each of which is finite-state. Thus, we get infinite behaviour, essentially through the unbounded/unknown number of processes, not through any infinite behaviour of each process. Such systems are quite popular for describing sophisticated communication and interaction protocols, where we want the protocols to work *regardless* of the number of other participants/processes involved. These are common (and often necessary) in *networks*, *service-oriented architectures*, *multi-agent systems*, and *open* systems in general.

7.4.1 Single process → PTL

We have already seen that simple finite-state machines can be straightforwardly[3] described using *propositional* TL, for example:

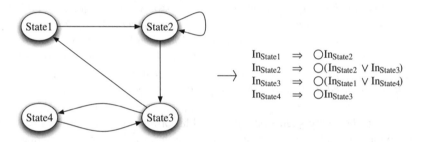

This then gives us a PTL characterization of the behaviour of one such machine, where any model for the set of formulae describes a run of the finite-state machine. Then we can establish properties of the state machine simply by employing deduction on the corresponding temporal formulae.

7.4.2 Finite number of processes → PTL

Now, if we have any fixed, finite number (say N) of such machines, working together in parallel, then we can still use PTL. Simply, we can just have a formula describing each state machine and then conjoin all the N formulae together, taking care with any interactions [158].

7.4.3 Arbitrary number of processes → monodic FOTL

Now, what if we consider an arbitrary, or even *infinite*, number of such identical finite-state machines? We cannot use PTL in a straightforward way, but we can use *first-order* TL. Since we know each state machine is identical, then we can use the formula

---

[3] However, additional formulae are needed to ensure that only one '*In*' proposition is true at any one moment. As we noted in Section 4.2.2, the PTL variation whereby certain propositions are disallowed together [158, 161] makes this a more concise description.

description of a state machine but then parameterize this with the number/name of the particular state machine being considered [211], for example:

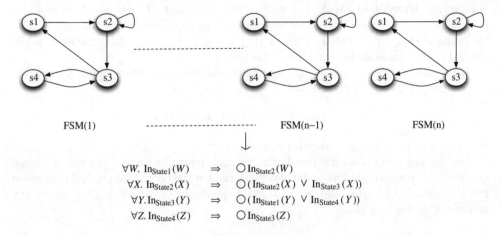

$$\forall W. \text{In}_{\text{State1}}(W) \quad \Rightarrow \quad \bigcirc \text{In}_{\text{State2}}(W)$$
$$\forall X. \text{In}_{\text{State2}}(X) \quad \Rightarrow \quad \bigcirc (\text{In}_{\text{State2}}(X) \vee \text{In}_{\text{State3}}(X))$$
$$\forall Y. \text{In}_{\text{State3}}(Y) \quad \Rightarrow \quad \bigcirc (\text{In}_{\text{State1}}(Y) \vee \text{In}_{\text{State4}}(Y))$$
$$\forall Z. \text{In}_{\text{State4}}(Z) \quad \Rightarrow \quad \bigcirc \text{In}_{\text{State3}}(Z)$$

For example, $\text{In}_{\text{State4}}(Z)$ is true if state machine 'Z' is in state 4 at this moment. Thus, for any particular instance, we have a PTL formula, but to describe the whole system we have a *monodic* (i.e. only one free variable crossing temporal operators) FOTL formula. So, now we have a monodic FOTL formula, *Sys*, describing the unbounded/infinite system of processes/automata/machines. If we wish to prove some property of such a system, say *Prop*, then we just need to prove

$$\vdash \; Sys \Rightarrow Prop \,.$$

Since this form of monodic FOTL is decidable then we can use a tool such as TSPASS to automatically establish this.

But, are there such protocols? And are we able to describe useful properties in FOTL? To answer these, we will next provide an overview of a simple case study from [211].

7.4.4   Case study [211]

This case study, derived from *cache coherence* protocols, involves an arbitrary number of identical automata, each of which can do one of several *actions*. In the full case study, the automata just have three states (I, S, and M), but have six actions that can potentially occur at any of these states. These action are W, R, T, together with their duals $\overline{W}$, $\overline{R}$, and $\overline{T}$. Thus, the basic automaton is described by a set of triples of the form

$$\langle StateFrom, Action, StateTo \rangle$$

For example, for state I the protocol allows the follwing triples (note that action T cannot be taken when in state I):

$$\langle I, W, M \rangle \qquad \langle I, R, S \rangle \qquad \langle I, \overline{W}, I \rangle \qquad \langle I, \overline{R}, I \rangle \qquad \langle I, \overline{T}, I \rangle$$

for the five actions W, R, $\overline{W}$, $\overline{R}$, and $\overline{T}$. It is a core part of such systems that when one automaton takes a transition executing an action such as W, R, or T, then all other automata must make a complementary transition (a *reaction*) $\overline{W}$, $\overline{R}$, or $\overline{T}$. (For further details, see [211].)

Automaton translation (*Sys*)   For each state, I, S, and M, we introduce monadic predicates In_I, In_S, and In_M. We also introduce monadic predicates for each action, for example Do_W for action W. Thus, $In_S(a)$ is true if automaton a is in state S, while $Do_R(a)$ is true if automaton a carries out action R.

It is clear now how to specify the automaton transitions in such a logical framework. For example, $\langle I, R, S \rangle$ is represented as

$$\Box \forall X.\ (In_I(X) \wedge Do_R(X))\ \Rightarrow\ \bigcirc In_S(X).$$

And so the state is changed after this action.

We can carry out a similar translation for each transition and so produce the basic specification. However, in this protocol we also need at least one *distinguished* automaton that is required to be active. So, we add another predicate, A, where $A(X)$ is true of any automaton X that is *active*.

Aside.   The basic pattern of computation is that one (usually) or more active automata must all take the same action at the same time as all other automata take the complementary action.

As well as the specification of the transitions within the automaton, there are a number of system-wide requirements that must be captured in order to provide *Sys*.

- There always exists at least one active automaton:

$$\Box \exists X.\ A(X)$$

- The state of any *active* automaton is changed after an action (as above).

- The state of any *other* automaton is changed after the re-action (as above).

- Any active automaton performs one of the actions available:

$$\forall X.\ \Box(A(X) \Rightarrow (Do_W(X) \vee Do_R(X) \vee Do_T(X)))$$

- Every automaton is in one of the prescribed states:

$$\forall X.\ \Box(In_I(X) \vee In_S(X) \vee In_M(X))$$

And so on. The formulae above, and others of this style, together provide the FOTL specification for the system, *Sys*.

Property translation (*Prop*)   Next we must develop a number of properties, in FOTL, to check. These, of course, depend on the system being specified and the requirements we have for it. Some sample properties, from [211], are given as follows:

- There is at least one process/automaton that will eventually reach the M state:

$$\exists Z.\ \Diamond In_M(Z)$$

- There is at least one process/automaton that will eventually remain in the S state:

$$\exists W. \; \Diamond \Box In_s(W)$$

- All processes/automata will eventually remain in the S state:

$$\forall Y. \; \Diamond \Box In_s(Y)$$

- There is a point after which no process/automaton enters the I state:

$$\Diamond \forall X. \; \Box \neg In_I(X)$$

- There should never be a moment when one process/automaton is in the S state while another is in the M state (also called *non-co-occurrence*):

$$\Box \; \forall X, Y. \; \neg(In_s(X) \wedge In_M(Y))$$

- At most one automaton is active:

$$\forall X, Y. \; (A(X) \wedge A(Y)) \Rightarrow (X = Y)$$

And so on. Whichever property, *Prop*, we choose we can then try to establish $\vdash Sys \Rightarrow Prop$.

Aside: Problems. Some properties might need equality, for example the 'at most one automaton is active' property above and the property stating 'at most one process/automaton can be in state M':

$$\Box \; \forall X, Y. \; (In_M(X) \wedge In_M(Y)) \Rightarrow (X = Y)$$

However, as mentioned in Section 2.8.7, monadic FOTL with equality is not *recursively enumerable* in general! So we need to be careful with the use of equality, and use alternatives where possible. The same problems occur if we wish to *count* how many processes/automata are in a certain state.

7.4.5 Failing processes

So, using the above approach we can verify properties of systems/protocols involving arbitrary numbers of participants. However, what if some processes *fail*? Can we still verify our properties of such systems? But how many failures are allowed? If a fixed number of failures are allowed, then we can re-cast this problem as above and verify. If arbitrary numbers of failures occur, then we likely have very little chance of verifying anything! However, what if we have arbitrary numbers of processes/automata, but allow only a *finite* (but unknown) number of failures? What then? In [212], this problem is tackled for the above form of FOTL specifications and it is shown that verification can, indeed, be carried out even if we do not know exactly how many failures will occur.

7.5   Reasoning with intervals

For practical temporal reasoning, it is often useful to work with various dif-
ferent granularities of time, for example minutes, hours, days and years. A
particularly influential class of temporal logics has tackled this aspect with
the logics being based on the notion of 'intervals' rather than 'points'. These
interval temporal logics have been devised and analysed both in Computer Sci-
ence and Artificial Intelligence and have been shown to be useful in as distinct
areas as temporal planning, ecological simulation and hardware verification.
We will here outline the basic idea of using intervals within a temporal logic
and then introduce the two main themes in this area: Allen's Interval Algebra
and Moszkowski's Interval Temporal Logic.

For practical temporal reasoning, it is often useful to work with various *different* granularities of time, for example minutes, hours, days, and years. As we saw in Section 2.8, the
temporal logic we have been examining can be adapted to have an underlying *interval*,
rather than point, basis. Although work in Philosophy, Linguistics and Logic had earlier
considered time periods, for example [101], interval temporal representations came to
prominence in Computer Science and Artificial Intelligence via two important routes:

- the development, in the early 1980s, of interval temporal logics for the description
  of computer systems, typically hardware and protocols [257, 381, 386, 449]; and

- the development, by Allen, of interval representations within Artificial Intelligence,
  primarily for use in planning systems [14, 15, 17].

Such temporal logics have been particularly influential both in Computer Science and
Artificial Intelligence, and have been shown to be useful in diverse areas such as
temporal planning, ecological simulation and hardware verification. We have already
seen Moszkowski's Interval Temporal Logic (ITL) in Sections 2.8.10 and Section 6.4.4,
directly executing ITL statements in the latter [253, 382], so we will instead outline
Allen's Interval Algebra [14, 15, 17], one of the most influential works within practical
Artificial Intelligence.

7.5.1   From points to intervals

Throughout this book we have been concerned with point-based temporal logic. Here, the
truth/falsity of a logical statement is assessed at a specific moment/point in time. Thus,
if $\langle \mathcal{M}, i \rangle \models \varphi$, then we say that φ is satisfied *at* moment 'i'. However, many aspects
are not so naturally modelled as happening at distinct points; for example, real-world
processes typically have more complex, continuous dynamics. Even within English we
often consider properties over some duration. Thus,

> "I will be writing this Chapter throughout the day"

does not easily fit within a point-based approach. For example, if we consider each minute
to represent a different 'moment', then we can model the statement above, but what about
properties over 'seconds'? Or over 'daylight hours'?

Consequently, an alternative approach is to consider the truth/falsity of temporal statements over a temporal *interval*. An interval can be thought of as 'section' of time with some duration. Thus, if i now refers to an interval, for example an hour, then $\langle \mathcal{M}, i \rangle \models \varphi$ might now mean that φ is true during that hour. There are, of course, many variations here. Rather than interpreting the above as

'φ is true during hour i'

we might equally well interpret this as any of

'φ is true throughout hour i'

'φ is true in hour i'

'φ is true for an hour i'

The choice of semantics here is clearly important.

Before moving on to different varieties of interval descriptions, we note two aspects relevant to our discussion earlier:

- interval models generally still have a finite past – thus there is some concept of an *initial interval*; and

- the underlying domain can be viewed as the Real Numbers ($\mathbb{R}$) rather than the Natural Numbers ($\mathbb{N}$).

The latter provides some intuition about why we cannot easily describe interval-approaches in discrete, point-based formalisms. However, effectively using a dense model can also be complex, potentially allowing us to keep on examining smaller and smaller intervals forever.

7.5.2 Granularity

In mentioning the possibility of an underlying dense domain above we can begin to see some of the complexity; between any two time points there are an infinite number of other time points. Thus, time intervals can be described at arbitrary *granularities*. Since we can typically take a sub-interval of any existing interval then we can generate an increasingly fine granularity; since we can concatenate intervals to form new ones, then we can generate increasingly coarse granularities.

This can be both dangerous and useful. In an example from [196], consider a discussion between participants who agree to organize a meeting *every month*. They must agree to either a date (e.g. the 25th) or to a particular day (e.g. the last Tuesday in the month). Later, they will consider times within that day. Then they might possibly consider more detailed times within the meeting itself, and so on. In the first case, the participants wish to represent the possibilities without having to deal with minutes or even hours. Later, hours, minutes and seconds may be needed. Consequently, many different, interlinked, granularities may be required. This has led to both granularity hierarchies [90, 217, 378, 422] and specific logics such as *calendar logic* [395]. (For the interested reader, excellent descriptions of many of these varieties can be found in [62, 173].)

7.5.3 Defining intervals

As mentioned above, an *interval* captures some duration of time over which certain properties hold. As in the earlier point-based approaches, there are many different varieties of definition for intervals. For simplicity, let us consider an interval to be a structure representing all the period of time between two 'moments'. But, now, many possibilities still remain. Does the interval include the end points? Can we have intervals where the start point and end point are the same? Can we have infinite length intervals? And so on.

Typical questions we might want to frame in a temporal language concern the relationship of points to intervals and the relationship between intervals. It is also important to define, and distinguish between, *open* and *closed* intervals. For example, if we consider just intervals over some numerical domain such as $\mathbb{R}$, where $\{begin, end, i\} \subseteq \mathbb{R}$, then i is within the *open* interval $(begin, end)$ if $begin < i$ and $i < end$, and is within the *closed* interval $[begin, end]$ if $begin \leq i$ and $i \leq end$. Then when we move on to the properties of formulae over intervals and the calculus for manipulating intervals, there are many possibilities. Next, we will just mention the most famous.

7.5.4 Allen's interval algebra

Allen produced an interval algebra that was not only able to manipulate intervals but could also describe temporal properties and their evolution over such intervals [14, 15]. This was particularly useful in *planning* systems, and remains so today. The essence of planning some sequence of actions is that the plan is active over an interval and, within that interval, subplans/actions are undertaken. How the subplans/actions are brought together is the cornerstone of planning and describing how their intervals interact is a central theme developed within Allen's algebra.

Thus, once we have some interval, I, we might ask:

- does a particular time point a occur *within* the interval I?

- is a particular time point b *adjacent* to (i.e. immediately before or immediately after) the interval I (and what interval, if any, do we get if we add b to I)?

- does another interval, I_2, *overlap* with I?

- is the interval I_2 a strict *sub-interval* of I?

- what interval represents the temporal *intersection* of the intervals I and I_2?

- ...and so on...

As well as providing such logical apparatus, Allen's algebra allows a logical language with a variety of useful operations. These are typically relations on intervals and examples include:

- I_1 *overlaps* I_2 is true if the intervals I_1 and I_2 overlap;

- I_3 *during* I_4 is true if the interval I_3 is completely contained within I_4;

- I_5 *before* I_6 is true if I_5 occurs before I_6;

- etc.

Allen analysed this interval model, together with the potential relations, and produced a core Interval Algebra containing 13 such binary relations between intervals [14, 15]. This allows us then to ask logical questions such as

- does the proposition φ hold *throughout* the interval I?

- does the proposition φ hold anywhere *within* the interval I?

- does the proposition φ hold by the *end* of interval I?

- does the proposition φ hold immediately *before* the interval I?

- given an interval I_1 where φ holds, is there another interval, I_2, occurring in the future (i.e. strictly *after* I_1), on which φ also holds?

- can we split up an interval I into two sub-intervals, I_1 and I_2 such that φ holds continuously throughout I_1 but not at the beginning of I_2, and where joining I_1 and I_2 back together gives I?

- ...and many more...

As mentioned above, this work has been *very* influential and has led to broad areas of research within Artificial Intelligence, particularly in planning and natural language representation systems. For example, further work on the formalization and checking of Allen's interval relations can be found in [16, 255, 332, 333, 349, 350] with the algebraic aspects being explored further in [272, 273]. In addition, the basic interval algebra has been extended and improved in many different ways; see [232] for some of these aspects and [163] for a thorough analysis of the computational problems associated with such interval reasoning. These last two references also bring in the work on representing such problems as temporal constraint networks [140, 448] and solving them via constraint satisfaction techniques [439].

7.5.5 Beyond IA/ITL

With the success of both Allen's Interval Algebra (IA) and Moszkowski's ITL it is no surprise that considerable work has been carried out examining the variety of interval temporal logics and their potential extensions. Important work by Halpern and Shoham [255] showed how a powerful logic (HS) over intervals (not just of linear orders) essentially subsumes Allen's algebra. Indeed, the HS language with unary modal operators captures entirely Allen's algebra, while binary operators are needed to capture the '*chop*' operator within ITL [242], reflecting its additional complexity. Significant work concerns extending this still further [91, 92].

There are many potential extensions of the above interval approaches. One is to consider intervals, not just over linear orders, but also over arbitrary relations. This moves us towards spatial and spatio-temporal logics, see [227] or [127]. Alternatively, one might bring real-time aspects into interval temporal logics. This has been developed within the work on *duration calculi* [107, 508]. Another interesting extension to interval temporal logic is to add operators that allow endpoints to be moved, thus giving *compass logic* [368].

Finally in this section, we note that there are a number of excellent articles covering much more than we can here: introductory articles, such as [358, 491]; surveys of interval problems in Artificial Intelligence, such as [163, 232]; and the comprehensive survey of interval and duration calculi by Goranko, Montanari and Sciavicco [242].

7.6 Planning

Temporal planning plays an important role in many areas, particularly Artificial Intelligence. We will here outline the basic idea of planning, how this evolved to planning over temporal events, and how the idea of using temporal logic to control the planning process is now leading to the direct planning of temporal logic statements. A further advance is the utilization of model-checking technology to provide efficient planning systems.

Planning is a fundamental issue in Artificial Intelligence and, consequently, has been investigated widely [18]. Of the many research areas concerned with planning, several intersect with topics in this book; we will mention just a few of these below.

7.6.1   What is planning?

Assuming that we have a set of actions, *Act*, a typical *planning problem* (over *Act*) is to transform an *initial* state into a *goal* state through a series of state-transition operations derived from the actions. Here, both the initial and goal states are described logically, and the operations follow the traditional STRIPS [348] format:

- *preconditions* for the operator to be applied;

- *deletions* of facts from the state following the operation; and

- *additions* of facts to the state following the operation.

Example.   A typical (and traditional) planning problem might be to transform from the initial state

$$\{on(blockA, table), on(blockB, table), on(blockC, table)\}$$

to the goal state

$$\{on(blockA, blockB), on(blockB, blockC), on(blockC, table)\}$$

using the *move(X, From, To)* operation:

   preconditions: $on(X, From), \neg\exists Y.on(Y, To)$
     deletions: $on(X, From)$
     additions: $on(X, To)$

So, a typical automated planning system takes three inputs: precise description of the initial state; precise description of the goal state; and a set of actions, represented as transformation operations in standard notations such as STRIPS (above) or PDDL (below). If successful, the planning system returns with a sequence of actions that will lead from the initial state to the goal state.

7.6.2   Planning as model checking

The success of model-checking technologies led to their application in unexpected areas. One such is that of planning. The efficient BDD and SAT-based technologies used in contemporary model-checkers mean that the search they perform is, indeed, very fast. Since model checking is, in essence, *search* the idea behind the 'planning as model checking' approach is to formalize the planning domain and describe the desired behaviours of plans in the model. The planning problem is then solved by searching through the state space, checking that there exists a plan that satisfies the goal [112, 237]. The highly engineered model-checking technology has ensured that this approach has been very successful in the planning community.

7.6.3   Temporal planning

Much of *temporal planning* [216] is concerned with real-time aspects such as ensuring that plans complete before certain deadlines. This is necessary because real-world planning does, indeed, require such temporal constraints. These constraints concern not just deadlines but durations, as we have seen in Section 7.5. Consequently, plan/domain description languages require a variety of temporal aspects. In particular, the *planning domain definition language* (PDDL) used in international planning competitions aims to standardize planning/problem descriptions and itself incorporates strong temporal aspects [215]. Indeed, most (if not all) practical planning systems incorporate mechanisms for handling temporal constraints and so significant work on both the computational complexity of such problems [163] and sophisticated techniques for handling them in practice [232] have been developed.

7.6.4   Controlling planning

While real-time temporal logics are used in the temporal planners described above, we might ask if there is any role for standard temporal logics such as PTL? It turns out that there is! In considering standard planning we can see that there are numerous possible routes from a given configuration; some are beneficial, leading to the goal, while others lead us to failure. A key part of the planning process is to control which choices to make and which routes to take from the initial state towards our goal. Bacchus and colleagues recognized that simple temporal logics could be usefully used for *meta-level* control of the planning process [32], leading on to the development of the TLPlan system.

This was very successful and has led to many other researchers using temporal control within planning. Indeed, more recent work has increased the temporal aspect until planning can be seen as temporal deduction [104, 369]. This has a link not only to Chapter 4 in this book, since the view of planning as deduction, that is proving a theorem, is well known, going back to the work of Green [246], but also to Chapter 6 since the distinction between planning and execution is *very* narrow, and the use of temporal logic for heuristic control of execution is very appealing [48].

7.6.5   Planning control

In complex Engineering scenarios such as robotic movement, sensor manipulation or process management, control systems are used to guide the evolution of the subsystems.

But how can we be sure that the control systems developed are appropriate for the task in hand, and can guarantee that they will achieve what is required? This is a difficult problem, but for certain classes of system it has been solved using model-checking/planning techniques. For example, in [317], a temporal requirement is given (in PTL) and this is used to partition the state space of the underlying hybrid system. This partitioning is finite, so a model-checking type procedure is then used to find a suitable path through the partitions that guarantee the required area is reached. From such a path an appropriate control strategy is devised. This, together with related work on using temporal logic for guidance and analysis has shown how very useful temporal techniques are in the field of control systems [175, 176, 316, 315, 324, 465, 466].

8

Summary

I never think of the future; it comes soon enough.
   – Albert Einstein

So, we are at the end. But, what have we seen?

It turns out that temporal logic (certainly the variety we have looked at here) is essentially quite a simple formalism. However, in spite of this simplicity, we have seen how temporal logics have very many potential applications, beginning with the formal specification of concurrent, interacting and communicating systems. Describing such *reactive* systems in a formal way clarifies their behaviour and formalizes our intuitive view of what this behaviour should comprise.

Although carrying out such formal specification is both an informative and useful process, if we cannot manipulate the resulting specifications in interesting and productive ways, then surely the logical framework itself would not be so appealing. Yet, as we have seen, this is not a problem with temporal specifications. A wide range of techniques has been developed for manipulating these descriptions, together with very many practical tools that implement these techniques.

Deductive techniques allow us to establish whether temporal statements are true or false, with the *clausal temporal resolution* approach specifically being the focus of our attention. *Algorithmic* techniques move towards analysing the detailed temporal properties of structures, something that is particularly important in *model checking* where the structures typically describe the runs of a particular system. Techniques for the *direct execution* of temporal specifications, as characterized by the METATEM approach, then allow us to animate and prototype temporal descriptions. This also provides the basic concepts of a dynamic and flexible multi-agent programming language. Beyond these aspects, we have also highlighted a great number of areas where temporal logics are being, or can be, applied.

An Introduction to Practical Formal Methods Using Temporal Logic, First Edition. Michael Fisher.
© 2011 John Wiley & Sons, Ltd. Published 2011 by John Wiley & Sons, Ltd.

Throughout this book we have endeavoured to provide a *practical* focus to our exploration of temporal logics. This has led to the introduction of tools such as `Concurrent MetateM`, `Spin`, and `TSPASS`, together with our use of many examples and exercises. While these tools are too complex to fully learn by reading several chapters within a book, we hope to have given the reader sufficient insight to begin to use them in their own practical scenarios.

As will have become clear, there are *many* more aspects that we could have covered. There are different logics, alternative techniques, distinct tools and a plethora of applications, all of which we have mentioned, albeit briefly. In addition, more new aspects relating to temporal formalisms are appearing all the time. While we cannot hope to describe all these in detail, this book at least provides the basics from which you can explore, assess and apply these tools and techniques based on temporal logics.

A

Review of classical logic

Logic, n: *The art of thinking and reasoning in strict accordance with the limitations and incapacities of human misunderstanding.*
  – Ambrose Bierce

A.1   Introduction

A.1.1   What is logic?

The study of logic can be seen from two viewpoints. The first is where logical systems are seen as being self-contained and not meant to model any real-world objects or actions. Philosophers and logicians have typically studied such systems, developing and extending various different notions of truth, provability etc.

An alternative view of logic, and one that we tend to follow, is as a formalization of some aspect of the real world. Here, a logical system describes, or models, some abstraction of objects or actions that either do, or may, occur in real life. But logic is not the only language that can be used to describe such aspects. For example, English is a language that has been widely used to describe and reason about situations and problems. However, to discover the validity or consistency of a statement in English, the ambiguities and inconsistencies of natural language must be overcome; lawyers often make a living from arguing about the meaning of statements in English!

Thus, a more 'formal' language with well-defined semantics and a consistent reasoning mechanism is required; logical systems fit into this category. However, if logic is used to model properties of the real world, we must recognize that a logic represents a particular abstraction. Naturally, if we want to represent a more complex or more detailed view of the world, a more complex logical system must be used.

An analogy of this occurs in the field of mechanics. For several centuries, Newtonian Mechanics was believed to describe all the mechanical properties of real-world objects. Indeed, it seemed to work easily for any of the practical examples that people

An Introduction to Practical Formal Methods Using Temporal Logic, First Edition. Michael Fisher.
© 2011 John Wiley & Sons, Ltd. Published 2011 by John Wiley & Sons, Ltd.

tried. However, in the 20th century physicists discovered situations where Newtonion Mechanics did not give the correct answers (e.g., objects moving *very* quickly). Thus, a more detailed model of the world was required and Relativistic Mechanics was invented. So, in order to describe more aspects of the real world, a simple system that was easily applicable to everyday situations was replaced by a complex system in which calculations about everyday situation were hopelessly intricate. Similarly, logics, such as *propositional logic*, are easily mechanisable but are not expressive enough to model many situations. More complex logics, such as *first-order logic*, are able to model a much wider range of properties but are much harder to mechanize. To summarize, a logic just represents an abstraction of the world. If we want a more precise view, a more complex logic must, in general, be used.

Thus, there are a wide variety of logical systems that have been developed. Fortunately, all have several things in common, for example notions of truth, falsity, logical structure, validity, proof, consistency etc. In general, logic consists of two parts: the model theory, which is concerned with what statements in the logic describe and gives a semantics for such statements; and the proof theory, which is concerned with how statements in the language can be combined.

A.1.2   What is proof?

The proof theory of a logic consists of axioms and inference rules (alternatively called proof rules). Axioms describe 'universal truths' for the logic and inference rules transform true statements into other true statements (theorems). The process involved in a proof is generally that of transforming such true statements in the logic using only the inference rules and axioms. Technically, a proof is a sequence of formulae, each of which is either an axiom or follows from earlier formulae by a rule of inference.

The two main approaches to the construction of proofs are *forwards deduction* and *backwards deduction*. In the case of forwards deduction, we start with a set of axioms, apply rules of inference to some elements of this set to generate further valid statements (i.e., theorems) and add these to the original set of axioms. This process is then repeated on the new set of theorems and axioms. Eventually (hopefully!) the hypothesis that we wanted to prove will appear in this set of theorems. The skill of the logician or the automated theorem-prover lies in the choice of appropriate rules of inference and axioms to apply in order to reach the hypothesis as quickly as possible.

The backwards deduction method proceeds by continually decomposing the hypothesis into sub-problems until the sub-problems reached are axioms of the logic. Thus, a statement can be proved either if it is an axiom or if it can be decomposed into sub-statements that can be proved and, when suitable inference rules are applied, can validate the original statement.

Thus, proving a formula essentially consists of searching for a sequence of axioms and inference rule applications that form the steps of the proof.

A.1.3   Automating deduction

Logic gives us a framework in which we can express problems, propose theorems and develop proofs. However, the process of proving things in logic is often very long and tedious, requiring little, if any, 'intelligence'. Consequently, there has been a general move to automate, as far as possible, the proof process for various logics.

For some logics, the whole deduction mechanism can be automated. In such a logic (e.g., propositional calculus) we can formulate a hypothesis (i.e., a question) and can completely automate the process of giving a *true* or *false* interpretation (i.e., answer) to the hypothesis. Such logics are called *decidable*. In other, more expressive, logics such a decision process might never terminate! Logics with this property are called *undecidable*.

It may seem a rather hopeless task to attempt to automate undecidable logics. However, there are several ways round these problems. For example, we can *semi-automate* the decision process. This means that the decision process can be automated, but it may need to be given 'hints' every so often to make sure it finds the correct answer. Alternatively, we can restrict ourselves to reason in the subset of an undecidable logic that is decidable. Finally, some logics are *semi-decidable*, which means that *if* the decision process gives an answer it will be the correct one, though it may not give an answer at all!

A.2   Propositional logic

We begin with a discussion of propositional logic. This is a very simple logic, and is often insufficient to provide the basis for a reasonably powerful logical language. However, by introducing propositional logic here the basic notions (e.g., of proof, refutation, resolution) can be described relatively simply and we can extend these notions to more powerful logics elsewhere.

In propositional logic we can make sentences only out of the constructs such as 'and', 'if....then', 'or', 'not', the constant symbols 'true' and 'false', and a set of propositional symbols representing atomic assertions. In *classical* propositional logic, a propositional symbol can only be either 'true' or 'false'.

Example A.1 *Suppose we have propositions 'it is Monday', 'it is Manchester', and 'it is raining', then one particular sentence of propositional logic is*

if 'it is Manchester' and 'it is Monday' then 'it is raining'

Rather than using 'it is Manchester', 'it is Monday', and 'it is raining', we use propositional symbols as abbreviations for these statements. Similarly, we use the symbols $\land$, $\Rightarrow$, *etc., to represent 'and', 'if...then', etc. Thus, with the following mapping*

Statement	Represented by
it is Manchester	p
it is Monday	q
it is raining	r

the original sentence becomes $(p \land q) \Rightarrow r$.

A.2.1   The language of propositional logic

Formally, the language of propositional logic consists of the following symbols

- A set, PROP, of *propositional symbols*, p, q, r, etc., representing the *atomic* propositions of the logic.

- A set of propositional connectives, as follows, categorized by arity.
  - Nullary connectives: **true, false**.
  - Unary connective: $\neg$.
  - Binary connectives: $\wedge, \vee, \Rightarrow, \Leftrightarrow$.

- The symbols "(" and ")" which are used simply to avoid ambiguity.

The set of *well-formed propositional formulae* (WFF_p) is defined by the following rules.

- Any propositional symbol is in WFF_p.

- The nullary connectives, **true** and **false** are both in WFF_p.

- If A is in WFF_p, then so is $\neg A$.

- If A and B are in WFF_p, then so are $(A \vee B)$, $(A \wedge B)$, $(A \Leftrightarrow B)$, and $(A \Rightarrow B)$.

We will call elements of WFF_p propositional formulae. To avoid confusion, we will represent propositional symbols by lowercase letters, such as p, q, and r, and *meta-variables* ranging over propositional formulae by uppercase letters, such as A, B, and C. In addition, when the meaning is unambiguous, we will omit the parentheses '(' and ')'.

Example A.2 *The following strings are well-formed propositional formulae*

$$p \qquad (p) \qquad (p \vee q) \qquad (((p \Rightarrow q) \wedge \neg q) \Rightarrow \neg p)$$

whereas

$$(()) \qquad (p \vee q) \wedge \qquad (p\neg \Rightarrow q)$$

are not propositional formulae as they cannot be constructed using the rules given above.

A *literal* is either a propositional symbol or a negated propositional symbol. In the former case, it is a *positive literal* and in the latter it is a *negative literal*. For example, p and $\neg q$ are literals whereas $p \wedge q$ and $\neg(p \vee q)$ are not.

A.2.2   The semantics of propositional logic

In classical propositional logic, an *interpretation* for a well-formed formula is an assignment of either T or F (representing truth and falsehood respectively) to each statement of the language. Associated with each interpretation, $\mathcal{I}$, is a function, $\mathcal{I}_{atomic}$ from propositional symbols to elements of $\{T, F\}$. Thus, a propositional symbol p is true in an interpretation $\mathcal{I}$ if, and only if, $\mathcal{I}_{atomic}(p) = T$. An interpretation relation, $\models$, can then be used to give the semantics of propositional formulae in such interpretations. This infix binary relation is defined as follows for an interpretation $\mathcal{I}$ and a propositional formula.

$$
\begin{array}{lll}
\mathcal{I} \models & p & \text{iff} \quad \mathcal{I}_{atomic}(p) = T \quad \text{for } p \in \text{PROP} \\
\mathcal{I} \models & \neg A & \text{iff} \quad \text{not } \mathcal{I} \models A \quad (\text{or } \mathcal{I} \not\models A) \\
\mathcal{I} \models & A \wedge B & \text{iff} \quad \mathcal{I} \models A \text{ and } \mathcal{I} \models B \\
\mathcal{I} \models & A \vee B & \text{iff} \quad \mathcal{I} \models A \text{ or } \mathcal{I} \models B \\
\mathcal{I} \models & A \Leftrightarrow B & \text{iff} \quad \mathcal{I} \models (A \Rightarrow B) \text{ and } \mathcal{I} \models (B \Rightarrow A) \\
\mathcal{I} \models & A \Rightarrow B & \text{iff} \quad \text{if } \mathcal{I} \models A, \text{ then } \mathcal{I} \models B
\end{array}
$$

We need not have used all the connectives $\wedge$, $\vee$, $\neg$ and $\Rightarrow$ because certain connectives can be simulated by combinations of the others.

Example A.3 *We can do without disjunction ('$\vee$') because any formula of the form $A \vee B$ is equivalent to $(\neg A) \wedge (\neg B)$.*

This leads to the question of what, if it exists, is the minimal set of connectives needed to express all the others. It turns out that several pairs of connectives are able to simulate all the others, for example:

$$\{\vee, \neg\} \quad \{\wedge, \neg\} \quad \{\Rightarrow, \neg\} \quad \{\Rightarrow, \textbf{false}\}$$

Such sets of connectives are called *functionally complete* sets of connectives.

Exercise A.1 *Simulate the connectives $\vee$, $\wedge$, and **true** using only $\Rightarrow$ and **false** (and variables).*

A.2.3 Satisfaction and validity

For a propositional formula A, if $\mathcal{I} \models A$, then $\mathcal{I}$ is said to *satisfy* A. Similarly, A is *satisfied* by $\mathcal{I}$.

Example A.4 *The formula '$(p \wedge q) \Rightarrow r$' is satisfied by the interpretation $\mathcal{I}$, such that $\mathcal{I}_{atomic}(p) = F$, $\mathcal{I}_{atomic}(q) = T$ and $\mathcal{I}_{atomic}(r) = T$.*

A formula, B, is *satisfiable* if there is some interpretation, $\mathcal{I}$, in which $\mathcal{I}$ satisfies B. If there are no interpretations that satisfy B, then B is said to be *unsatisfiable*.

Example A.5 *The formula '$(p \wedge q) \Rightarrow r$' is satisfiable, whereas the formula '$p \wedge$ **false**' is unsatisfiable.*

> *For example, if we find an interpretation $\mathcal{I}$ in which p, q and r are all true then $(p \wedge q) \Rightarrow r$ is true in $\mathcal{I}$. However, no matter what truth values we find for p, then $p \wedge$ **false** can never be true in any interpretation.*

By using the above semantics, the interpretation function, $\mathcal{I}$, can also be seen to incorporate a function, $\mathcal{I}_f$, from propositional formulae to truth values. In this case, the following hold

$$\mathcal{I} \models A \quad \text{iff} \quad \mathcal{I}_f(A) = T$$
$$\mathcal{I} \not\models A \quad \text{iff} \quad \mathcal{I}_f(A) = F$$

Thus, given two propositional symbols, p and q, and an interpretation for these symbols, we can represent the truth or falsehood of their basic combinations by the following *truth table*.

p	q	$\neg p$	$\neg q$	$p \wedge q$	$p \vee q$	$p \Rightarrow q$
T	T	F	F	T	T	T
T	F	F	T	F	T	F
F	T	T	F	F	T	T
F	F	T	T	F	F	T

A formula, C, is said to be *valid* if for all interpretations, $\mathcal{I}$, then $\mathcal{I}$ satisfies C. An alternative term for a formula that is valid is a *tautology*. Valid formulae are satisfiable no matter what interpretation is put on the propositional symbols. They may be considered as *universal truths*.

> **Example A.6** *The formulae '$p \vee$ **true**' and '$p \Rightarrow p$' are both valid.*

If a formula, C, is valid, it is written as $\models C$. From the definitions of satisfiability and validity, we can derive the following relation

$$\models A \quad \text{iff} \quad \forall \mathcal{I}. \, \mathcal{I} \models A$$
$$\text{iff} \quad \neg \exists \mathcal{I}. \, \mathcal{I} \models \neg A$$

Thus, a formula is valid if there is no interpretation that satisfies the negation of the formula. This is the basis for many decision methods because if we are interested in checking the validity of a formula, we can try to construct an interpretation for the negation of the formula. If this construction fails (and the method is sound, i.e., you would have been able to construct an interpretation satisfying the negation if one existed), then we know that the original formula is valid.

A.2.4 Provability

The notions of satisfiability and validity relate to the model theory of the logic. Specifically, they relate to the interpretation of formulae within the logic. However, checking the validity of a statement is not the only way to ascertain its truth or falsehood. We would like a way of reasoning, syntactically, about statements of the logic without having to have any particular interpretation in mind. For this reason, a proof theory for the logic has been developed. This gives a way of reasoning with universal truths (*axioms*) and assumptions. Using *inference rules*, true statements may generate further true statements (*theorems*).

Axioms are generally propositional statements prefixed by a '⊢' symbol, meaning 'is provable' or 'is true'. 'True statements' are usually referred to as 'theorems'.

Example A.7 *The following could be axioms for propositional logic.*

$$\vdash\ \neg\neg A\ \Leftrightarrow\ A$$
$$\vdash\ \neg(A \wedge B)\ \Leftrightarrow\ (\neg A) \vee (\neg B)$$

Inference rules are generally represented by diagrams such as

$$\frac{\begin{array}{c}\vdash\ A\\ \vdash\ B\end{array}}{\vdash\ C}$$

meaning that if A is a *theorem* (a true statement) and B is a *theorem*, then it can be inferred that C is a *theorem*.

Two very common (and typical) inference rules are

Modus Ponens $\qquad \dfrac{\begin{array}{c}\vdash\ A\\ \vdash\ A \Rightarrow B\end{array}}{\vdash\ B}$

$\wedge$-*Introduction* $\qquad \dfrac{\begin{array}{c}\vdash\ A\\ \vdash\ B\end{array}}{\vdash\ A \wedge B}$

Aside. Although '(' and ')' are generally only used to avoid ambiguity, we could also have an inference rule such as

$$\frac{\vdash A}{\vdash (A)}$$

Example A.8 *Suppose the following statement was to be treated as a theorem.*

if 'it is Monday' and 'it is November' then 'it is a Formal Methods lecture'

Representing 'it is Monday', 'it is November', and 'it is a Formal Methods lecture' by m, n, and l, respectively, we generate the theorem $(m \wedge n) \Rightarrow l$.
Now if we know that it is Monday (i.e., m is true) and it is November (i.e., n is true), then we should be able to infer that it is a Formal Methods lecture (i.e., l is true). The proof that l is true proceeds as follows:

(a) Assume that m and n are true (i.e., both $\vdash m$ and $\vdash n$).

> (b) *Replacing A by m and B by n in the ∧-introduction rule above, then ⊢ m ∧ n can be inferred.*
>
> (c) *Replacing A by m ∧ n and B by (m ∧ n) ⇒ l in the modus ponens rule above, then ⊢ l can be inferred.*

A.2.5   Relating proof theory and model theory

As we have seen, the model theoretic definition of validity is not the only way to define truth. It is possible in propositional logic to write down a set of axioms and inference rules that also define a notion of truth known as provability. The development of proof theory and model theory for a logic leads to the question of whether the same statements can be expressed and inferred in both frameworks.

Soundness.   Given a proof theory for a logic, we must at least ensure that all statements that are treated as axioms are valid in the model theory. Also, we must ensure that all inference rules used are valid. Such a property of a proof system is called its *soundness* (or *consistency*) with respect to the model theory.

Theorem (Soundness) *If a statement is provable, then it is valid.*

In more detail the soundness theorem is

Theorem (Soundness) *For any $A \in \text{WFF}_p$,   if   ⊢ A then   ⊨ A.*

Completeness.   The dual of soundness is *completeness*. Completeness says that any formula that can be shown to be valid using the model theory is also provable using the proof theory. Again two different forms of the theorem can be given.

Theorem (Completeness) *If a statement is valid, then it is provable.*

Theorem (Completeness) *For any $A \in \text{WFF}_p$,   if   ⊨ A then   ⊢ A.*
If the logic we deal with is complete, then we can talk about the truth of logical statements in a purely syntactic way – we need never refer to particular interpretations of symbols!

A.3   Normal forms

During proofs, particularly the resolution proofs that we will be using later, it is often necessary to rewrite propositional formulae into specific forms, called *normal forms*, or into canonical/minimal formulae. The next two sections deal with these aspects of propositional logic.

A.3.1   Simplification of propositional formulae

The following is a list of the basic equivalences that can be used to rewrite propositional formulae from which the implications ($\Rightarrow$) and equivalences ($\Leftrightarrow$) have been removed. (Note that formulae of the form $A \Rightarrow B$ can be rewritten as $\neg A \lor B$ and equivalences such as $A \Leftrightarrow B$ can be rewritten as $(\neg A \vee B) \wedge (\neg B \vee A)$.) Below, the relation '$\equiv$' is taken to mean 'is logically equivalent to'.

- Associativity rules:

$$((A \vee B) \vee C) \quad \equiv \quad (A \vee (B \vee C))$$
$$((A \wedge B) \wedge C) \quad \equiv \quad (A \wedge (B \wedge C))$$

- Commutativity rules:

$$(A \vee B) \quad \equiv \quad (B \vee A)$$
$$(A \wedge B) \quad \equiv \quad (B \wedge A)$$

- Distributivity rules:

$$(A \vee (B \wedge C)) \quad \equiv \quad ((A \vee B) \wedge (A \vee C))$$
$$(A \wedge (B \vee C)) \quad \equiv \quad ((A \wedge B) \vee (A \wedge C))$$

- De Morgan's rules:

$$\neg(A \vee B) \quad \equiv \quad \neg A \wedge \neg B$$
$$\neg(A \wedge B) \quad \equiv \quad \neg A \vee \neg B$$

- Idempotency rules:

$$(A \vee A) \quad \equiv \quad A$$
$$(A \wedge A) \quad \equiv \quad A$$

- Double negation rule:

$$\neg\neg A \quad \equiv \quad A$$

- Absorption rules:

$$(A \vee (A \wedge B)) \quad \equiv \quad A$$
$$(A \wedge (A \vee B)) \quad \equiv \quad A$$

- Laws of zero and one:

$$(A \vee \mathbf{false}) \quad \equiv \quad A$$
$$(A \wedge \mathbf{false}) \quad \equiv \quad \mathbf{false}$$
$$(A \vee \mathbf{true}) \quad \equiv \quad \mathbf{true}$$
$$(A \wedge \mathbf{true}) \quad \equiv \quad A$$
$$(A \vee \neg A) \quad \equiv \quad \mathbf{true}$$
$$(A \wedge \neg A) \quad \equiv \quad \mathbf{false}$$

A.3.2   Normal forms

It is often useful to transform propositional formulae into certain *normal forms*. Thus, formulae are translated into logically equivalent but syntactically different forms that are more appropriate for reasoning with.

Literal normal form

Literal normal form (alternatively called negation normal form) is really an intermediate normal form used as part of the translation to both conjunctive and disjunctive normal forms.

Basically, a formula is translated to literal normal form by 'pushing' all negations in the formula to the literals (i.e., propositions) and removing double negations. Typical transformation rules for translating a propositional formula into literal normal form are given below.

Formula	Translates to	Formula
$\neg\neg A$	$\longrightarrow$	A
$\neg(A \vee B)$	$\longrightarrow$	$\neg A \wedge \neg B$
$\neg(A \wedge B)$	$\longrightarrow$	$\neg A \vee \neg B$

Note that both implications ($\Rightarrow$) and equivalences ($\Leftrightarrow$) may be transformed into disjunctions, conjunctions and negations before the above process is carried out.

Conjunctive normal form

Once a formula is in literal normal form, it can be translated into *Conjunctive Normal Form* (CNF) using the following transformation rules.

Formula	Translates to	Formula
$A \vee (B \wedge C)$	$\longrightarrow$	$(A \vee B) \wedge (A \vee C)$
$(B \wedge C) \vee A$	$\longrightarrow$	$(B \vee A) \wedge (C \vee A)$

Disjunctive normal form

Alternatively, a formula in literal normal form can be transformed into one in *Disjunctive Normal Form* (DNF) using the following rules (the duals of the translation to CNF).

Formula	Translates to	Formula
$A \wedge (B \vee C)$	$\longrightarrow$	$(A \wedge B) \vee (A \wedge C)$
$(B \vee C) \wedge A$	$\longrightarrow$	$(B \wedge A) \vee (C \wedge A)$

Exercise A.2 *Given the formula* $((\neg p \vee q) \wedge \neg r) \vee p$,

(a) *put it into conjunctive normal form, and,*

(b) *put it into disjunctive normal form.*

It can be shown that a propositional formula can always be transformed into any of the above normal forms and that the new formula will be semantically equivalent to the original one.

Note that these translations do not produce unique normal form. For example, both '$(a \wedge b \wedge c) \vee (b \wedge c)$' and '$b \wedge c$' are in DNF and are semantically equivalent, yet are not syntactically equivalent. If necessary, the rules given in Section A.3.1 can be applied further to generate an unique formula. In the discussion of proofs by resolution provided later, a given formula must be converted into conjunctive normal form. During this process, removal of tautologies and *subsumption* are both carried out. Subsumption is, in the propositional case, simply a process of applying the absorption rules described in Section A.3.1, that is

$$(A \vee (A \wedge B)) \quad \equiv \quad A$$
$$(A \wedge (A \vee B)) \quad \equiv \quad A$$

Once a formula has been processed into conjunctive normal form, it can be represented by

$$C_1 \wedge C_2 \wedge \ldots \wedge C_m$$

where each C_i is a disjunct. The C_i are called *clauses* and the whole formula can alternatively be regarded as a set of clauses, that is

$$\{C_1, C_2, \ldots, C_m\}.$$

A.4   Propositional resolution

The *resolution* inference rule is a generalization of modus ponens that is particularly useful in many logics. Recall that the modus ponens inference rule is

$$\frac{\vdash \quad P}{\vdash \quad P \Rightarrow B}$$
$$\frac{}{\vdash \quad B}$$

The resolution rule, applied to propositional logic, can be seen as a generalization of modus ponens and is given as

$$\frac{\vdash \quad \neg A \Rightarrow p}{\vdash \quad p \Rightarrow B}$$
$$\frac{}{\vdash \quad \neg A \Rightarrow B}$$

where p is a proposition symbol and A and B are arbitrary clauses. Thus, when A is false, resolution collapses to an instance of modus ponens.

The *resolution method* is a proof procedure based on the use of the resolution rule. It is a *refutation procedure*.

Refutation

If we want to prove that C follows from H, written as $\vdash H \Rightarrow C$, then one way to proceed is to attempt to use the proof rules of the form described above to derive $\vdash C$, starting from just $\vdash H$. This is called *forward deduction* or *forward chaining*.

An alternative approach, and one that is widely used in both automatic theorem-provers and Logic Programming, is to use *refutation*. The idea here is that, if we wish to show $\vdash H \Rightarrow C$, then if we can show that H and $\neg C$ together are inconsistent then, by completeness, it must be the case that C is a logical consequence of H. This method of proof, also termed *reductio ad absurdum*, can be characterized by the following inference rule[1]

$$\frac{\neg A \vdash \textbf{false}}{\vdash A}$$

---

[1] Note that $X \vdash Y$ means 'if X is a theorem, then Y is a theorem'.

In our particular example,

$$\frac{H, \neg C \vdash \textbf{false}}{\vdash H \Rightarrow C}$$

This refutation method, combined with the particular resolution proof rule described below, provides a *backward deduction*, or *backward chaining*, approach.

Thus the first step in the use of the resolution procedure is to negate the formula to be proved valid. It is often the case that the formula to be checked for validity is in the form of a set of axioms implying a theorem, that is

$$A_1 \wedge A_2 \wedge \ldots \wedge A_n \Rightarrow B.$$

In this case, the negation of the formula is simply

$$A_1 \wedge A_2 \wedge \ldots \wedge A_n \wedge \neg B.$$

The formula must be rewritten into conjunctive normal form as described in Section A.3.2. Thus, the new formula is a set of clauses (i.e., disjuncts)

$$\{C_1, C_2, \ldots, C_m\}.$$

Now, the central idea behind resolution is that one of the clauses should be reduced, if possible, to **false** (or, equivalently, the empty clause). If this occurs, then the conjunction of clauses is inconsistent (since $C_i \wedge \textbf{false} \equiv \textbf{false}$) and thus the negated formula is inconsistent. The negation of the original formula is unsatisfiable, so the original formula must be a theorem!

Aside: A resolution proof procedure is *refutation complete* if, given an inconsistent set of clauses, the procedure is *guaranteed* to produce **false** (or the empty clause).

A.4.1   Binary resolution

Although the resolution rule is presented above in an implicational form, it is more usually represented in a disjunctive form, that is

$$\frac{\begin{array}{ll}\vdash & A \vee p \\ \vdash & B \vee \neg p\end{array}}{\vdash \quad A \vee B}$$

Some terminology:

- $A \vee B$ is called the *resolvent*;

- $A \vee p$ and $B \vee \neg p$ are called the *parents* of the resolvent;

- p and $\neg p$ are called *complementary literals*.

Thus, if a literal appears in one parent while its negation appears in the other, both these complementary literals are removed from the combined formula. The combined formula is the resolvent. The above rule is usually called *binary resolution* because two literals are complementary.

Example A.9 *The following is a proof of the inconsistency (and hence unsatisfiability) of the set of clauses*

$$\{p \vee q, \ p \vee \neg q, \ \neg p \vee q, \ \neg p \vee \neg q\}$$

given using the binary resolution rule. Such proofs are called refutations.

1.	$p \vee q$	
2.	$p \vee \neg q$	
3.	$\neg p \vee q$	
4.	$\neg p \vee \neg q$	
5.	p	*(resolve 1 and 2)*
6.	$\neg p$	*(resolve 3 and 4)*
7.	**false**	*(resolve 5 and 6)*

Note that we implicitly use some simplification rules in such proofs, for example $(A \vee A) \equiv A$ in generating formulae (5) and (6).

A.4.2 Strategies

Although the basic resolution method is complete (i.e. it will generate **false** if the original set of clauses is unsatisfiable) it is not very efficient when measured in terms of the size of the search space for a resolution refutation. For example, consider the following refutation:

1.	$\neg p \vee \neg q \vee \neg r$	
2.	p	
3.	q	
4.	r	
5a.	$\neg q \vee \neg r$	(resolve 1 and 2)
5b.	$\neg p \vee \neg r$	(resolve 1 and 3)
5c.	$\neg p \vee \neg q$	(resolve 1 and 4)
6a.	$\neg r$	(resolve 3 and 5a)
6b.	$\neg q$	(resolve 4 and 5a)
6c.	$\neg r$	(resolve 2 and 5b)
6d.	$\neg p$	(resolve 4 and 5b)
6e.	$\neg q$	(resolve 2 and 5c)
6f.	$\neg p$	(resolve 3 and 5c)
7.	**false**	(resolve 4 and 6c)

Here, several redundant clauses are produced.

Since the original development of resolution [437], many refinements have improved its efficiency. Some, such as subsumption and the elimination of tautologies, discard useless or redundant results (these were described earlier). Many forms of resolution restrict which pairs of clauses are allowed to resolve with each other. Some of these restrictions, such as the *set of support* strategy, preserve completeness, whilst others, such as *unit resolution*, are complete only for some sets of clauses.

Before considering some of these varieties of resolution, we will look at various strategies or *heuristics* that can be used to guide the search. As we have seen, the unrestricted application of the basic resolution rule will produce far too much new, and possibly redundant, information. Therefore various strategies can be used to guide the search for a refutation.

A.4.3   Weighting strategies

A *weighting* strategy is a way for the user of the (semi-) automated reasoning system to give the system hints about which clauses are most likely to be useful in a refutation. Thus, all clauses are weighted. The lighter the weight, the sooner the program will look at the clause. Also, the program will concentrate on trying to use the lightest clauses in its resolutions.

Note that this use of weightings can also be used to enforce some order on the use of clauses (see also Section A.4.5).

One problem with this is that care must be taken to avoid the situation where new clauses are continually being generated that have lighter weights than one of the original clauses that *really* needs to be used in the refutation because such behaviour can interfere with the refutation completeness of the resolution procedure.

A.4.4   The set of support strategy

One of the most widely used strategies is the *set of support* strategy. Basically, a subset of the set of clauses is identified which is considered to be essential to the refutation proof. The set of identified clauses is called the *set of support*, and the members of this set are said to *have support* or be *supported*. The system is not allowed to perform resolution between two clauses unless at least one is supported or has been derived from a supported clause.

For example, consider the set of clauses, where clauses 1 to 4 are the original ones and where the set of support is simply $\{\neg r\}$.

1. $p \vee q$
2. $\neg p \vee r$
3. $\neg q \vee r$
4. $\neg r$
5. $\neg p$ (resolve 2 and 4)
6. $\neg q$ (resolve 3 and 4)
7. q (resolve 1 and 5)
8. p (resolve 1 and 6)
9. **false**   (resolve 6 and 7)

Notice that each new clause is derived only through resolution of a clause with clause 4 or with one of its descendants. Notice also that without the set of support restriction, several redundant clauses could have been generated, for example

5a. $q \vee r$ (resolve 1 and 2)
6a. $p \vee r$ (resolve 1 and 3)

Thus, such a strategy is useful if we can identify a set of literals that is *essential* to the proof. We will not explore this further here although it should be stated that the set of support strategy is one of the most widely used and successful resolution strategies developed.

A.4.5   Ordered resolution and selection

See also Section 4.3.2.

The introduction of *ordered* resolution [33, 178] provided a better basis for deciding on clauses and literals, and strengthened termination properties in first-order versions of resolution. Essentially, an ordering is provided on literals which aids the choice of clauses to be resolved. As we saw in Section 4.3, TSPASS uses ordered resolution with *selection*, where this selection is a complementary mechanism for directing the search for a refutation [356]. Ordered resolution with selection is now the predominant approach used in first-order resolution-based systems (such as SPASS).

A.4.6   Varieties of resolution rule

We will now go through several traditional varieties of resolution rule, most of which are designed to cut down the redundancy in standard resolution.

Hyperresolution

Hyperresolution is a more efficient version of standard binary resolution that was introduced in order to reduce the number of clauses generated. Although ordinary resolution operations take two clauses as arguments, hyperresolution is the first of several rules that will be shown that may require an arbitrary number of arguments.

Each hyperresolution operation takes a single mixed or negative clause, termed the *nucleus*, as one of its arguments together with as many positive clauses, called *electrons*, because there are negative literals in the nucleus. Each negative literal of the nucleus is resolved with a literal in one of the electrons. The hyperresolvent consists of all the positive literals of the nucleus disjoined with the unresolved literals of the electrons. The hyperresolution inference rule thus looks like

$$
\begin{array}{l}
\vdash \quad A \vee \neg p_1 \vee \neg p_2 \vee \ldots \vee \neg p_n \\
\vdash \quad B_1 \vee p_1 \\
\vdash \quad B_2 \vee p_2 \\
\vdash \quad \ldots \ldots \\
\hline
\vdash \quad B_n \vee p_n \\
\vdash \quad A \vee B_1 \vee B_2 \vee \ldots \vee B_n
\end{array}
$$

Hyperresolution can be seen as the simultaneous application of a set of binary resolution steps. *Negative hyperresolution* is exactly the same as hyperresolution except that all the electrons are *negative* literals. Hyperresolution (and thus negative hyperresolution) is complete because it can be seen as a combination of several binary resolution steps, but without the production of redundant intermediate clauses.

Example A.10 *A simple example of the use of hyperresolution is given by the following refutation*

    1. $\neg p \vee \neg q \vee \neg r$
    2. p
    3. q
    4. r
    5. $\neg r$ *(hyperresolve 1, 2 and 3)*
    6. **false**   *(resolve 4 and 5)*

Note that if binary resolution had been used, redundant clauses, such as

    5a. $\neg q \vee \neg r$ *(resolve 1 and 2)*
    5b. $\neg p \vee \neg r$ *(resolve 1 and 3)*
    5c. $\neg p \vee \neg q$ *(resolve 1 and 4)*
    6a. $\neg r$ *(resolve 3 and 5a)*
    6b. $\neg q$ *(resolve 4 and 5a)*
    6c. $\neg r$ *(resolve 2 and 5b)*
    6d. $\neg p$ *(resolve 4 and 5b)*
    6e. $\neg q$ *(resolve 2 and 5c)*
    6f. $\neg p$ *(resolve 3 and 5c)*

might have been generated.

Unit resolution

A *unit resolvent* is one in which at least one of the parent clauses is a *unit clause* (that is, one containing a single literal). The example refutation given above for the set of support strategy can also be seen as a *unit refutation* because, in all the resolutions, at least one of the resolvents is a literal ($\neg r$, $\neg p$, $\neg q$, q).

Inference procedures based on unit resolution are easy to implement and are usually quite efficient. It is worth noting that, whenever a clause is resolved with a unit clause, the conclusion has fewer literals than the parent does. This helps to focus the search towards producing the empty clause (i.e., a contradiction) and thereby improves efficiency.

Unfortunately, inference procedures based on unit resolution generally are not complete. For example, the set of clauses

$$\{p \vee q,\ \neg p \vee q,\ p \vee \neg q,\ \neg p \vee \neg q\}$$

is unsatisfiable, and using general resolution a contradiction can easily be derived. However, unit resolution fails in this case since none of the initial clauses is a unit clause.

Unit-resulting resolution

Unit-resulting resolution (UR-resolution) is a more efficient version of unit resolution. The UR-resolution operation, like the hyperresolution operation, takes an arbitrary number of arguments. Where hyperresolution acts on a single mixed or negative clause and a set of positive clauses and produces a positive or empty clause as its output, UR-resolution operates on a single non-unit clause and a set of unit clauses and produces a unit or empty clause as its output. Thus UR-resolution can be seen as the simultaneous application of several unit resolutions, where the intermediate non-unit clauses are not produced. Also, a UR-resolution operation produces only unit clauses.

Input resolution

An *input resolvent* is one in which at least one of the two parent clauses is a member of the original clause set. It can be shown that input resolution is equivalent to unit resolution in power. A consequence of this is that input resolution is again incomplete. For example, the clauses

$$\{p \vee q, \neg p \vee q, p \vee \neg q, \neg p \vee \neg q\}$$

are unsatisfiable, but, as with unit resolution, there is no input refutation for these clauses. This is because there is no way of generating the empty clause (or **false**) by a resolution operation on two clauses, one of which is an initial one. We need to do resolution on p and $\neg p$ or q and $\neg q$, but none of these clauses are initial.

Although such a restricted resolution rule might not seem useful, we will see later that a particular variety of input resolution is used as the basis of `Prolog`.

Linear resolution

Linear resolution is a slight generalization of input resolution. A *linear resolvent* is one in which at least one of the parents is either in the initial clause set or is an ancestor of the other parent.

Linear resolution takes its name from the linear forms of refutation that it generates. A linear deduction starts with a clause in the initial clause set (called the *top clause*) and produces a linear chain of resolutions. Each resolvent after the first one is obtained from the last resolvent (called the *near parent*) and some other clause (called the *far parent*). In linear resolution, the far parent must either be in the initial clause set or be an ancestor of the near parent.

Much of the redundancy in unconstrained resolution derives from the resolution of intermediate conclusions with other intermediate conclusions. The advantage of linear resolution is that it avoids many useless inferences by focusing deduction at each point on the ancestors of each clause and on the elements of the initial clause set.

Linear resolution is known to be refutation complete. Furthermore, it is not necessary to try every clause from the initial clause set as the top clause. As with the set of support strategy, a subset of the original clauses can be used and the top clause can be chosen from this set.

Example A.11

1. $p \vee q$
2. $\neg p \vee q$

> 3. $p \vee \neg q$
> 4. $\neg p \vee \neg q$
> 5. q *(resolve 1 and 2)*
> 6. p *(resolve 1 and 3)*
> 7. $\neg q$ *(resolve 1 and 4)*
> 8. **false**   *(resolve 5 and 7)*
>
> Notice that at least one of the parents of each resolution operation is always either clause 1 or a descendant of clause 1.

A.5   Horn clauses

Although resolution is, in general, inefficient, it turns out that for a particular type of clause (Horn Clauses), then resolution can be made *much* more efficient. In fact so efficient that deduction within Horn Clauses can be used as the basis for a programming paradigm – Logic Programming.

Definition A.5.1 *A Horn Clause is a clause containing at most one positive literal.*

Definition A.5.2 *A Horn Clause is* definite *if it contains a positive literal; otherwise it is a* negative *Horn Clause.*

> **Example A.12** *Clauses can be categorized into different varieties, for example*
>
> $a \vee \neg b \vee \neg c$ *is a definite Horn Clause.*
> $\neg a \vee \neg b \vee \neg c$ *is a negative Horn Clause.*
> a *is a definite Horn Clause, but it is also a unit clause.*
> $a \vee \neg b \vee c$ *is not a Horn Clause.*

A.5.1   Resolution methods for horn clauses

One of the important aspects of Horn Clauses is that both input and unit resolution are complete for sets of Horn Clauses.

> **Example A.13** *The earlier example showing that neither input nor unit resolution are complete, namely,*
> $$\{p \vee q, \ p \vee \neg q, \ \neg p \vee q, \ \neg p \vee \neg q\}$$
> *is not a set of Horn Clauses. In particular, $p \vee q$ is not a Horn Clause.*

In particular, when using Horn Clauses, linear and input resolution become equivalent.

A.5.2   Rule interpretation of horn clauses

One useful way to represent Horn Clauses is as rules, facts and goals. The analogy between this rule form and the clausal form is as follows:

- A definite clause, such as

$$a \vee \neg b \vee \neg c$$

  can be represented by the implication

$$(b \wedge c) \Rightarrow a.$$

  Thus, the negative literals represent some *hypothesis*, while the positive literal can be considered as the conclusion.
  We might also write such an implication as a *rule*, that is in the form

$$a \leftarrow b, c.$$

- A positive Horn Clause (and, hence, a unit clause), such as

$$d$$

  can be represented as a *fact*.

- A negative Horn Clause, such as

$$\neg a$$

  can be represented as the implication

$$a \Rightarrow \textbf{false}$$

  and then as the *goal*

$$\leftarrow a.$$

  Finally, the resolution rule,

$$\frac{\begin{array}{ccc} A & \vee & p \\ B & \vee & \neg p \end{array}}{A \quad \vee \quad B}$$

  can be rewritten as the rule

$$\frac{\begin{array}{ll} p & \leftarrow A \\ q & \leftarrow p, B \end{array}}{q \quad \leftarrow A, B}$$

  In particular, when we start with a goal, the rule will be

$$\frac{\begin{array}{ll} & \leftarrow p, B \\ p & \leftarrow A \end{array}}{\leftarrow A, B}$$

This is the essential form of clause structure and deduction mechanism that underlies `Prolog`.

A.5.3 From refutation to execution

We can now give an example of resolution within Horn Clauses, as follows:

1.	$a \vee \neg b \vee \neg c$	
2.	$b \vee \neg d$	
3.	c	
4.	d	
5.	$\neg a$	
6.	$a \vee \neg c \vee \neg d$	(resolve 1 and 5)
7.	$a \vee \neg c$	(resolve 4 and 6)
8.	a	(resolve 3 and 7)
9.	**false**	(resolve 5 and 8)

We can also give an analogous computation in rule form. First, imagine that the initial set of rules, facts and goals was as follows.

1.	$a \leftarrow b, c$	(rule)
2.	$b \leftarrow d$	(rule)
3.	c	(fact)
4.	d	(fact)
5.	$\leftarrow a$	(goal)

Now, in the Logic Programming framework, the computation can proceed as follows:

$\leftarrow a$	(goal)
$\leftarrow b, c$	(by rule 1)
$\leftarrow d, c$	(by rule 2)
$\leftarrow c$	(by fact 4)
$\square$	(by fact 3)

Thus, the process of resolution corresponds to reducing the goal clause to the empty clause by use of the rule and fact clauses.

A.6 First-order logic

One of the main limitations of propositional logic is that it can only reason over very simple structures. Within *first-order* predicate logic increased expressiveness is achieved by allowing propositional symbols to have arguments ranging over the elements of these structures (and calling such propositions *predicates*). For example, in the previous section, we introduced the proposition 'it is Manchester'. If we wish to extend our statements to refer to other towns in England, we have to add several new propositions, one for each town, that is 'it is London', 'it is Liverpool' etc. If we had a structure representing the set of all towns in England and a predicate 'is', then the statement 'is(Manchester)' in predicate logic would be equivalent to 'it is Manchester' in propositional logic. At first sight it seems that the two languages are equivalent in power. However, if we want to describe a new property of towns, such as 'is a city', we can just add a predicate 'city' to first-order predicate logic that applies to elements of the set of towns, whereas in propositional logic we would have to add a whole new set of propositions such as 'Manchester is a city', 'London is a city' etc. The predicate version is clearly more concise and, in cases when the structures being considered are infinite, then the propositional version might not even be appropriate. *First-order logic* allows such predicate symbols but also provides expressive power through the availability of quantifiers and variables. These allow us to state facts about elements of the structure without enumerating the particular elements.

Example A.14 *If we again consider the statement involving the propositions 'it is Monday', 'it is Manchester' and 'it is raining', where*

> *if 'it is Manchester' and 'it is Monday' then 'it is raining'*

is represented by

$$(p \wedge q) \Rightarrow r$$

then we cannot express the statement

> *for all cities, if 'it is Monday' then 'it is raining in the city'*

in propositional logic without enumerating all the cities.

In first-order logic, we can express such a statement. If we take the proposition m to mean 'it is Monday' and the predicate r with argument c to mean 'it is raining in city c', then the above statement might be expressed as

$$\forall c. \; m \Rightarrow r(c)$$

Note that the symbol c in the above formula is a variable, and the '$\forall c$.' prefix means 'for all c in the domain, then it is true that'.

A.6.1   The language of first-order logic

The language of first-order logic consists of the symbols from propositional logic, together with

- a set, VAR, of *variables*, x, y, z etc., ranging over a particular domain[2]
- the quantifiers, $\forall$ (for all) and $\exists$ (there exists)
- a set of *non-logical symbols*[3] consisting of
  - a set, CONS, of constant symbols, c_1, c_2, etc.
  - a set, FUNC, of function symbols, f_1, f_2 etc., together with an *arity* function, which gives the number of arguments for each function symbol

---

[2] For simplicity we will assume just one domain.

[3] These symbols are termed non-logical because they are associated with the actual structure reasoned about in the problem domain, rather than with abstract reasoning.

    – a set, PRED, of predicate symbols, p_1, p_2 etc., together with an *arity* function, which gives the number of arguments for each predicate symbol.

Note that proposition symbols can be seen as predicate symbols of arity 0.
The set of *terms*, TERM, is defined by the following rules.

- Any constant is in TERM.

- Any variable is in TERM.

- If $t_1, \ldots, t_n$ are in TERM, and f is a function symbol of arity $n > 0$, then $f(t_1, \ldots, t_n)$ is in TERM.

Terms are thus names of objects in the universe of discourse.
Next we define the notion of an *atomic formula*.

- Any propositional symbol is an atomic formula.

- **true** and **false** are atomic formulae.

- If $t_1, \ldots, t_n$ are in TERM, and p is a predicate symbol of arity $n > 0$, then $p(t_1, \ldots, t_n)$ is an atomic formula.

The set of *well-formed first-order formulae*, WFF$_f$, is now defined by the following rules:

- Any atomic first-order formula is in WFF$_f$.

- If A is in WFF$_f$, then so is $\neg A$.

- If A and B are in WFF$_f$, then so are $(A \vee B)$, $(A \wedge B)$, $(A \Leftrightarrow B)$, and $(A \Rightarrow B)$.

- If A is in WFF$_f$, and x is a variable, then both $\forall x.A$ and $\exists x.A$ are in WFF$_f$.

We will call elements of WFF$_f$ first-order formulae.

    Again, we will represent first-order predicates and propositions as lowercase letters, such as p, q, and r, and *meta-variables* ranging over first-order formulae by uppercase letters, such as A, B, and C. We will also represent function and constant symbols by lowercase letters, such as f, g, and h for function symbols, and a, b, and c for constant symbols. We also omit parentheses when we can.

Example A.15 *The following strings are well-formed first-order formulae.*

$$\forall x.\ p(x) \qquad (\exists y.q) \qquad (p(x))$$

whereas

$$\forall \exists x.p(x) \qquad \qquad \exists q(x)$$

are not well-formed first-order formulae because they cannot be constructed using the rules given above.

Note that a variable can occur in a formula without an enclosing quantifier. When used in this way, a variable is said to be *free*, whereas a variable that occurs in a formula and in the scope of an enclosing quantifier is said to be *bound*.

Example A.16 *In the following formulae, x is free in the first, bound in the second and occurs both free and bound in the third:*

$$p(x) \wedge q(x)$$
$$\forall x.\ (p(x) \Rightarrow q(x))$$
$$p(x) \wedge \exists x.\ q(x)$$

Terminology: ground formulae.   If a formula has no free variables, it is called a *closed* formula. If it has neither free nor bound variables, it is called a *ground* formula. We will usually only deal with closed first-order formulae.

Monadic logic.   As we have seen above, each predicate symbol has an *arity* specifying how many arguments it can take. *Monadic* predicates have arity ≤ 1, for example $p(x)$, $q(y)$ or $r(22)$. The simplicity of this fragment of first-order logic is often useful, both in extending to the temporal framework and in providing decidable fragments.

A.6.2   The semantics of first-order logic

The interpretations defined for propositional logic must be extended to include the domain of discourse D and assignments of T or F to predicates and quantified statements.

Thus, associated with each interpretation for first-order formulae, $\mathcal{I}$, are

- D, the domain of discourse[4],

- $\mathcal{I}_c$, a mapping from every constant symbol c_i to an element of D,

- $\mathcal{I}_f$, a mapping associating with every n-ary function symbol f a function from D^n to D,

- $\mathcal{I}_{atomic}$, a mapping associating with every n-ary predicate symbol p a function from D^n to $\{T,\ F\}$.

Note that, as nullary predicates are propositions, $\mathcal{I}_{atomic}$ here is a natural extension of the $\mathcal{I}_{atomic}$ mapping given for propositional interpretations.

Given a domain, D, a *variable assignment*, $\mathcal{A}$, is a mapping from variables to elements of D. Given a first-order interpretation $\mathcal{I}$ and a variable assignment $\mathcal{A}$, the *term assignment*, $T_{\mathcal{I}\mathcal{A}}$, is a mapping from terms to elements of D. Now, $T_{\mathcal{I}\mathcal{A}}$ is defined as follows:

- *If c is a constant (i.e., $c \in$ CONS)*
  then $T_{\mathcal{I}\mathcal{A}}(c) = \mathcal{I}_c(c)$

- *If v is a variable (i.e., $v \in$ VAR)*
  then $T_{\mathcal{I}\mathcal{A}}(v) = \mathcal{A}(v)$

- *If f is an n-ary function symbol (i.e., $f \in$ FUNC) and $t_1, \ldots, t_n$ are terms*
  then $T_{\mathcal{I}\mathcal{A}}(f(t_1, \ldots, t_n)) = \mathcal{I}_f(f)(T_{\mathcal{I}\mathcal{A}}(t_1), \ldots, T_{\mathcal{I}\mathcal{A}}(t_n))$

---

[4] This can be thought of as another element of $\mathcal{I}$, i.e., $\mathcal{I}_D$.

Thus, we give the semantics of first-order formulae, not just with respect to an interpretation that gives meaning to the constant symbols in the language, but also with respect to a variable assignment that gives meaning to the variables in the language.

We again use the interpretation relation, $\models$, relating the semantics of first-order formulae to an interpretation, $\mathcal{I}$, and a variable assignment, $\mathcal{A}$. (Below, the '$\dagger$' operator adds a new mapping to our variable assignment, overwriting any previous assignment to that variable.)

$$
\begin{array}{lll}
\mathcal{I}, \mathcal{A} \models & \neg B & \text{iff} \quad \text{not } \mathcal{I}, \mathcal{A} \models B \quad (\text{or } \mathcal{I}, \mathcal{A} \not\models B) \\
\mathcal{I}, \mathcal{A} \models & B \wedge C & \text{iff} \quad \mathcal{I}, \mathcal{A} \models B \text{ and } \mathcal{I}, \mathcal{A} \models C \\
\mathcal{I}, \mathcal{A} \models & B \vee C & \text{iff} \quad \mathcal{I}, \mathcal{A} \models B \text{ or } \mathcal{I}, \mathcal{A} \models C \\
\mathcal{I}, \mathcal{A} \models & B \Leftrightarrow C & \text{iff} \quad \mathcal{I}, \mathcal{A} \models (B \Rightarrow C) \text{ and } \mathcal{I}, \mathcal{A} \models (C \Rightarrow B) \\
\mathcal{I}, \mathcal{A} \models & B \Rightarrow C & \text{iff} \quad \text{if } \mathcal{I}, \mathcal{A} \models B \text{ then } \mathcal{I}, \mathcal{A} \models C \\
\mathcal{I}, \mathcal{A} \models & q(t_1, \ldots, t_n) & \text{iff} \quad \mathcal{I}_{atomic}(q)(T_{\mathcal{I}\mathcal{A}}(t_1), \ldots, T_{\mathcal{I}\mathcal{A}}(t_n)) = T \\
\mathcal{I}, \mathcal{A} \models & \forall x.\ B & \text{iff} \quad \text{for all } d \in D,\ \mathcal{I}, \mathcal{A} \dagger [x \mapsto d] \models B \\
\mathcal{I}, \mathcal{A} \models & \exists x.\ B & \text{iff} \quad \text{there is a } d \in D, \text{ such that } \mathcal{I}, \mathcal{A} \dagger [x \mapsto d] \models B
\end{array}
$$

We will usually deal with closed formulae. In this case, the initial variable assignment is unimportant and we will usually use an empty mapping ($[\,]$) as the variable assignment.

Example A.17 *Consider the interpretation, $\mathcal{I}$, containing*

$D = \mathbb{N}$ *the set of all natural numbers*
$\mathcal{I}_c = [a \mapsto 0,\ b \mapsto 1,\ c \mapsto 3]$
$\mathcal{I}_f = [f \mapsto +,\ g \mapsto -]$
$\mathcal{I}_{atomic} = [p \mapsto\ >]$

and the variable assignment $\mathcal{A}$ given by

$$[x \mapsto 1,\ y \mapsto 2]$$

Now, the term assignment associated with $\mathcal{I}$ and $\mathcal{A}$, that is $T_{\mathcal{I}\mathcal{A}}$, maps the term

$$f(g(f(y, b), a), x)$$

onto the element of D represented by

$$+(-(+(2, 1), 0), 1)$$

Since
$$
\begin{array}{rcl}
+(2, 1) & = & 3 \\
-(3, 0) & = & 3 \\
+(3, 1) & = & 4
\end{array}
$$

then
$$T_{\mathcal{I}\mathcal{A}}(f(g(f(y, b), a), x)) = 4.$$

Example A.18 *Assume the interpretation, $\mathcal{I}$, as described in Example A.17, and now consider the satisfaction of the formula*

$$\exists x.\; p(x, b) \wedge p(f(b, c), x)$$

given an empty variable assignment (because the formula is closed). Thus, we want to find out if the following is true:

$$\mathcal{I}, [\,] \models \exists x.\; p(x, b) \wedge p(f(b, c), x)$$

It is satisfied if there exists a 'd' in D (i.e., in the natural numbers) such that

$$\mathcal{I}, [x \mapsto d] \models p(x, b) \wedge p(f(b, c), x)$$

In this case, there must be a $d \in \mathbb{N}$ such that both

$$\mathcal{I}, [x \mapsto d] \models p(x, b)$$

and

$$\mathcal{I}, [x \mapsto d] \models p(f(b, c), x)$$

are satisfied. This is equivalent to finding some d such that

$$>(d, 2) \quad \text{and} \quad >(+(1, 3), d)$$

that is $>(d, 2)$ and $>(4, d)$. Thus, we can make this statement true by choosing $d = 3$.

A.6.3   Provability

The inference rules of first-order logic are basically those of propositional logic, augmented by the following

Universal Instantiation
$$\frac{\vdash\ \forall x.B}{\vdash\ B_{x/t}}$$
where t is free for x in B

and

Existential Instantiation
$$\frac{\vdash\ \exists x.B}{\vdash\ B_{x/h(x_1,\ldots x_n)}}$$

where h is a new n-ary function symbol and where $x_1, \ldots, x_n$ are the free variables in B.

Universal Instantiation (UI) allows us to reason from the general to the particular. It states that whenever we believe a universally quantified sentence, we can infer an instance of that sentence in which the universally quantified variable is replaced by any appropriate term.

Example A.19 *Again taking the interpretation described in previous examples, consider the statement*

$$\forall x.\ p(x, a) \lor p(b, x) \tag{1}$$

which says that any element of the domain of discourse is either greater than 0 or less than 1. Using the UI rule, we can deduce from (1) above that, for any term t,

$$p(t, a) \lor p(b, t)$$

In particular, we can infer $p(d, a) \lor p(b, d)$ where d is a constant. Note that we can also infer, from (1),

$$p(f(a, b), a) \lor p(b, f(a, b))$$

because $f(a, b)$ is a term.

Note that the reason for the side condition on the UI inference rule is that if we have

$$\forall x.\ \exists y.\ p(y, x)$$

and use UI to generate a conclusion with a free variable such as

$$\exists y.\ p(y, f(y, b))$$

then we get an invalid theorem! The original one said that for any x there is a y such that y is greater than x. However, we have derived a theorem that says that there exists a y such that y is greater than $y + 1$! The restriction on the term that can be substituted avoids this problem.

Existential Instantiation (EI) allows us to eliminate existential quantifiers. Like Universal Instantiation (UI), this rule states that we can infer an instance of the quantified sentence in which the existentially quantified variable is replaced by a suitable term. This process is called *Skolemization*.

Example A.20 *If we have the theorem discussed in the last example, that is*

$$\forall x.\ \exists y.\ p(y, x)$$

and use UI to generate

$$\exists y.\ p(y, t)$$

where t is some suitable term, then because there are no free variables in the sentence, we can infer

$$p(d, t)$$

where $d > t$. The new constant d is called a Skolem constant.
If we had used the EI rule before the UI rule we would, from

$$\forall x.\ \exists y.\ p(y, x)$$

have derived some theorem of the form

$$\forall x.\ p(h(x), x)$$

where h is a new function symbol (called a Skolem function*). We can give an interpretation to the new function h that satisfies this statement by making $h(x)$ equivalent to $f(x, 1)$.*
Thus, we can derive the theorem

$$\forall x.\ p(f(x, 1), x)$$

which says that for any number, x, $x + 1$ is always greater than x!

Intuitively, the reason why the UI rule introduces a constant yet the EI rule introduces a function is that if we choose an instantiation of a universally quantified variable, we can choose *any* appropriate term – the theorem must be true for any term chosen. However, if we choose an instantiation for an existentially quantified variable, not just any term will do. Also, such a term may depend on the (universally quantified) free variables.

Example A.21 *In the previous example, we used the EI rule to generate*

$$\forall x.\ p(h(x), x)$$

and then defined h to have the same meaning as '+1'. However, if we had instantiated the variable with a Skolem constant, rather than a Skolem function, we would have produced

$$\forall x.\ p(t, x)$$

This is only satisfiable if the set of Natural Numbers has a maximal value, which it does not.

A.6.4   Soundness and completeness

First-order logic still retains the useful properties of soundness and completeness (second-order logic is incomplete) and finite axiomatizations of the logic exist. Unfortunately, general first-order logic is undecidable. The reduced complexity of certain restricted forms of first-order logic is one of the reasons that these logics have become popular. Examples of such restrictions are Horn Clauses and propositional Modal and Temporal Logics.

A.6.5   First-order resolution

As with propositional formulae, first-order formulae can be simplified and rewritten into normal forms. We first consider the additional simplification rules that can be used in first-order logic.

Simplification of first-order formulae

- Pushing negations:

$$\neg\forall x.\ A \quad \equiv \quad \exists x.\ \neg A$$
$$\neg\exists x.\ A \quad \equiv \quad \forall x.\ \neg A$$

- Distributivity rules:

$$(\forall x.\ A) \wedge (\forall x.\ B) \quad \equiv \quad \forall x.\ (A \wedge B)$$
$$(\exists x.\ A) \vee (\exists x.\ B) \quad \equiv \quad \exists x.\ (A \vee B)$$

- Subsumption:

$$(\forall x.\ p(x)) \wedge p(a) \quad \equiv \quad \forall x.\ p(x)$$
$$(\exists x.\ p(x)) \vee p(a) \quad \equiv \quad \exists x.\ p(x)$$

- Laws of zero and one:

$$(\forall x.\ \mathbf{false}) \quad \equiv \quad \mathbf{false}$$
$$(\exists x.\ \mathbf{false}) \quad \equiv \quad \mathbf{false}$$
$$(\forall x.\ \mathbf{true}) \quad \equiv \quad \mathbf{true}$$
$$(\exists x.\ \mathbf{true}) \quad \equiv \quad \mathbf{true}$$

- Variable renaming:

$$\forall x.\ A \quad \equiv \quad \forall y.\ A_{x/y}$$
$$\exists x.\ A \quad \equiv \quad \exists y.\ A_{x/y}$$

Here $A_{x/y}$ means the formula A with all occurrences of x replaced by y. Note that this transformation can only take place if the new variable y does not already occur in A.

- Quantifier movement:

$$(\forall x.\ A) \wedge B \quad \equiv \quad \forall x.(A \wedge B)$$
$$(\forall x.\ A) \vee B \quad \equiv \quad \forall x.(A \vee B)$$
$$(\exists x.\ A) \wedge B \quad \equiv \quad \exists x.(A \wedge B)$$
$$(\exists x.\ A) \vee B \quad \equiv \quad \exists x.(A \vee B)$$

These transformations are allowed only if x is not free in B.

As with propositional formulae, first-order formulae can be rewritten into standard clausal forms for use in automated reasoning procedures. We will concentrate on the translation of general first-order formulae to clausal form for use in resolution.

Example A.22 *If quantification occurs over a* finite *set, then a first-order formula can effectively be reduced to a propositional formula. Consider $\forall x.p(x)$ where x can only take its value from the finite set, $S = \{s_1, s_2, \ldots, s_n\}$. This formula can be characterized by*

$$\bigwedge_{i=1}^{n} p(s_i).$$

But now, since each $p(s_i)$ is a ground formula, then we can introduce n new propositions, $p_{s_1}, p_{s_2}, \ldots, p_{s_n}$, and re-cast the above formula as

$$p_{s_1} \wedge p_{s_2} \wedge \cdots \wedge p_{s_n}.$$

Rewriting first-order formulae into clausal form

The clausal form used is basically CNF with all quantifiers and existential variables resoved and all remaining variables being (implicitly) universally quantified.

The steps involved in rewriting a formula into clausal form are as follows [231]:

1. Removal of implications and equivalences: rewrite $A \Leftrightarrow B$ as $A \Rightarrow B$ and $B \Rightarrow A$, and then rewrite any $C \Rightarrow D$ as $(\neg C) \vee D$.

2. Rewrite into literal normal form: use the following (augmented) list of rewrite rules to perform this rewriting.

Formula	Translates to	Formula
$\neg\neg A$	$\longrightarrow$	A
$\neg(A \vee B)$	$\longrightarrow$	$\neg A \wedge \neg B$
$\neg(A \wedge B)$	$\longrightarrow$	$\neg A \vee \neg B$
$\neg(\forall x.A)$	$\longrightarrow$	$\exists x.(\neg A)$
$\neg(\exists x.A)$	$\longrightarrow$	$\forall x.(\neg A)$

3. Standardize variables: here, we rename all the variables so that each quantifier has a different variable.

   For example, if we had $(\forall x.p(x)) \wedge (\exists x.q(x))$ we can rewrite this to $(\forall x.p(x)) \wedge (\exists y.q(y))$ using the simplification rules described in the last section. Again, the new variable introduced must not already occur anywhere inside the scope of the quantifier.

4. Skolemization: this is the process of removing existential quantifiers and replacing existentially quantified variables by either Skolem constants or Skolem functions. For example, the formula

$$\exists x.\ p(x) \wedge (\forall y.\ (\exists z.\ q(z)) \wedge (\forall v.\ \exists w.\ r(w)))$$

could be skolemized as follows (where c is a Skolem constant and h_1 and h_2 are unary and binary Skolem functions respectively)

$$p(c) \wedge (\forall y.\ q(h_1(y)) \wedge (\forall v.\ r(h_2(y, v))))$$

5. Extraction of quantifiers: the remaining universal quantifiers are moved to the 'outside' of the formula. Note that because all the variable names are different, this is no problem.

   Once all the universal quantifiers are at the outermost part of the formula, they are removed. This is because all the remaining variables are universally quantified.

6. Rewriting into CNF: this is more or less the same as in the propositional case.

7. Conjunction removal: as in the propositional case, the formula in CNF is rewritten as a set of disjunctions. For example,

$$(p(a, b) \vee q) \wedge r(b, c)$$

is rewritten as

$$\{p(a, b) \vee q, \ r(b, c)\}.$$

8. Standardizing the variables apart: the variables are renamed so that no variable appears in more than one clause. For example

$$\{p(a, x) \vee q, \ r(x, c)\}$$

is rewritten as

$$\{p(a, x) \vee q, \ r(y, c)\}.$$

Example A.23 *Let us follow through this process in detail, for one particular example.*

initial: $\forall x.(\forall y.p(x, y)) \Rightarrow \neg(\forall y.q(x, y)) \Rightarrow r(x, y))$

step 1: $\forall x.\neg(\forall y.p(x, y)) \vee \neg(\forall y.\neg q(x, y) \vee r(x, y))$

step 2: $\forall x.(\exists y.\neg p(x, y)) \vee (\exists y.q(x, y) \wedge \neg r(x, y))$

step 3: $\forall x.(\exists y.\neg p(x, y)) \vee (\exists z.q(x, z) \wedge \neg r(x, z))$

step 4: $\forall x.\neg p(x, h_1(x)) \vee (q(x, h_2(x)) \wedge \neg r(x, h_2(x)))$

step 5: $\neg p(x, h_1(x)) \vee (q(x, h_2(x)) \wedge \neg r(x, h_2(x)))$

step 6: $(\neg p(x, h_1(x)) \vee q(x, h_2(x))) \wedge (\neg p(x, h_1(x)) \vee \neg r(x, h_2(x)))$

step 7: $\{ \ \neg p(x, h_1(x)) \vee q(x, h_2(x)), \neg p(x, h_1(x)) \vee \neg r(x, h_2(x)) \ \}$

step 8: $\{ \ \neg p(x_1, h_1(x_1)) \vee q(x_1, h_2(x_1)), \neg p(x_2, h_1(x_2)) \vee \neg r(x_2, h_2(x_2)) \ \}$

Variable-free resolution

If a set of clauses contains no variables, then resolution can be applied to the clauses just as it is applied to propositional clauses. This is because ground terms can be treated as propositions.

Example A.24 *Consider the set of clauses*

$$\{p(a) \vee q(b, c), \quad p(a) \vee \neg q(b, c), \quad \neg p(a) \vee q(b, c), \quad \neg p(a) \vee \neg q(b, c)\}$$

A refutation can be produced using binary resolution as follows

1. $p(a) \vee q(b, c)$
2. $p(a) \vee \neg q(b, c)$
3. $\neg p(a) \vee q(b, c)$
4. $\neg p(a) \vee \neg q(b, c)$
5. $p(a)$ *(resolve 1 and 2)*
6. $\neg p(a)$ *(resolve 3 and 4)*
7.   **false**   *(resolve 5 and 6)*

Substitutions

Resolution for terms that include variables is more complicated than resolution either in the propositional or variable-free cases. For example, if we have the clauses

$$p(a)$$
$$\neg p(x)$$

where a is a constant and x is a variable, we would like to apply binary resolution to generate false[5]. Thus, we need some way to generate $p(a)$ from $p(x)$. This is achieved by applying a *substitution* to the formula $p(x)$.

A substitution is a mapping from variables to terms in which no variable maps to a term that contains that variable within any of its associated expressions. A substitution is another name for the variable assignment introduced earlier.

Example A.25 *The substitution*

$$[x \mapsto a, \, y \mapsto f(a), \, z \mapsto h(x)]$$

is a mapping that maps variables x, y, and z to terms a, $f(a)$, and $h(x)$ respectively. However, neither of the following mappings are legitimate substitutions

$$[x \mapsto f(y), \, y \mapsto g(x)]$$
$$[x \mapsto h(a, g(x))]$$

The terms associated with each variable in a substitution are often called the *bindings* of that substitution. A substitution mapping can be *applied* to a first-order formula to produce a new formula with the variables replaced in the appropriate way.

Example A.26 *The substitution*

$$[x \mapsto h(z), \, y \mapsto a, \, z \mapsto b]$$

applied to the formula

$$f(y, g(z))$$

gives

$$f(a, g(b)).$$

However, when applied to

$$f(x, g(z))$$

it gives

$$f(h(z), g(b)).$$

---

[5] Remember that all variables that remain in the formula, in this case x, are implicitly universally quantified.

> *This is because all the variable replacements are carried out at once on the original term.*

Thus, if we had the clauses

$$p(a)$$
$$\neg p(x)$$

we would need to generate and apply the substitution

$$[x \mapsto a]$$

to get

$$p(a)$$
$$\neg p(a)$$

and then use (variable-free) binary resolution.

The process of generating such substitutions from sets of formulae is called *unification*.

Unification

A set of formulae, $\{A_1, A_2, \ldots, A_n\}$, is *unifiable* if there exists a substitution, θ, that, when applied to all the formulae in the set, generates identical formulae. In this case, θ is said to be the unifier for the set of formulae.

Example A.27 *The substitution*

$$[x \mapsto a, \ y \mapsto f(b), \ z \mapsto g(a, f(b))]$$

is a unifier for the set

$$\{h(z), \ h(g(x, f(b))), \ h(g(a, y))\}$$

In many cases there is more than one unifier for a particular set of formulae. For example, both the substitutions

$$[x \mapsto f(y)]$$
$$[x \mapsto f(a), \ y \mapsto a]$$

are unifiers for the formulae

$$\{g(x), \ g(f(y))\}.$$

The *most general unifier* is the substitution that reduces the number of variables in the formulae the least. We can consider this as the substitution that unifies with maximum generality. Note that if a set of formulae is unifiable, then there always exists an unique most general unifier (*mgu*).

Why do we choose the most general unifier? Well, consider the clauses

$$h(f(x)) \wedge g(x)$$
$$\neg h(f(y))$$
$$\neg g(b)$$

Now, we would like to resolve the first two clauses together. If we take the *mgu* for the formulae $h(f(x))$ and $h(f(y))$, we get[6] $[x \mapsto y]$. Applying this *mgu*, we get the clauses

$$h(f(y)) \wedge g(y)$$
$$\neg h(f(y))$$
$$\neg g(b)$$

and can resolve the first two clauses to get

$$g(y)$$
$$\neg g(b)$$

and then apply the *mgu* $[y \mapsto b]$ to generate

$$g(b)$$
$$\neg g(b)$$

However, if we had not chosen the *mgu* for $h(f(x))$ and $h(f(y))$ in the first place, we might not have derived a contradiction. For example, the substitution

$$[x \mapsto a, \ y \mapsto a]$$

is a unifier for $h(f(x))$ and $h(f(y))$. Applying this, we get

$$h(f(a)) \wedge g(a)$$
$$\neg h(f(a))$$
$$\neg g(b)$$

and resolving the first two clauses gives us

$$g(a)$$
$$\neg g(b)$$

From this we cannot derive false!

Deriving the MGU from a pair of formulae

The algorithm for generating an *mgu* from a pair of formulae is given below[7].

```
MGU(A, B)
{
    if (A == B)                              return {};
    if is_variable(A)                        return MGU-var(A,B);
    if is_variable(B)                        return MGU-var(B,A);
    if (is_constant(A) || is_constant(B))    return {FALSE};
    if (length(A) != length(B))              return {FALSE};

    i = 0;
    g = {};
```

---

[6] Alternatively, we could have generated the *mgu* $[y \mapsto x]$ because the most general unifier is unique up to variable renaming.

[7] This algorithm is derived from the one given in the excellent [231].

```
    while (i != length(A))
    {
        s = MGU( part(A,i), part(B,i) );

        if (s == {FALSE})                    return {FALSE};

        g = compose(g,s);
        A = substitute(A,g);
        B = substitute(B,g);
        i = i + 1;
    }
    return g;
}

MGU-var(A, B)
{
    if occurs(A,B)        return {FALSE};

    return { x --> y };
}
```

If two expressions are unifiable, the MGU function returns the *mgu* (i.e., a set of variable mappings), otherwise it returns the set containing FALSE.

The length() function gives the number of arguments of the function symbol, predicate symbol or logical connective to which it is applied. For example,

$$\text{length}(p(f(a,b,c))) == 1$$

The part() function splits a function or predicate application or a compound formula as follows: the 0th part is the function symbol, predicate symbol or connective; the arguments are the other parts. part(A,i) returns the ith part of the formula A. For example, the length of

$$f(a, h(b), g(x, a))$$

is 3. The 0th part is f, the first part is a, the second $h(b)$, and the third $g(x, a)$.

The composition of two substitutions is the substitution obtained by applying each of the substitutions in turn. compose(g,s) returns a substitution that is equivalent to first applying g, then applying s. The function substitute() applies a substitution to a formula, returning the new formula.

Finally, occurs(A,B) returns true if the variable A occurs in the formula B. This process is called the *occurs check*.

Full first-order resolution

The resolution rules introduced earlier for use with propositional logic can now be extended for use with first-order logic. For example, the binary resolution rule becomes

$$\frac{\vdash \quad A \vee C_1}{\vdash \quad \theta(A \vee B)}$$
$$\vdash \quad B \vee \neg C_2$$

where θ is the *mgu* of C_1 and C_2. The result of such a resolution application is the formula $A \vee B$ with the substitution θ applied.

Example A.28 *Consider the statement*

    if

        tom is the father of jane, and
        bob is the father of anne, and
        everyone who is the father of someone is a parent of that person,

    then

        tom is a parent of jane.

We can rewrite this statement in first order logic as

$$(father(tom, jane) \wedge father(bob, anne) \wedge$$
$$(\forall x. \forall y. father(x, y) \Rightarrow parent(x, y)))$$
$$\Rightarrow$$
$$parent(tom, jane)$$

Transforming this into clausal form, we get

        father(tom, jane)
        father(bob, anne)
        $\neg$*father(x, y)* $\vee$ *parent(x, y)*
        $\neg$*parent(tom, jane)*

and we can use binary resolution to generate a refutation as follows.

1.	*father(tom, jane)*	
2.	*father(bob, anne)*	
3.	$\neg$*father(x, y)* $\vee$ *parent(x, y)*	
4.	$\neg$*parent(tom, jane)*	
5.	*parent(tom, jane)*	*(resolve 1 and 3)*
6.	*parent(bob, anne)*	*(resolve 2 and 3)*
7.	**false**	*(resolve 4 and 5)*

Example A.29 *Consider the statement*

if

    tom is the father of jane, and
    bob is the father of anne, and
    everyone who is the father of someone is a parent of that person,

then

    there is a person who is a parent of jane.

We can rewrite this statement in first order logic as

$$(father(tom, jane) \; \wedge \; father(bob, anne) \; \wedge$$
$$(\forall x. \, \forall y. \, father(x, y) \Rightarrow parent(x, y)))$$
$$\Rightarrow$$
$$(\exists z. \, parent(z, jane))$$

Transforming this into clausal form, we get (where 'z' is a universally quantified variable)

$$father(tom, jane)$$
$$father(bob, anne)$$
$$\neg father(x, y) \vee parent(x, y)$$
$$\neg parent(z, jane)$$

again we can use binary resolution to generate a refutation:

1. *father(tom, jane)*
2. *father(bob, anne)*
3. *¬father(x, y) ∨ parent(x, y)*
4. *¬parent(z, jane)*
5. *parent(tom, jane)*   *(resolve 1 and 2)*
6. *parent(bob, anne)*   *(resolve 2 and 3)*
7. **false**   *(resolve 4 and 5)*

The other forms of resolution rule introduced for propositional logic generalize to first-order logic in the obvious way. Also, the results about the refutation completeness of various resolution rules carry over from propositional logic. For example, input resolution is not refutation complete for first-order formulae.

However, as we have seen, there are restricted versions of classical logic, in particular Horn Clauses, that have better properties with respect to refutation completeness.

A.6.6   Onwards to logic programming

As we saw, in the propositional case, `Prolog` (and Logic Programming languages in general) is based on the search for a refutation. Specifically, `Prolog` programs consist of first-order Horn Clauses and computation involves the search for a refutation within these clauses using a variety of linear resolution. In first-order logic this requires unification and the binding of variables to values. Such values are then used by the computation to compute answers. For example, recall that the last resolution operation in Example A.29 uses the *mgu* $[z \mapsto tom]$. Note that, in some cases, we require to know what the actual instantiation is that derives a refutation. For example, if we had asked if

    tom is the father of jane, and
    bob is the father of anne, and
    everyone who is the father of someone is a parent of that person,

then

    who is a parent of jane.

we would have had to record the instantiation of z from which a refutation was derived, namely $[z \mapsto tom]$. (This is exactly how Prolog works – we will not explain further here but can recommend many useful and informative works, such as [320, 321, 459, 391].)

B

Solutions to exercises

These Things Take Time.
  – The Smiths

Solutions: Chapter 2

[2.1] *Which of the following are* not *legal WFF of PTL, and why?*

  (a) *april* ∨ ○*U may* .... **not** legal
      – left-hand side of '*U*' is not a WFF

  (b) *may* ∨ ○((*april W may*)) .... **legal**

  (c) ○*july* ∧ *august*(◇*september*) .... **not** legal
      – binary operator expected between the propositional symbol '*august*'
      and '('

[2.2] *How might we represent the following statements in PTL?*

  (a) *"In the next moment in time, 'running' will be true and, at some time after
      that, 'terminated' will be true."* .... ○(*running* ∧ ○◇*terminated*)

  (b) *"There is a moment in the future where either 'pink' is always true, or
      'brown' is true in the next moment in time."* .... ◇(□*pink* ∨ ○*brown*)

  (c) *"In the* second *moment in time, 'hot' will be true."* .... **start** ⇒ ○*hot*

[2.3] *By examining the formal semantics of* □, *show that* □*p* ⇒ □□*p*.
  In model $\mathcal{M}$ at moment i we know that '□*p*' has semantics

$$\text{for all } j, \text{ if } (j \geq i), \text{ then } \langle \mathcal{M}, j \rangle \models p$$

An Introduction to Practical Formal Methods Using Temporal Logic, First Edition. Michael Fisher.
© 2011 John Wiley & Sons, Ltd. Published 2011 by John Wiley & Sons, Ltd.

Essentially, since ∀ implies ∀∀ in classical logic, then the above implies

for all j, if $(j \geq i)$, then for all k, if $(k \geq j)$, then $\langle M, k \rangle \models p$

which corresponds directly to the semantics of $\square \square p$.

[2.4] (a) $\varphi U \psi \Leftrightarrow (\varphi W \psi \wedge \Diamond \psi)$

(b) $(l \wedge a)U$ ... means that l must be true until ....
$(l \wedge b)U c$ means that also lUc.
So, $(l \wedge a)U(l \wedge (l \wedge b)U c)$ implies l until $(l$ and l until $c)$, and so l until c.

[2.5] Consider a temporal sequence on which q is true at all even numbered moments in time (i.e. 0, 2, 4, ...) but false at all odd numbered moments (i.e. 1, 3, 5, ...). Now:

$\square \Diamond q$ is satisfied on such a sequence since there is always another future (even) moment at which q is true; but

$\square q$ is *not* satisfied on this sequence.

Thus, this sequence shows that $\square \Diamond q \Rightarrow \square q$ is *not* valid.

[2.6] $\Diamond attempt \Rightarrow \Diamond \square succeed$ essentially means that we need only *attempt* once in order to get persistent success. Note that this is quite a strange constraint because not only does this mean one attempt can lead to continual success, but also success can, potentially, occur *before* an attempt is even made!

[2.7] Below is a Büchi Automaton corresponding to 'start $\Rightarrow aUb$':

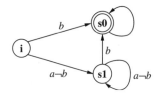

Alternatively:

$A = \mathbf{P}(\{a, b\})$

$S = \{\mathbf{i}, \mathbf{s0}, \mathbf{s1}\}$

$\delta = \{(\mathbf{i}, b, \mathbf{s0}), (\mathbf{i}, a\neg b, \mathbf{s1}), (\mathbf{s0}, true, \mathbf{s0}), (\mathbf{s1}, a\neg b, \mathbf{s1}), (\mathbf{s1}, b, \mathbf{s0})\}$

$I = \{\mathbf{i}\}$

$F = \{\mathbf{s0}\}$

[2.8] (a) $\bigcirc \bigcirc \bigcirc ((x = 0) \wedge \square (y > 1))$

(b) $\langle M, i \rangle \models \blacklozenge \varphi$ iff there exists j such that $(0 \leq j < i)$ and $\langle M, j \rangle \models \varphi$

(c) An infinite number, that is $p \wedge \bigcirc p \wedge \bigcirc \bigcirc p \wedge \bigcirc \bigcirc \bigcirc p \wedge \bigcirc \bigcirc \bigcirc \bigcirc p \wedge \ldots$

Alternatively,

$$\bigwedge_{i=0}^{\infty} \bigcirc^i p$$

[2.9] (a) $\Diamond \varphi$ says that, at some point in the future, φ will occur at least once. From $\Box(\varphi \Rightarrow \Diamond \psi)$ if φ occurs anywhere at least once, then $\Diamond \psi$ will occur at least once.

This, in turn, forces ψ to occur at least once. If we instead have $\Box \varphi$, then we will require $\Box \Diamond \psi$, that is ψ occurs infinitely often.

(b) $\Box(p \Rightarrow \Diamond q) \wedge \Box p$ is satisfied (for example) on a sequence where p is true throughout, but where q is only true at *even* numbered states. However, $\Box q$ is *not* satisfied on such a sequence. Consequently, $\Box(p \Rightarrow \Diamond q) \wedge \Box p$ does not imply $\Box q$.

(c) The formula is true. We can justify this, for example, with the transformations:

$$\bigcirc \Box \bigcirc \varphi \Rightarrow \Box \bigcirc \bigcirc \varphi$$

$$\Rightarrow \Diamond \bigcirc \bigcirc \varphi$$

$$\Rightarrow \Diamond \Diamond \bigcirc \varphi$$

$$\Rightarrow \Diamond \bigcirc \Diamond \varphi$$

(d) The formula is true. We can justify this, for example, with the transformations:

$$\Box \Diamond \varphi \Rightarrow \bigcirc \Diamond \varphi$$

$$\Rightarrow \Diamond \bigcirc \varphi$$

[2.10] (a) $\langle \mathbb{Z}, \pi \rangle \longrightarrow$ since the past now goes on forever, there is clearly no finite past and no need for the '**start**' operator. Other future-time operators remain and we can add their past-time counterparts.

(b) $\langle \mathbb{R}, \pi \rangle \longrightarrow$ as above, we have an infinite past, but the main difference now is that there is no concept of distinct/discrete next/last states. Consequently, next-time and last-time operators are irrelevant.

Solutions: Chapter 3

[3.1] A typical pattern of behaviour corresponding to the formulae is

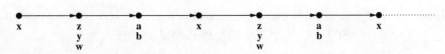

[3.2] (a) *Give a temporal semantics (i.e. a temporal formula characterizing the execution sequences) of this program:*

```
z:=3; z:=z+5; z:=z/4; end
```

Assuming we can interpret arithmetic functions within our semantics:

$$\bigcirc(z = 3) \wedge \bigcirc\bigcirc(z = 8) \wedge \bigcirc\bigcirc\bigcirc(z = 2)$$

If we cannot, then:

$$\bigcirc(z = 3) \wedge \bigcirc\bigcirc(z = 3 + 5) \wedge \bigcirc\bigcirc\bigcirc(z = (3 + 5)/4)$$

(b) *Give a temporal formula capturing a semantics of the statement in a standard imperative programming language:*

```
(if (y>2) then y:=1 else y:=y+2); end
```

$$((y > 2) \Rightarrow \bigcirc(y = 1)) \wedge$$
$$((y \leq 2) \Rightarrow \exists w.y = w \wedge \bigcirc(y = w + 2))$$

(c) *What temporal formula would we write to describe the property that, at some point in the execution of the program in part (b) the variable y is guaranteed to have a value less than 5? Is this a liveness property, a safety property or neither?*

$$\Diamond(y < 5) \dots\dots\dots\dots\dots\dots\dots\dots\dots\dots\dots\dots\dots$$**liveness property**

[3.3] (a) **start** $\Rightarrow a \wedge \bigcirc b \wedge \bigcirc\bigcirc c$

(b) **start** $\Rightarrow a \wedge \bigcirc b \wedge \bigcirc\bigcirc c \wedge \bigcirc\Diamond x \wedge \bigcirc\Diamond\bigcirc y \wedge \square(b \Rightarrow \Diamond x)$

(c) **start** $\Rightarrow a \wedge \bigcirc b \wedge \bigcirc\bigcirc c \wedge \bigcirc\Diamond x \wedge \bigcirc\Diamond\bigcirc y \wedge \square(b \Rightarrow \Diamond x) \wedge$
$$\bigcirc\Diamond\bigcirc\Diamond w \wedge \bigcirc\Diamond\bigcirc\Diamond\Diamond d \wedge \bigcirc\Diamond\bigcirc\Diamond\Diamond\bigcirc e$$

(d) $a \Rightarrow \bigcirc b$
$$\Rightarrow \bigcirc\Diamond x$$
$$\Rightarrow \bigcirc\Diamond\bigcirc y$$
$$\Rightarrow \bigcirc\Diamond\bigcirc\Diamond w$$
$$\Rightarrow \bigcirc\Diamond\bigcirc\Diamond\Diamond a$$

So, $\square(a \Rightarrow \Diamond a)$ and, given that **start** $\Rightarrow a$ then **start** $\Rightarrow \square\Diamond a$. This, together with $a \Rightarrow \bigcirc\bigcirc c$, implies **start** $\Rightarrow \square\Diamond c$.

[3.4] (a) *send_msg* is made true at the second step of A's execution, that is we have 'p' at step 0, 'q' at step 1, and '*send_msg*' at step 2.

(b) Here, p is true at every even numbered state, while q is true at every odd numbered state. This requires that '*send_msg*' is also true at every even numbered state.

(c) Since '*send_msg*' is true at every even numbered state then $\square\Diamond$*send_msg* is, indeed, true.

[3.5] (a) If *rcv_msg* is false, then *Spec$_B$* is not required to do anything, so *Spec$_A$* ∧ *Spec$_B$* effectively gives us just *Spec$_A$*. The specification of *Spec$_B$* does not prescribe the values of its propositions too much, so f, g etc. may well become true without *rcv_msg* being true.

(b) $Comms(A, B) = \Box(send\_msg \Rightarrow rcv\_msg)$

(c) $Comms(A, B) = \Box(send\_msg \Rightarrow \Diamond rcv\_msg)$

(d) $Comms(A, B) = \Box(send\_msg \Leftrightarrow rcv\_msg)$

(e) Add $rcv\_msg \Rightarrow \Diamond send\_rcpt$ to *Spec$_B$*.
And add $\Box(send\_rcpt \Rightarrow \Diamond rcv\_rcpt)$ to $Comms(A, B)$ allowing us to use *rcv_rcpt* in *Spec$_A$* if required.

Solutions: Chapter 4

[4.1] *Translate '$fU(\bigcirc g)$' into SNF.*
We go through a series of steps, as follows

(*a*)	**start** $\Rightarrow$	$fU(\bigcirc g)$	from initial formula
(*b*)	**start** $\Rightarrow$	x_2	from (*a*)
(*c*)	$x_2 \Rightarrow$	$fU x_1$	from (*a*)
(*d*)	$x_1 \Rightarrow$	$\bigcirc g$	from (*a*)
(*c$_1$*)	$x_2 \Rightarrow$	$\Diamond x_1$	from (*c*)
(*c$_2$*)	$x_2 \Rightarrow$	x_3	from (*c*)
(*c$_3$*)	$x_3 \Rightarrow$	$x_1 \lor f$	from (*c*)
(*c$_4$*)	$x_3 \Rightarrow$	$x_1 \lor \bigcirc x_3$	from (*c*)
(*c$_{41}$*)	$(x_3 \land \neg x_1) \Rightarrow$	$\bigcirc x_3$	from (*c$_4$*)

The SNF output consists of the clauses (*b*), (*c$_1$*), (*c$_2$*), (*c$_3$*), (*c$_{41}$*) and (*d*). However, (*c$_2$*) and (*c$_3$*) may have to be rewritten further if *present/global* rules are not allowed.

[4.2] *Rewrite $\bigcirc(a \Rightarrow b) \land \bigcirc a \land \bigcirc \neg b$ into SNF.* For example:

(1)	**start** $\Rightarrow$	x_1
(2)	$x_1 \Rightarrow$	$\bigcirc x_2$
(3)	**start** $\Rightarrow$	$(\neg x_2 \lor \neg a \lor b)$
(4)	**true** $\Rightarrow$	$\bigcirc(\neg x_2 \lor \neg a \lor b)$
(5)	$x_1 \Rightarrow$	$\bigcirc a$
(6)	$x_1 \Rightarrow$	$\bigcirc \neg b$

Alternatively, just

(1)	**start** $\Rightarrow$	x_1
(2)	$x_1 \Rightarrow$	$\bigcirc(\neg a \lor b)$
(3)	$x_1 \Rightarrow$	$\bigcirc a$
(4)	$x_1 \Rightarrow$	$\bigcirc \neg b$

[4.3] *Apply temporal resolution rule as in Example 4.10.*
Given

$$gambler \Rightarrow \bigcirc\square\neg rich$$
$$ambitious \Rightarrow \Diamond rich$$

then applying the temporal resolution rule gives

$$ambitious \Rightarrow (\neg gambler)W rich$$

which can be rewritten to

(1) **start** $\Rightarrow$ $\neg ambitious \lor rich \lor \neg gambler$
(2) **start** $\Rightarrow$ $\neg ambitious \lor rich \lor x$
(3) **true** $\Rightarrow$ $\bigcirc(\neg ambitious \lor rich \lor \neg gambler)$
(4) **true** $\Rightarrow$ $\bigcirc(\neg ambitious \lor rich \lor x)$
(5) x $\Rightarrow$ $\bigcirc(rich \lor \neg gambler)$
(6) x $\Rightarrow$ $\bigcirc(rich \lor x)$

[4.4] *Explain TSPASS translation output of* **start** $\Rightarrow (aU b)$.
Generates new predicate symbols, _P, _S, _U, _U1, _U2, and _waitforb.
And then the clauses:

```
formula(_P).
formula(always(implies(_P,or(a,b)))).
formula(always(implies(_P,or(_S,b)))).
formula(always(implies(and(_P,not(b)),_waitforb))).
formula(always(sometime(not(_waitforb)))).
formula(always(implies(_waitforb,next(_U)))).
formula(always(implies(_S,next(_U1)))).
formula(always(implies(_S,next(_U2)))).
formula(always(implies(_U,or(_waitforb,b)))).
formula(always(implies(_U1,or(a,b)))).
formula(always(implies(_U2,or(_S,b)))).
```

We can write the output here in our more recognizable form, as follows.

$$\begin{aligned}
\textbf{start} &\Rightarrow \text{\_P} \\
\text{\_P} &\Rightarrow a \lor b \\
\text{\_P} &\Rightarrow \text{\_S} \lor b \\
(\text{\_P} \land \neg b) &\Rightarrow \text{\_waitforb} \\
\textbf{true} &\Rightarrow \Diamond\neg\text{\_waitforb} \\
\text{\_waitforb} &\Rightarrow \bigcirc\text{\_U} \\
\text{\_S} &\Rightarrow \bigcirc\text{\_U1} \\
\text{\_S} &\Rightarrow \bigcirc\text{\_U2} \\
\text{\_U} &\Rightarrow \text{\_waitforb} \lor b \\
\text{\_U1} &\Rightarrow a \lor b \\
\text{\_U2} &\Rightarrow \text{\_S} \lor b
\end{aligned}$$

Given that **true** $\Rightarrow \Diamond\neg$_waitforb, we would like to make _waitforb false infinitely often. If *b* occurs immediately, then we can indeed make _waitforb false all the time. However, if *b* is false at the start, then **start** $\Rightarrow$ _P and

(_P ∧ ¬*b*) ⇒ _waitforb together ensure that _waitforb is true. This, in turn, ensures that _U is true in the next state and this causes *b* to again be checked, ensuring that at least one of *a* or *b* must be true. And so on. If *b* is continually false, the _waitforb continues to be true and so **true** ⇒ ◇¬_waitforb cannot be satisfied.

[4.5] *We can produce the TSPASS translation of Example 4.5, using the '--extendedstepclauses' flag:*

```
list_of_symbols.
predicates[(_P,0),(_waitforp,0),(p,0)].
end_of_list.

list_of_formulae(axioms).
formula(always(sometime(not(p)))).
formula(always(implies(p,next(p)))).
formula(_P).
formula(always(implies(and(_P,not(p)),_waitforp))).
formula(always(implies(_waitforp,next(or(_waitforp,p))))).
formula(always(sometime(not(_waitforp)))).
end_of_list.
```

Here,

$$\texttt{formula(always(sometime(not(p))))}$$

and

$$\texttt{formula(always(implies(p,next(p))))}$$

are straight-forward translations of $\Box \Diamond \neg p$ and $\Box(p \Rightarrow \bigcirc p)$, respectively. The remaining clauses, however, provide a (slightly contorted) translation of $\Diamond p$ into a formula of the form '$\Box \Diamond \ldots$'.

[4.6] *Try proving*

$$(\Diamond p \land \Box(p \Rightarrow \bigcirc \Diamond p)) \Rightarrow \Box \Diamond p$$

that is negate, giving

$$\Diamond p \land \Box(p \Rightarrow \bigcirc \Diamond p) \land \Diamond \Box \neg p$$

and input this to TSPASS.
If we invoke fotl-translate, we get

```
list_of_symbols.
predicates[(_P,0),(_P1,0),(_P2,0),(_P3,0),(_R,0),
           (_waitfor_P2_1,0),(_waitforp,0),(p,0)].
end_of_list.

list_of_formulae(axioms).
formula(_P).
formula(always(implies(p,_P1))).
```

```
formula(_P3).
formula(always(implies(_P1,next(_P)))).
formula(always(implies(_P2,_R))).
formula(always(implies(_R,next(_R)))).
formula(always(implies(_R,not(p)))).
formula(always(implies(and(_P,not(p)),_waitforp))).
formula(always(implies(_waitforp,next(or(_waitforp,p))))).
formula(always(sometime(not(_waitforp)))).
formula(always(implies(and(_P3,not(_P2)),_waitfor_P2_1))).
formula(always(implies(_waitfor_P2_1,
        next(or(_waitfor_P2_1,_P2))))).
formula(always(sometime(not(_waitfor_P2_1)))).
end_of_list.
```

Then, invoking the TSAPSS resolution procedure gives us

```
1[0:Inp:LS] || ->   _P3(temp_zero)*.
2[0:Inp:LS] || ->   _P(temp_zero)*.
3[0:Inp:LS] || _P2(U)* ->   _R(U).
4[0:Inp:LS] || p(U) ->   _P1(U)*.
7[0:Inp:LS] || _R(U)* p(U) -> .
8[0:Inp:LS] || _P3(U) ->   _P2(U)* _waitfor_P2_1(U).
9[0:Inp:LS] || _P(U)* ->   p(U) _waitforp(U).
15[0:Res:8.1,3.0:LS] || _P3(U)* ->   _waitfor_P2_1(U) _R(U).
16[0:Res:2.0,9.0] || ->   p(temp_zero) _waitforp(temp_zero)*.
18[0:Res:1.0,15.0:LS] || ->
    _waitfor_P2_1(temp_zero)_R(temp_zero)*.
20[0:Res:18.1,7.0] || p(temp_zero) ->
    _waitfor_P2_1(temp_zero)*.
43[0:LoopSearch::LS] || _waitforp(U) _R(U)* -> .
47[0:Res:18.1,43.1] || _waitforp(temp_zero) ->
    _waitfor_P2_1(temp_zero)*.
90[0:LoopSearch::LS] || _P1(U)* _waitfor_P2_1(U) -> .
91[0:LoopSearch::LS] || _waitforp(U) _waitfor_P2_1(U)* -> .
92[0:Res:4.1,90.0:LS] || p(U) _waitfor_P2_1(U)* -> .
95[0:Res:47.1,91.1] || _waitforp(temp_zero)*
    _waitforp(temp_zero)* -> .
97[0:Obv:95.0:LS] || _waitforp(temp_zero)* -> .
99[0:Res:16.1,97.0:LS] || ->   p(temp_zero)*.
101[0:Res:20.1,92.1] || p(temp_zero)* p(temp_zero)* -> .
102[0:Obv:101.0:LS] || p(temp_zero)* -> .
104[0:Res:99.0,102.0] || -> .
```

[4.7] *Use* fotl-translate *to translate* **start** $\Rightarrow ((\lozenge g) \wedge (\lozenge h))$.
If we do this, we generate:

```
list_of_formulae(axioms).
formula(_P).
formula(_P1).
formula(always(implies(and(_P,not(g)),_waitforg))).
formula(always(implies(_waitforg,next(or(_waitforg,g))))).
formula(always(sometime(not(_waitforg)))).
```

```
formula(always(implies(and(_P1,not(h)),_waitforh_1))).
formula(always(implies(_waitforh_1,
       next(or(_waitforh_1,h))))).
formula(always(sometime(not(_waitforh_1)))).
end_of_list.
```

Here, the new propositions _P and _P1 rename $\Diamond g$ and $\Diamond h$, respectively. The clauses relating to each of _P and _P1 are then essentially identical. Thus, for $\Diamond g$ we get

$$(\_P \land \neg g) \;\Rightarrow\; \_\texttt{waitforg}$$
$$\_\texttt{waitforg} \;\Rightarrow\; \bigcirc(g \lor \_\texttt{waitforg})$$
$$\textbf{true} \;\Rightarrow\; \Diamond\neg\_\texttt{waitforg}$$

[4.8] Translating the 'printing' example gives us the output in 'ex_printing.snf'. Running TSPASS on this gives a straightforward proof; see 'ex_printing.out'.

[4.9] It would be a useful exercise to try this with various flags in order to see if a single proposition can be achieved. In essence, though, fotl-translate does not carry out very sophisticated matching and so typically uses two new propositions.

Solutions: Chapter 5

[5.1] Since we need to check $\langle M, 0 \rangle \models \textbf{start} \Rightarrow \Box(a \Rightarrow \bigcirc b)$ on the particular model specified, we first expand the semantic definitions, as follows:

$$\forall i \geq 0.\ \langle M, i \rangle \models (a \Rightarrow \bigcirc b)$$

$$\forall i \geq 0.\ \textbf{if}\langle M, i \rangle \models a \ \textbf{then}\langle M, i \rangle \models \bigcirc b$$

$$\forall i \geq 0.\ \textbf{if}\langle M, i \rangle \models a \ \textbf{then}\langle M, i+1 \rangle \models b$$

Now, look at all i where $\langle M, i \rangle \models a$:

$i = 1$ check $\langle M, i+1 \rangle \models b$ . . . . i.e. $\langle M, 2 \rangle \models b$ is true

$i = 4$ check $\langle M, i+1 \rangle \models b$ . . . . i.e. $\langle M, 5 \rangle \models b$ is true

So, formula is satisfied on this particular model.

[5.2] Keep on checking b on each state. If b is true, mark the state and move on to the next state (because $\Box b \Leftrightarrow (b \land \bigcirc \Box b)$).

Once we reach a state that is already marked, we stop, knowing that $\Box b$ is satisfied on this structure.

For the second example, we again start checking b on each state. When we reach the '$\neg b$' state this fails and we know that $\Box b$ is *not* satisfied on this structure.

[5.3] *In the algorithm for checking $\Box$ given in Section 5.1.2, if the next state is already marked, we stop knowing that $\Box\varphi$ is true on this sequence. Why?*

Since every time we use the $\Box\varphi \Leftrightarrow (\varphi \land \bigcirc \Box\varphi)$ equivalence we check φ, mark the state and move on, then if we again reach a marked state, all the intervening states have φ satisfied. We have now found a cycle, so we have successfully found an infinite path on which φ is always true.

[5.4] Keep on checking b on each state. If b is ever true, we stop with success. If not, we mark the state and move on to the next state (because $\Diamond b \Leftrightarrow (b \vee \bigcirc \Diamond b)$).

In the first example, we reach such a state where b is true and succeed.

In the second example, we fail to find such a state and keep traversing the graph. Once we reach a state that is already marked, we stop, knowing that $\Diamond b$ cannot be satisfied on this structure.

[5.5] *In the algorithm for checking* $\Diamond$ *given in Section* 5.1.2, *if the* next *state is already marked, we stop knowing that* $\Diamond \varphi$ *is false on this sequence. Why?*
Since every time we use the $\Diamond \varphi \Leftrightarrow (\varphi \vee \bigcirc \Diamond \varphi)$ equivalence we check φ. If it is false, mark the state and move on, then if we again reach a marked state, all the intervening states have φ false. We have now found a cycle, so we have found an infinite path on which φ is never satisfied.

[5.6] Sample Büchi Automaton for the *Comms* formula $\Box(x \Rightarrow \Diamond^+ y)$ is

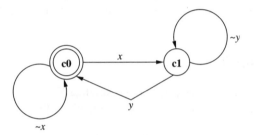

Product construction will be carried out as expected. In particular combined states involving both **a2** and **c0** can step to states involving both **a1** and **c1**. Now the combined automaton cannot remain in **c1** forever and so **c1** and **b1** must transform to **c0** and **b2**. And so on.

[5.7] *Run this yourself*

[5.8] *Again, try this yourself!*

[5.9] *And again!!*

[5.10] *What happens if we comment out the second message send in* MyProcess *in the* rendezvous.pml *example?*
In the basic rendezvous.pml execution, an error occurs with a timeout waiting for the other process to (synchronously) read the value that MyProcess wants to send.

When the second message is commented out, no error occurs and the processes just exchange '909'.

[5.11] Let us examine three choices of value for **N**.

N=0:
```
MyProcess sent 909.
YourProcess received 909.
timeout
```

N=1:
```
MyProcess sent 909.
YourProcess received 909.
MyProcess sent 693.
```

N=2:
```
MyProcess sent 909.
YourProcess received 909.
MyProcess sent 693.
```

[5.12] (a) Producer and Consumer share two channels, one in each direction, while Producer and Sink share one. Thus data goes

- from Producer to Consumer via p2c,

- from Consumer to Producer via c2p, and

- from Producer to Sink via d.

(b) The assertion in the Consumer process will succeed as only values less than 8 are sent from Producer to Consumer.

(c) Similarly, the assertion in the Sink process will succeed as Producer eventually sends 9 to Sink.

[5.13] *Run this yourself.*

[5.14] *And again!*

[5.15] XX successively increments total through 0, 3, 6, 9 and then sends this to YY. So, YY assigns 9 to dumped.
If never claim represents <> (dumped == 10) then no error is found. However, if it represents <> (dumped == 9), then the never claim is satisfied and so the property is violated.

[5.16] • spin -f "[]p":

• spin -f "X p":

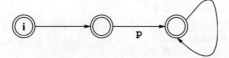

• spin -f "p U q":

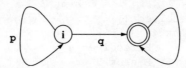

- `spin -f " XXq":`

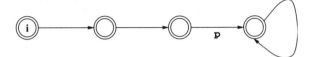

- `spin -f "[]<>r":`

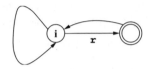

[5.17] (a) Data passes between processes as follows.

- process A sends information to process B via channel a2b,

- process B sends information to process C via channel b2c, and

- process C sends information to process A via channel c2a.

(b) • Set `total` variable to be zero.

- Increment the `total` variable (modulo 8).

- Output this value on the `out` channel.

- Print out the value of `total`.

- Read in a new value for `total`.

- Print out the new value of `total`.

- If `total` is zero, output the value, then terminate.

- Otherwise, go back to incrementing `total`.

(c) • An integer value is read from channel `in` and stored in `total`.

- This is printed out.

- If the `total` is zero, write the value on the output channel and terminate.

- If the `total` is non-zero, increment it (modulo 8), output the new `total`, write it out, and go back to reading a new value.

(d) A sample trace would be:

```
A sent 1
        B received 1
        B sent 2
                C received 2
                C sent 3
A received 3
A sent 4
        B received 4
        B sent 5
```

```
                             C received 5
                             C sent  6
             A received 6
             A sent  7
                      B received 7
                      B sent  0
                             C received 0
             A received 0
                      B received 0
```

See `promela3ex.pml` and try running this.

(e) Assertion will succeed, as A never sees a '1' after it has been sent it the first time.

(f) Now, since A receives a '3' from C during computation, the assertion fails.

(g) `! [] <> (total != 0)`

Solutions: Chapter 6

[6.1] *Assuming the 'oldest eventuality first' strategy, what two sequences of states represent possible* METATEM *executions for the following set of rules?*

$$
\begin{aligned}
\textbf{start} &\Rightarrow \neg f \\
\textbf{start} &\Rightarrow \neg g \\
\textbf{true} &\Rightarrow \Diamond f \\
\textbf{true} &\Rightarrow \Diamond g \\
\textbf{true} &\Rightarrow \bigcirc(\neg f \lor \neg g)
\end{aligned}
$$

We first note that at most one of f or g can occur at any moment in time. In addition, both $\Diamond f$ and $\Diamond g$ are continually regenerated whenever they are satisfied. Thus we have two possible executions, as follows.

(a) We choose the make 'f' true at $t = 0$, then 'g' must be made true at $t = 1$ since $\Diamond g$ has been outstanding longer than $\Diamond f$. At $t = f$, 'f' is made true for similar reasons, and so on. Essentially, 'f' and 'g' alternate throughout the whole execution.

(b) The second possible execution again alternates 'f' and 'g' alternate through the whole execution, but this time begins with 'g' at $t = 0$.

[6.2] *Run this yourself.*

[6.3] Try running `Concurrent MetateM` on both

$$
\textbf{true} \Rightarrow \bigcirc(\exists X.p(X) \lor q(X))
$$

and on

$$
\begin{aligned}
\textbf{true} &\Rightarrow \bigcirc(\exists X.p(X) \lor q(X)) \\
\textbf{true} &\Rightarrow \bigcirc p(c1)
\end{aligned}
$$

See `exist_start.sys`. A sample exection has `free_agent` generating

```
state 0:  [true]
state 1:  [p(a)]
state 2:  [p(c)]
state 3:  [p(e)]
.....
```

so producing new constants ar every state, while `freeish_agent` produces

```
state 0:  [true]
state 1:  [p(c1)]
.....
```

essentially re-using the `c1` constant.

Solutions: Appendix A

[**A.1**] *Simulate the connectives* $\vee$, $\wedge$, *and* **true** *using only* $\Rightarrow$ *and* **false** *(and variables).*

$$
\begin{aligned}
A \vee B &\equiv (A \Rightarrow \textbf{false}) \Rightarrow B \\
A \wedge B &\equiv (A \Rightarrow (B \Rightarrow \textbf{false})) \Rightarrow \textbf{false} \\
\textbf{true} &\equiv \textbf{false} \Rightarrow A \\
\neg A &\equiv A \Rightarrow \textbf{false}
\end{aligned}
$$

[**A.2**] *Given the formula* $((\neg p \vee q) \wedge \neg r) \vee p$, *(a) put it into conjunctive normal form, and, (b) put it into disjunctive normal form.*

(a) $((\neg p \vee q) \wedge \neg r) \vee p$ becomes $(\neg p \vee q \vee p) \wedge (\neg r \vee p)$ in CNF. This, in turn, could be simplified just to $(\neg r \vee p)$

(b) $((\neg p \vee q) \wedge \neg r) \vee p$ becomes $(\neg p \wedge \neg r) \vee (q \wedge \neg r) \vee p$ in DNF.

References

[1] M. Abadi. The Power of Temporal Proofs. *Theoretical Computer Science*, 65(1):35–83, 1989.

[2] M. Abadi and L. Lamport. The Existence of Refinement Mappings. In *Proc. 3rd Annual Symposium on Logic in Computer Science (LICS)*, pages 165–175. IEEE Computer Society Press, 1988.

[3] M. Abadi and L. Lamport. An Old-Fashioned Recipe for Real Time. *ACM Transactions on Programming Languages and Systems*, 16(5):1543–1571, 1994.

[4] M. Abadi and Z. Manna. Nonclausal Temporal Deduction. In *Proc. International Conference on Logics of Programs*, volume 193 of *LNCS*, pages 1–15. Springer, 1985.

[5] M. Abadi and Z. Manna. Temporal Logic Programming. *Journal of Symbolic Computation*, 8:277–295, 1989.

[6] M. Abadi and Z. Manna. Nonclausal Deduction in First-Order Temporal Logic. *Journal of the ACM*, 37(2):279–317, 1990.

[7] J.-R. Abrial. *The B-Book: Assigning Programs to Meanings*. Cambridge University Press, 1996. ISBN 0-521-49619-5.

[8] Ada'83 Language Reference Manual. `http://www.adaic.org/standards/ada83.html`.

[9] C. Aggarwal, editor. *Data Streams: Models and Algorithms*, volume 31 of *Advances in Database Systems*. Springer, 2007. ISBN 0-387-28759-0.

[10] J. Agrawal, Y. Diao, D. Gyllstrom and N. Immerman. Efficient Pattern Matching over Event Streams. In *Proc. ACM SIGMOD International Conference on Management of Data*, pages 147–160. ACM Press, 2008. ISBN 978-1-60558-102-6.

[11] R. Agrawal and R. Srikant. Mining Sequential Patterns. In *Proc. 11th International Conference on Data Engineering (ICDE)*, pages 3–14. IEEE Computer Society, 1995. ISBN 0-818-66910-1.

[12] F. Aguado, P. Cabalar, G. Pérez and C. Vidal. Strongly Equivalent Temporal Logic Programs. In *Proc. 11th European Conference on Logics in Artificial Intelligence (JELIA)*, volume 5293 of *LNCS*, pages 8–20. Springer, 2008.

[13] A. V. Aho, J. E. Hopcroft and J. D. Ullman. *The Design and Analysis of Computer Algorithms*. Addison-Wesley, 1974. ISBN 0-201-00029-6.

[14] J. F. Allen. Maintaining Knowledge about Temporal Intervals. *Communications of the ACM*, 26(11):832–843, 1983.

An Introduction to Practical Formal Methods Using Temporal Logic, First Edition. Michael Fisher.
© 2011 John Wiley & Sons, Ltd. Published 2011 by John Wiley & Sons, Ltd.

[15] J. F. Allen. Towards a General Theory of Action and Time. *Artificial Intelligence*, 23(2): 123–154, 1984.

[16] J. F. Allen and P. J. Hayes. A Common Sense Theory of Time. In *Proc. 9th International Joint Conference on Artificial Intelligence (IJCAI)*, pages 528–531. Morgan Kaufmann, 1985.

[17] J. F. Allen and J. A. Koomen. Planning Using a Temporal World Model. In *Proc. 8th International Joint Conference on Artificial Intelligence (IJCAI)*, pages 741–747. Morgan Kaufmann, 1983.

[18] J. F. Allen, J. Hendler and A. Tate, editors. *Readings in Planning*. Morgan Kaufmann Publishers, 1990. ISBN 1-558-60130-9.

[19] R. Alur and D. L. Dill. The Theory of Timed Automata. In de Bakkar *et al.* [137], pages 45–73.

[20] R. Alur and D. L. Dill. A Theory of Timed Automata. *Theoretical Computer Science*, 126(1): 183–235, 1994.

[21] R. Alur and T. A. Henzinger. Logics and Models of Real Time: A Survey. In de Bakker *et al.* [137], pages 74–106.

[22] R. Alur and T. A. Henzinger. Real-Time Logics: Complexity and Expressiveness. *Information and Computation*, 104(1):35–77, 1993.

[23] R. Alur and P. Madhusudan. Decision Problems for Timed Automata: A Survey. In *Formal Methods for the Design of Real-Time Systems*, volume 3185 of *LNCS*, pages 1–24. Springer, 2004.

[24] R. Alur, C. Courcoubetis and D. L. Dill. Model-Checking in Dense Real-time. *Information and Computation*, 104(1):2–34, 1993.

[25] R. Alur, C. Courcoubetis, N. Halbwachs, T. A. Henzinger, P.-H. Ho, X. Nicollin, A. Olivero, J. Sifakis and S. Yovine. The Algorithmic Analysis of Hybrid Systems. *Theoretical Computer Science*, 138(1):3–34, 1995.

[26] R. Alur, T. A. Henzinger and O. Kupferman. Alternating-Time Temporal Logic. *Journal of the ACM*, 49(5):672–713, 2002.

[27] C. M. Antunes and A. L. Oliviera. Temporal Data Mining: An Overview. SIGKDD Workshop on Temporal Data Mining, San Francisco, USA, 2001.

[28] R. Armoni, L. Fix, A. Flaisher, R. Gerth, B. Ginsburg, T. Kanza, A. Landver, S. Mador-Haim, E. Singerman, A. Tiemeyer, M. Y. Vardi and Y. Zbar. The ForSpec Temporal Logic: A New Temporal Property-Specification Language. In *Proc. International Conference on Tools and Algorithms for Construction and Analysis of Systems (TACAS)*, volume 2280 of *LNCS*, pages 296–211. Springer, 2002.

[29] L. Astefanoaei, M. Dastani, J.-J. Meyer and F. S. de Boer. On the Semantics and Verification of Normative Multi-Agent Systems. *Journal of Universal Computer Science*, 15(13):2629–2652, 2009.

[30] G. Audemard, A. Cimatti, A. Kornilowicz and R. Sebastiani. Bounded Model Checking for Timed Systems. In *Proc. 22nd IFIP WG 6.1 International Conference on Formal Techniques for Networked and Distributed Systems (FORTE)*, volume 2529 of *LNCS*, pages 243–259. Springer, 2002.

[31] M. Baaz, A. Leitsch and R. Zach. Completeness of a First-Order Temporal Logic with Time-Gaps. *Theoretical Computer Science*, 160(1&2):241–270, 1996.

[32] F. Bacchus and F. Kabanza. Using Temporal Logics to Express Search Control Knowledge for Planning. *Artificial Intelligence*, 116(1–2):123–191, 2000.

[33] L. Bachmair and H. Ganzinger. Ordered Chaining Calculi for First-Order Theories of Transitive Relations. *ACM Journal*, 45(6):1007–1049, 1998.

[34] C. Baier and J.-P. Katoen. *Principles of Model Checking*. MIT Press, 2008. ISBN 0-262-02649-X.

[35] P. Balbiani, A. Herzig and M. Lima-Marques. TIM: The Toulouse Inference Machine for Non-Classical Logic Programming. In *Proc. International Workshop on Processing Declarative Knowledge (PDK)*, volume 567 of *LNCS*, pages 366–382, 1991.

[36] T. Ball and S. K. Rajamani. The SLAM Toolkit. In *Proc. 13th International Conference on Computer Aided Verification (CAV)*, volume 2102 of *LNCS*, pages 260–264. Springer, 2001.

[37] T. Ball, A. Podelski and S. K. Rajamani. Relative Completeness of Abstraction Refinement for Software Model Checking. In *Proc. International Conference on Tools and Algorithms for the Construction and Analysis of Systems (TACAS)*, volume 2280 of *LNCS*. Springer, 2002.

[38] P. Balsiger, A. Heuerding and S. Schwendimann. Logics Workbench 1.0. In *Proc. International Conference on Automated Reasoning with Analytic Tableaux and Related Methods (TABLEAUX)*, volume 1397 of *LNCS*, pages 35–37. Springer, 1998.

[39] B. Banieqbal and H. Barringer. A Study of an Extended Temporal Language and a Temporal Fixed Point Calculus. Technical Report UMCS-86-10-2, Department of Computer Science, University of Manchester, UK, 1986.

[40] B. Banieqbal and H. Barringer. Temporal Logic with Fixed Points. In *Proc. International Colloquium on Temporal Logic in Specification*, volume 398 of *LNCS*, pages 62–74. Springer, 1987.

[41] R. S. Barga, J. Goldstein, M. H. Ali and M. Hong. Consistent Streaming Through Time: A Vision for Event Stream Processing. In *Proc. 3rd Biennial Conference on Innovative Data Systems Research (CIDR)*, pages 363–374. http://www.cidrdb.org, 2007.

[42] H. Barringer and R. Kuiper. Towards the Hierarchical, Temporal Logic, Specification. In *Proc. Workshop on the Analysis of Concurrent Systems*, volume 207 of *LNCS*, pages 157–183. Springer, 1983.

[43] H. Barringer and R. Kuiper. Hierarchical Development of Concurrent Systems in a Temporal Logic Framework. In *Proc. NSF/SERC Seminar on Concurrency*, volume 197 of *LNCS*, pages 35–61. Springer, 1985.

[44] H. Barringer, R. Kuiper and A. Pnueli. Now You May Compose Temporal Logic Specifications. In *Proc. 16th ACM Symposium on Theory of Computing (STOC)*, pages 51–63. ACM Press, 1984.

[45] H. Barringer, R. Kuiper and A. Pnueli. A Compositional Temporal Approach to a CSP-like Language. In *Proc. IFIP Working Conference "The Role of Abstract Models in Information Processing"*, pages 207–227, Vienna, Austria, 1985.

[46] H. Barringer, R. Kuiper and A. Pnueli. A Really Abstract Concurrent Model and its Temporal Logic. In *Proc. 13th ACM Symposium in Principles of Programming Languages (POPL)*, pages 173–183. ACM Press, 1986.

[47] H. Barringer, M. Fisher, D. Gabbay, G. Gough and R. Owens. METATEM: A Framework for Programming in Temporal Logic. In *Proc. REX Workshop on Stepwise Refinement of Distributed Systems: Models, Formalisms, Correctness*, volume 430 of *LNCS*, pages 94–129. Springer, 1989.

[48] H. Barringer, M. Fisher, D. Gabbay and A. Hunter. Meta-Reasoning in Executable Temporal Logic. In *Proc. International Conference on Principles of Knowledge Representation and Reasoning (KR)*, pages 40–49. Morgan Kaufmann, 1991.

[49] H. Barringer, M. Fisher, D. Gabbay, G. Gough and R. Owens. METATEM: An Introduction. *Formal Aspects of Computing*, 7(5):533–549, 1995.

[50] H. Barringer, M. Fisher, D. Gabbay, R. Owens and M. Reynolds, editors. *The Imperative Future: Principles of Executable Temporal Logics*. Research Studies Press, 1996.

[51] H. Barringer, M. Fisher, D. Gabbay and G. Gough, editors. *Advances in Temporal Logic*. Kluwer Academic Publishers, 2000.

[52] H. Barringer, D. E. Rydeheard and K. Havelund. Rule Systems for Run-Time Monitoring: From Eagle to RuleR. In *Proc. 7th International Workshop on Runtime Verification (RV)*, volume 4839 of *LNCS*, pages 111–125. Springer, 2007.

[53] H. Barringer, K. Havelund, D. E. Rydeheard and A. Groce. Rule Systems for Runtime Verification: A Short Tutorial. In *Proc. 9th International Workshop on Runtime Verification (RV)*, volume 5779 of *LNCS*, pages 1–24. Springer, 2009.

[54] M. Baudinet. Temporal Logic Programming is Complete and Expressive. In *Proc. 16th Annual ACM Symposium on Principles of Programming Languages (POPL)*, pages 267–280. ACM Press, 1989.

[55] M. Baudinet. A Simple Proof of the Completeness of Temporal Logic Programming. In L. F. del Cerro and M. Penttonen, editors, *Intensional Logics for Programming*. Oxford University Press, 1992.

[56] M. Baudinet. On the Expressiveness of Temporal Logic Programming. *Information and Computation*, 117(2):157–180, 1995.

[57] A. Bauer, M. Leucker and C. Schallhart. Monitoring of Real-time Properties. In *Proc. 26th Conference on Foundations of Software Technology and Theoretical Computer Science (FSTTCS)*, volume 4337 of *LNCS*. Springer, 2006.

[58] A. Behdenna, C. Dixon and M. Fisher. Deductive Verification of Simple Foraging Robotic Behaviours. *International Journal of Intelligent Computing and Cybernetics*, 2(4):604–643, 2009.

[59] G. Behrmann, A. David, K. G. Larsen, O. Möller, P. Pettersson and W. Yi. UPPAAL – Present and Future. In *Proc. 40th IEEE Conference on Decision and Control (CDC)*, pages 2881–2886. IEEE Computer Society Press, 2001.

[60] J. Bengtsson and W. Yi. Timed Automata: Semantics, Algorithms and Tools. In *Lecture Notes on Concurrency and Petri Nets*, volume 3098 of *LNCS*, pages 87–124. Springer, 2004.

[61] B. Bennett, C. Dixon, M. Fisher, E. Franconi, I. Horrocks and M. de Rijke. Combinations of Modal Logics. *AI Review*, 17(1):1–20, 2002.

[62] C. Bettini, S. Jajodia and X. Wang. *Time Granularities in Databases, Data Mining, and Temporal Reasoning*. Springer, 2000.

[63] D. Beyer, C. Lewerentz and A. Noack. Rabbit: A Tool for BDD-Based Verification of Real-Time Systems. In *Proc. 15th International Conference on Computer Aided Verification, (CAV)*, volume 2725 of *LNCS*, pages 122–125. Springer, 2003.

[64] A. Biere, A. Cimatti, E. M. Clarke, M. Fujita and Y. Zhu. Symbolic Model Checking using SAT Procedures instead of BDDs. In *Proc. 36th Annual ACM/IEEE Design Automation Conference (DAC)*, pages 317–320. ACM Press, 1999.

[65] A. Biere, A. Cimatti, E. M. Clarke and Y. Zhu. Symbolic Model Checking without BDDs. In *Proc. 5th International Conference on Tools and Algorithms for Construction and Analysis of Systems (TACAS)*, volume 1579 of *LNCS*, pages 193–207. Springer, 1999.

[66] N. Bjorner, A. Browne, E. Chang, M. Colón, A. Kapur, Z. Manna, H. B. Sipma and T. E. Uribe. *STeP: The Stanford Temporal Prover (Educational Release Version 1.0) User's Manual*. Computer Science Department, Stanford University, USA, 1995.

[67] P. Blackburn, M. de Rijke, and Y. Venema. *Modal Logic*, volume 53 of *Cambridge Tracts in Theoretical Computer Science*. Cambridge University Press, 2002. ISBN 0-521-52714-7.

[68] P. Blackburn, J. van Benthem, and F. Wolter, editors. *Handbook of Modal Logic*. Elsevier, 2006. ISBN 0-444-51690-5.

[69] A. Blass and Y. Gurevich. Existential Fixed-Point Logic. In *Computation Theory and Logic, In Memory of Dieter Rodding*, volume 270 of *LNCS*, pages 20–36. Springer, 1987.

[70] E. Bodden. *J-LO – A Tool for Runtime-Checking Temporal Assertions*. Diploma thesis, RWTH Aachen University, Germany, 2005.

[71] A. Bolotov. *Clausal Resolution for Branching-Time Temporal Logic*. PhD thesis, Department of Computing and Mathematics, Manchester Metropolitan University, UK, 2000.

[72] A. Bolotov and A. Basukoski. A Clausal Resolution Method for Extended Computation Tree Logic ECTL. *Journal of Applied Logic*, 4(2):141–167, 2006.

[73] A. Bolotov and A. Basukoski. A Clausal Resolution Method for Branching-Time Logic ECTL$^+$. *Annals of Mathematics and Artificial Intelligence*, 46(3):235–263, 2006.

[74] A. Bolotov and C. Dixon. Resolution for Branching Time Temporal Logics: Applying the Temporal Resolution Rule. In *Proc. 7th International Workshop on Temporal Representation and Reasoning (TIME)*, pages 163–172. IEEE Press, 2000.

[75] A. Bolotov and M. Fisher. A Clausal Resolution Method for CTL Branching-Time Temporal Logic. *Journal of Experimental and Theoretical Artificial Intelligence*, 11:77–93, 1999.

[76] A. Bolotov, M. Fisher and C. Dixon. On the Relationship between ω-Automata and Temporal Logic Normal Forms. *Journal of Logic and Computation*, 12(4):561–581, 2002.

[77] A. Bolotov, A. Basukoski, O. Grigoriev and V. Shangin. Natural Deduction Calculus for Linear-Time Temporal Logic. In *Proc. 10th European Conference on Logics in Artificial Intelligence (JELIA)*, volume 4160 of *LNCS*, pages 56–68. Springer, 2006.

[78] A. Bolotov, O. Grigoriev and V. Shangin. Automated Natural Deduction for Propositional Linear-Time Temporal Logic. In *Proc, 14th International Symposium on Temporal Representation and Reasoning (TIME)*, pages 47–58. IEEE Computer Society Press, 2007.

[79] A. H. Bond and L. Gasser, editors. *Readings in Distributed Artificial Intelligence*. Morgan Kaufmann, 1988.

[80] R. H. Bordini, M. Fisher, W. Visser and M. Wooldridge. Model Checking Rational Agents. *IEEE Intelligent Systems*, 19(5):46–52, 2004.

[81] R. H. Bordini, M. Dastani, J. Dix and A. El Fallah Seghrouchni, editors. *Multi-Agent Programming: Languages, Platforms and Applications*. Springer, 2005. ISBN 0-387-24568-5.

[82] R. H. Bordini, M. Fisher, W. Visser and M. Wooldridge. Verifying Multi-Agent Programs by Model Checking. *Journal of Autonomous Agents and Multi-Agent Systems*, 12(2):239–256, 2006.

[83] R. H. Bordini, L. A. Dennis, B. Farwer and M. Fisher. Automated Verification of Multi-Agent Programs. In *Proc. 23rd IEEE/ACM International Conference on Automated Software Engineering (ASE)*, pages 69–78. IEEE Computer Society Press, 2008.

[84] R. H. Bordini, M. Dastani, J. Dix and A. El Fallah-Seghrouchni, editors. *Multi-Agent Programming: Languages, Tools and Applications*. Springer, 2009. ISBN 0-387-89298-2.

[85] R. H. Bordini, M. Fisher, M. Wooldridge and W. Visser. Property-based Slicing for Agent Verification. *Journal of Logic and Computation*, 19(6):1385–1425, 2009.

[86] T. Bosse, C. M. Jonker, L. van der Meij and J. Treur. A Language and Environment for Analysis of Dynamics by Simulation. *International Journal on Artificial Intelligence Tools*, 16(3):435–464, 2007.

[87] H. Bowman and S. Thompson. A Decision Procedure and Complete Axiomatization of Finite Interval Temporal Logic with Projection. *Journal of Logic and Computation*, 13(2):195–239, 2003.

[88] J. C. Bradfield, J. Esparza and A. Mader. An Effective Tableau System for the Linear Time μ-Calculus. In *Proc. 23rd International Colloquium on Automata, Languages and Programming (ICALP)*, volume 1099 of *LNCS*, pages 98–109. Springer, 1996.

[89] M. E. Bratman. *Intentions, Plans, and Practical Reason*. Harvard University Press, 1987. ISBN 1-575-86192-5.

[90] D. Bresolin, A. Montanari and G. Puppis. Time Granularities and Ultimately Periodic Automata. In *Proc. 9th European Conference on Logics in Artificial Intelligence (JELIA)*, volume 3229 of *LNCS*, pages 513–525. Springer, 2004.

[91] D. Bresolin, D. D. Monica, V. Goranko, A. Montanari and G. Sciavicco. Decidable and Undecidable Fragments of Halpern and Shoham's Interval Temporal Logic: Towards a Complete Classification. In *Proc. 15th International Conference on Logic for Programming, Artificial Intelligence, and Reasoning (LPAR)*, volume 5330 of *LNCS*, pages 590–604. Springer, 2008.

[92] D. Bresolin, D. D. Monica, V. Goranko, A. Montanari and G. Sciavicco. Metric Propositional Neighborhood Logics: Expressiveness, Decidability, and Undecidability. In *Proc. 19th European Conference on Artificial Intelligence (ECAI)*, volume 215 of *Frontiers in Artificial Intelligence and Applications*, pages 695–700. IOS Press, 2010.

[93] R. E. Bryant. Graph-Based Algorithms for Boolean Function Manipulation. *IEEE Transactions on Computers*, C-35(8):677–691, Aug. 1986.

[94] C. Brzoska. Temporal Logic Programming and its Relation to Constraint Logic Programming. In *Proc. International Symposium on Logic Programming (ILPS)*, pages 661–677. MIT Press, 1991.

[95] J. R. Büchi. On a Decision Method in Restricted Second Order Arithmetic. In *Proceedings of the International Congress on Logic, Methodology, and Philosophy of Science, Berkeley, 1960*, pages 1–11. Standford University Press, 1962. Republished in [359].

[96] Büchi Store. http://buchi.im.ntu.edu.tw.

[97] R. A. Bull and K. Segerberg. Basic Modal Logic. In D. Gabbay and F. Guenthner, editors, *Handbook of Philosophical Logic (II)*, volume 165 of *Synthese Library*, chapter II.1, pages 1–88. Reidel, 1984.

[98] J. R. Burch, E. M. Clarke, K. L. McMillan, D. L. Dill and L. J. Hwang. Symbolic Model Checking: 10^{20} States and Beyond. In *Proc. 5th IEEE Symposium on Logic in Computer Science (LICS)*, pages 1–33. IEEE Computer Society Press, 1990.

[99] J. R. Burch, E. M. Clarke, D. E. Long, K. L. McMillan and D. L. Dill. Symbolic Model Checking for Sequential Circuit Verification. *IEEE Transactions on Computer-Aided Design of Integrated Circuits and Systems*, 13(4):401–424, 1994.

[100] J. Burgess. Axioms for Tense Logic (1): 'Since' and 'Until'. *Notre Dame Journal of Formal Logic*, 23(4):367–374, 1982.

[101] J. Burgess. Axioms for Tense Logic (2): Time Periods. *Notre Dame Journal of Formal Logic*, 23(4):375–383, 1982.

[102] P. Cabalar. A Normal Form for Linear Temporal Equilibrium Logic. In *Proc. 12th European Conference on Logics in Artificial Intelligence (JELIA)*, volume 6341 of *LNCS*, pages 64–76. Springer, 2010.

[103] A. R. Cavalli and L. Fariñas del Cerro. A Decision Method for Linear Temporal Logic. In *Proc. 7th International Conference on Automated Deduction (CADE)*, volume 170 of *LNCS*, pages 113–127. Springer, 1984.

[104] S. Cerrito and M. C. Mayer. Bounded Model Search in Linear Temporal Logic and Its Application to Planning. In *Proc. International Conference on Automated Reasoning with Analytic Tableaux and Related Methods (TABLEAUX)*, volume 1397 of *LNCS*, pages 124–140, 1998.

[105] A. K. Chandra, D. C. Kozen and L. J. Stockmeyer. Alternation. *Journal of the ACM*, 28(1):114–133, 1981.

[106] K. M. Chandy and J. Misra. An Example of Stepwise Refinement of Distributed Programs: Quiescence Detection. *ACM Transactions on Programming Languages and Systems*, 8(3):326–343, 1986.

[107] Z. Chaochen and M. R. Hansen. *Duration Calculus – A Formal Approach to Real-Time Systems*. EATCS Monographs in Theoretical Computer Science. Springer, 2004.

[108] K. Chaudhuri, D. Doligez, L. Lamport and S. Merz. A TLA+ Proof System. In *Proc. LPAR 2008 Workshops ("Knowledge Exchange: Automated Provers and Proof Assistants" and "Implementation of Logics"*, volume 418 of *CEUR Workshop Proceedings*. CEUR-WS.org, 2008.

[109] B. Chellas. *Modal Logic: An Introduction*. Cambridge University Press, 1980. ISBN 0-521-29515-7.

[110] J. Chomicki and D. Toman. Temporal databases. In Fisher *et al.* [210], pages 429–468. ISBN 0-444-51493-7.

[111] Y. Choueka. Theories of Automata on ω-tapes: A Simplified Approach. *Journal of Computer and System Sciences*, 8:117–141, 1974.

[112] A. Cimatti, F. Giunchiglia, E. Giunchiglia and P. Traverso. Planning via Model Checking: A Decision Procedure for *AR*. In *Proc. 4th European Conference on Planning (ECP)*, volume 1348 of *LNCS*, pages 130–142. Springer, 1997.

[113] A. Cimatti, E. M. Clarke, F. Giunchiglia and M. Roveri. NuSMV: A New Symbolic Model Verifier. In *Proc. 11th International Conference on Computer Aided Verification (CAV)*, volume 1633 of *LNCS*, pages 495–499. Springer, 1999.

[114] A. Cimatti, E. Clarke, E. Giunchuglia, F. Giunchiglia, M. Pistore, M. Roveri, R. Sebastiani and A. Tacchella. NuSMV 2: An OpenSource Tool for Symbolic Model Checking. In *Proc. 14th International Conference on Computer Aided Verification (CAV)*, volume 2404 of *LNCS*, pages 359–364. Springer, 2002.

[115] E. M. Clarke and E. A. Emerson. Design and Synthesis of Synchronization Skeletons Using Branching-Time Temporal Logic. In *Proc. Workshop on Logics of Programs*, volume 131 of *LNCS*, pages 52–71. Springer, 1981.

[116] E. M. Clarke and E. A. Emerson. Using Branching Time Temporal Logic to Synthesise Synchronisation Skeletons. *Science of Computer Programming*, 2:241–266, 1982.

[117] E. M. Clarke and B.-H. Schlingloff. Model Checking. In A. Robinson and A. Voronkov, editors, *Handbook of Automated Reasoning*, pages 1635–1790. Elsevier and MIT Press, 2001.

[118] E. M. Clarke, E. A. Emerson and A. P. Sistla. Automatic Verification of Finite State Concurrent Systems Using Temporal Logic Specifications: A Practical Approach. In *Proc. 10th ACM Symposium on the Principles of Programming Languages (POPL)*, pages 117–126. ACM Press, 1983.

[119] E. M. Clarke, E. A. Emerson and A. P. Sistla. Automatic Verification of Finite-State Concurrent Systems Using Temporal Logic Specifications. *ACM Transactions on Programming Languages and Systems*, 8(2):244–263, 1986.

[120] E. M. Clarke, O. Grumberg and D. E. Long. Model Checking and Abstraction. *ACM Transactions on Programming Languages and Systems*, 16(5):1512–1542, 1994. ISSN 0164-0925. URL http://www.acm.org/pubs/toc/Abstracts/0164-0925/186051.html.

[121] E. M. Clarke, O. Grumberg, S. Jha, Y. Lu and H. Veith. Counterexample-Guided Abstraction Refinement. In *Proc. 12th International Conference on Computer Aided Verification (CAV)*, volume 1855 of *LNCS*, pages 154–169. Springer, 2000.

[122] E. M. Clarke, O. Grumberg and D. A. Peled. *Model Checking*. The MIT Press, 2000. ISBN 0-262-03270-8.

[123] E. M. Clarke, A. Biere, R. Raimi and Y. Zhu. Bounded Model Checking Using Satisfiability Solving. *Formal Methods in System Design*, 19(1):7–34, 2001.

[124] E. M. Clarke, O. Grumberg, S. Jha, Y. Lu and H. Veith. Counterexample-Guided Abstraction Refinement for Symbolic Model Checking. *Journal of the ACM*, 50(5):752–794, 2003.

[125] P. R. Cohen and H. J. Levesque. Intention Is Choice with Commitment. *Artificial Intelligence*, 42(2–3):213–261, 1990.

[126] P. R. Cohen and H. J. Levesque. Teamwork. *Nous*, 25(4):487–512, 1991.

[127] A. Cohn. Spatial Reasoning. In F. van Harmelen, B. Porter and V. Lifschitz, editors, *Handbook of Knowledge Representation*. Elsevier Press, 2007.

[128] C. Colombo, G. J. Pace and G. Schneider. LARVA – Safer Monitoring of Real-Time Java Programs (Tool Paper). In *Proc. 7th IEEE International Conference on Software Engineering and Formal Methods (SEFM)*, pages 33–37. IEEE Computer Society, 2009.

[129] C. Courcoubetis, M. Y. Vardi, P. Wolper and M. Yannakakis. Memory-Efficient Algorithm for the Verification of Temporal Properties. *Formal Methods in System Design*, 1:275–288, 1992.

[130] P. Cousot and R. Cousot. Comparing the Galois Connection and Widening/Narrowing Approaches to Abstract Interpretation. In *Proc. 4th International Symposium on Programming Language Implementation and Logic Programming (PLILP)*, volume 631 of *LNCS*, pages 269–295. Springer, 1992.

[131] P. Cousot and R. Cousot. Abstract Interpretation and Application to Logic Programs. *Journal of Logic Programming*, 13(1, 2, 3 and 4):103–179, 1992.

[132] J. N. Crossley, C. J. Ash, C. J. Brickhill, J. C. Stillwell and N. H. Williams. *What is Mathematical Logic*. Oxford University Press, 1972. ISBN 0-486-26404-1.

[133] M. D'Agostino, D. Gabbay, R. Hähnle and J. Posegga, editors. *Handbook of Tableau Methods*. Kluwer Academic Press, 1999. ISBN 9-048-15184-8.

[134] M. Daniele, F. Giunchiglia and M. Y. Vardi. Improved Automata Generation for Linear Temporal Logic. In *Proc. 11th Conference on Computer Aided Verification (CAV)*, volume 1633 of *LNCS*, pages 249–260. Springer, 1999.

[135] J. Davies and J. Woodcock. *Using Z: Specification, Refinement and Proof*. Prentice Hall International, 1996. ISBN 0-139-48472-8.

[136] C. Dax, M. Hofmann and M. Lange. A Proof System for the Linear Time μ-Calculus. In *Proc. 26th Conference on Foundations of Software Technology and Theoretical Computer Science (FSTTCS)*, volume 4337 of *LNCS*, pages 274–285. Springer, 2006.

[137] J. W. de Bakker, C. Huizing, W. P. de Roever and G. Rozenberg, editors. *Proc. REX Workshop on Real-Time: Theory in Practice*, volume 600 of *LNCS*, 1991. Springer.

[138] N. de Carvalho Ferreira, M. Fisher and W. van der Hoek. Logical Implementation of Uncertain Agents. In *Proc. 12th Portuguese Conference on Artificial Intelligence (EPIA)*, volume 3808 of *LNCS*, pages 536–547. Springer, 2005.

[139] N. de Carvalho Ferreira, M. Fisher and W. van der Hoek. Specifying and Reasoning about Uncertain Agents. *International Journal of Approximate Reasoning*, 49(1):35–51, 2008.

[140] R. Dechter, I. Meiri and J. Pearl. Temporal Constraint Networks. *Artificial Intelligence*, 49(1–3):61–95, 1991.

[141] A. Degtyarev, M. Fisher and B. Konev. A Simplified Clausal Resolution Procedure for Propositional Linear-Time Temporal Logic. In *Automated Reasoning with Analytic Tableaux and Related Methods*, volume 2381 of *LNCS*, pages 85–99. Springer, 2002. ISBN 3-540-43929-3.

[142] A. Degtyarev, M. Fisher and A. Lisitsa. Equality and Monodic First-Order Temporal Logic. *Studia Logica*, 72(2):147–156, 2002.

[143] A. Degtyarev, M. Fisher and B. Konev. Monodic Temporal Resolution. *ACM Transactions on Computational Logic*, 7(1):108–150, 2006.

[144] L. A. Dennis and M. Fisher. Programming Verifiable Heterogeneous Agent Systems. In *Proc. 6th International Workshop on Programming in Multi-Agent Systems (ProMAS)*, volume 5442 of *LNCS*, pages 40–55. Springer Verlag, 2008.

[145] L. A. Dennis, B. Farwer, R. H. Bordini, M. Fisher and M. Wooldridge. A Common Semantic Basis for BDI Languages. In *Proc. 7th International Workshop on Programming Multiagent Systems (ProMAS)*, volume 4908 of *LNAI*, pages 124–139. Springer, 2008.

[146] E. W. Dijkstra. Guarded Commands, Nondeterminacy and Formal Derivation of Programs. *Communications of the ACM*, 18(8):453–457, 1975.

[147] C. Dixon. *Strategies for Temporal Resolution*. PhD thesis, Department of Computer Science, University of Manchester, UK, 1995.

[148] C. Dixon. Search Strategies for Resolution in Temporal Logics. In *Proc. 13th International Conference on Automated Deduction (CADE)*, volume 1104 of *LNCS*, pages 673–687. Springer, 1996.

[149] C. Dixon. Temporal Resolution: Removing Irrelevant Information. In *Proc. 4th International Workshop on Temporal Representation and Reasoning (TIME)*, pages 4–11. IEEE Computer Society Press, 1997.

[150] C. Dixon. Temporal Resolution using a Breadth-First Search Algorithm. *Annals of Mathematics and Artificial Intelligence*, 22:87–115, 1998.

[151] C. Dixon. Using Otter for Temporal Resolution. In Barringer *et al.* [51], pages 149–166.

[152] C. Dixon and M. Fisher. The Set of Support Strategy in Temporal Resolution. In *Proc. 5th International Workshop on Temporal Representation and Reasoning (TIME)*, pages 113–120. IEEE Computer Society Press, 1998.

[153] C. Dixon and M. Fisher. Clausal Resolution for Logics of Time and Knowledge with Synchrony and Perfect Recall. In *Proc. Joint International Conference on Temporal Logic and Advances in Modal Logic (AiML-ICTL)*, pages 43–52, Leipzig, Germany, 2000.

[154] C. Dixon, M. Fisher and H. Barringer. A Graph-Based Approach to Resolution in Temporal Logic. In *Proc. 1st International Conference on Temporal Logic (ICTL)*, volume 827 of *LNCS*, pages 415–429. Springer, 1994.

[155] C. Dixon, M. Fisher and M. Wooldridge. Resolution for Temporal Logics of Knowledge. *Journal of Logic and Computation*, 8(3):345–372, 1998.

[156] C. Dixon, M. Fisher and M. Reynolds. Execution and Proof in a Horn-clause Temporal Logic. In Barringer *et al.* [51], pages 413–432.

[157] C. Dixon, M. Fisher and A. Bolotov. Resolution in a Logic of Rational Agency. *Artificial Intelligence*, 139(1):47–89, 2002.

[158] C. Dixon, M. Fisher and B. Konev. Is There a Future for Deductive Temporal Verification? In *Proc. 13th International Symposium on Temporal Representation and Reasoning (TIME)*, pages 11–18. IEEE Computer Society Press, 2006.

[159] C. Dixon, M. C. Fernández Gago, M. Fisher and W. van der Hoek. Temporal Logics of Knowledge and their Applications in Security. *Electronic Notes in Theoretical Computer Science*, 186:27–42, 2007.

[160] C. Dixon, M. Fisher and B. Konev. Temporal Logic with Capacity Constraints. In *Proc. 6th International Symposium on Frontiers of Combining Systems (FroCoS)*, volume 4720 of *LNCS*, pages 163–177. Springer, 2007.

[161] C. Dixon, M. Fisher and B. Konev. Tractable Temporal Reasoning. In *Proc. 20th International Joint Conference on Artificial Intelligence (IJCAI)*, pages 318–323. AAAI Press, 2007.

[162] C. Dixon, M. Fisher, B. Konev and A. Lisitsa. Practical First-Order Temporal Reasoning. In *Proc. 15th International Symposium on Temporal Representation and Reasoning (TIME)*, pages 156–163. IEEE Computer Society Press, 2008.

[163] T. Drakengren and P. Jonsson. Computational Complexity of Temporal Constraint Problems. In Fisher *et al.* [210], pages 197–218. ISBN 0-444-51493-7.

[164] M. Duflot, M. Kwiatkowska, G. Norman and D. Parker. A Formal Analysis of Bluetooth Device Discovery. *International Journal on Software Tools for Technology Transfer (STTT)*, 8(6):621–632, 2006.

[165] Duration Calculus. http://www.iist.unu.edu/dc.

[166] E. Emerson and A. Sistla. Deciding Full Branching Time Logic. *Information and Control*, 61(3):175–201, 1984.

[167] E. A. Emerson. The Role of Büchi's Automata in Computing Science. In S. MacLane and D. Siefkes, editors, *The Collected Works of J. Richard Büchi*. Springer, 1990.

[168] E. A. Emerson. Temporal and Modal Logic. In J. van Leeuwen, editor, *Handbook of Theoretical Computer Science*, pages 996–1072. Elsevier, 1990.

[169] E. A. Emerson and E. M. Clarke. Characterizing Correctness Properties of Parallel Programs as Fixed Points. In *Proc. 7th International Colloquium on Automata, Languages and Programming (ICALP)*, volume 85 of *LNCS*, pages 169–181, 1980.

[170] E. A. Emerson and J. Y. Halpern. "Sometimes" and "Not Never" Revisited: On Branching Versus Linear Time Temporal Logic. *Journal of the ACM*, 33(1):151–178, 1986.

[171] E. A. Emerson and C.-L. Lei. Efficient Model Checking in Fragments of the Propositional Mu-Calculus. In *Proc. Symposium on Logic in Computer Science (LICS)*, pages 267–278. IEEE Computer Society, 1986.

[172] E. A. Emerson and A. P. Sistla. Deciding Full Branching Time Logic. *Information and Control*, 61:175–201, 1984.

[173] J. Euzenat and A. Montanari. Time Granularity. In Fisher *et al.* [210], pages 59–118. ISBN 0-444-51493-7.

[174] R. Fagin, J. Halpern, Y. Moses and M. Vardi. *Reasoning About Knowledge*. MIT Press, 1996.

[175] G. E. Fainekos and G. J. Pappas. Robustness of Temporal Logic Specifications for Continuous-time Signals. *Theoretical Computer Science*, 410(42):4262–4291, 2009.

[176] G. E. Fainekos, A. Girard, H. Kress-Gazit and G. J. Pappas. Temporal Logic Motion Planning for Dynamic Robots. *Automatica*, 45(2):343–352, 2009.

[177] J. Ferber, O. Gutknecht and F. Michel. From Agents to Organizations: An Organizational View of Multi-Agent Systems. In *Proc. 4th International Workshop on Agent-Oriented Software Engineering (AOSE)*, volume 2935 of *LNCS*, pages 214–230. Springer, 2003.

[178] C. Fermüller, A. Leitsch, U. Hustadt and T. Tammet. Resolution Decision Procedures. In A. Robinson and A. Voronkov, editors, *Handbook of Automated Reasoning*, chapter 25, pages 1791–1850. Elsevier, 2001.

[179] M.-C. Fernández Gago. *Efficient Control of Temporal Reasoning*. PhD thesis, Department of Computer Science, University of Liverpool, UK, 2003.

[180] M.-C. Fernández Gago, M. Fisher and C. Dixon. Algorithms for Guiding Clausal Temporal Resolution. In *Advances in Artificial Intelligence (Proc. 25th Annual German Conference on AI)*, volume 2479, pages 235–252. Springer, 2002.

[181] M.-C. Fernández Gago, U. Hustadt, C. Dixon, M. Fisher and B. Konev. First-Order Temporal Verification in Practice. *Journal of Automated Reasoning*, 34(3):295–321, 2005.

[182] J. H. Fetzer. Program Verification: The Very Idea. *Communications of the ACM*, 31(9):1048–1063, 1988.

[183] M. Finger and D. M. Gabbay. Combining Temporal Logic Systems. *Notre Dame Journal of Formal Logic*, 37(2):204–232, 1996.

[184] M. Finger, M. Fisher and R. Owens. METATEM at Work: Modelling Reactive Systems Using Executable Temporal Logic. In *Proc. 6th International Conference on Industrial and Engineering Applications of Artificial Intelligence and Expert Systems (IEA/AIE)*. Gordon and Breach Publishers, 1993.

[185] M. Fisher. A Resolution Method for Temporal Logic. In *Proc. 12th International Joint Conference on Artificial Intelligence (IJCAI)*, pages 99–104. Morgan Kaufman, 1991.

[186] M. Fisher. A Model Checker for Linear Time Temporal Logic. *Formal Aspects of Computing*, 4(3):299–319, 1992.

[187] M. Fisher. A Normal Form for First-Order Temporal Formulae. In *Proc. 11th International Conference on Automated Deduction (CADE)*, volume 607 of *LNCS*, pages 370–384. Springer, 1992.

[188] M. Fisher. Concurrent METATEM – A Language for Modeling Reactive Systems. In *Proc. Conference on Parallel Architectures and Languages, Europe (PARLE)*, volume 694 of *LNCS*, pages 185–196. Springer, 1993.

[189] M. Fisher. A Survey of Concurrent METATEM – The Language and its Applications. In *Proc. 1st International Conference on Temporal Logic (ICTL)*, volume 827 of *LNCS*, pages 480–505. Springer, 1994.

[190] M. Fisher. Representing and Executing Agent-Based Systems. In *Intelligent Agents (Proc. Workshop on Agent Theories, Architectures, and Languages)*, volume 890 of *LNCS*, pages 307–323. Springer, 1995.

[191] M. Fisher. An Introduction to Executable Temporal Logics. *Knowledge Engineering Review*, 11(1):43–56, 1996.

[192] M. Fisher. A Temporal Semantics for Concurrent METATEM. *Journal of Symbolic Computation*, 22(5/6):627–648, 1996.

[193] M. Fisher. An Open Approach to Concurrent Theorem-Proving. In J. Geller, H. Kitano and C. Suttner, editors, *Parallel Processing for Artificial Intelligence*, volume 3. Elsevier/North-Holland, 1997.

[194] M. Fisher. Implementing BDI-like Systems by Direct Execution. In *Proc. 15th International Joint Conference on Artificial Intelligence (IJCAI)*, pages 316–321. Morgan Kaufmann, 1997.

[195] M. Fisher. A Normal Form for Temporal Logic and its Application in Theorem-Proving and Execution. *Journal of Logic and Computation*, 7(4):429–456, 1997.

[196] M. Fisher. Temporal Representation and Reasoning. In F. van Harmelen, B. Porter and V. Lifschitz, editors, *Handbook of Knowledge Representation*, pages 513–550. Elsevier Press, 2007.

[197] M. Fisher. Agent Deliberation in an Executable Temporal Framework. Technical Report ULCS-08-014, Department of Computer Science, University of Liverpool, UK, July 2008.

[198] M. Fisher and C. Dixon. Guiding Clausal Temporal Resolution. In Barringer *et al.* [51], pages 167–184.

[199] M. Fisher and C. Ghidini. Programming Resource-Bounded Deliberative Agents. In *Proc. 16th International Joint Conference on Artificial Intelligence (IJCAI)*, pages 200–205. Morgan Kaufmann, 1999.

[200] M. Fisher and C. Ghidini. The ABC of Rational Agent Programming. In *Proc. 1st International Conference on Autonomous Agents and Multi-Agent Systems (AAMAS)*, pages 849–856. ACM Press, 2002.

[201] M. Fisher and C. Ghidini. Exploring the Future with Resource-Bounded Agents. *Journal of Logic, Language and Information*, 18(1):3–21, 2009.

[202] M. Fisher and C. Ghidini. Executable Specifications of Resource-Bounded Agents. *Journal of Autonomous Agents and Multi-Agent Systems*, 21(3):368–396, 2010.

[203] M. Fisher and A. Hepple. Executing Logical Agent Specifications. In Bordini *et al.* [84], pages 1–27. ISBN 0-387-89298-2.

[204] M. Fisher and T. Kakoudakis. Flexible Agent Grouping in Executable Temporal Logic. In *Proc. 12th International Symposium on Languages for Intensional Programming (ISLIP)*. World Scientific Press, 1999.

[205] M. Fisher and P. Noël. Transformation and Synthesis in METATEM – Part I: Propositional METATEM. Technical Report UMCS-92-2-1, Department of Computer Science, University of Manchester, UK, 1992.

[206] M. Fisher and M. Wooldridge. Executable Temporal Logic for Distributed A.I. In *Proc. 12th International Workshop on Distributed A.I.*, Hidden Valley Resort, Pennsylvania, USA, 1993.

[207] M. Fisher and M. Wooldridge. Temporal Reasoning in Agent Based Systems. In Fisher *et al.* [210], pages 469–496. ISBN 0-444-51493-7.

[208] M. Fisher, C. Dixon and M. Peim. Clausal Temporal Resolution. *ACM Transactions on Computational Logic*, 2(1):12–56, 2001.

[209] M. Fisher, C. Ghidini and B. Hirsch. Organising Computation through Dynamic Grouping. In *Objects, Agents and Features*, volume 2975 of *LNCS*, pages 117–136. Springer-Verlag, 2004.

[210] M. Fisher, D. Gabbay and L. Vila, editors. *Handbook of Temporal Reasoning in Artificial Intelligence*, volume 1 of *Foundations of Artificial Intelligence*. Elsevier Press, 2005. ISBN 0-444-51493-7.

[211] M. Fisher, B. Konev and A. Lisitsa. Practical Infinite-State Verification with Temporal Reasoning. In *Verification of Infinite State Systems and Security*, volume 1 of *NATO Security through Science Series: Information and Communication*, pages 91–100. IOS Press, 2006.

[212] M. Fisher, B. Konev and A. Lisitsa. Temporal Verification of Fault-Tolerant Protocols. In *Methods Models and Tools for Fault Tolerance*, volume 5454 of *LNCS*, pages 44–56. Springer, 2009.

[213] M. Fitting. Tableau Methods of Proof for Modal Logics. *Notre Dame Journal of Formal Logic*, 13(2):237–247, 1972.

[214] R. W. Floyd. Assigning Meanings to Programs. In *Proc. American Mathematical Society Symposia on Applied Mathematics*, volume 19, pages 19–31, 1967.

[215] M. Fox and D. Long. PDDL2.1: An Extension to PDDL for Expressing Temporal Planning Domains. *Journal of Artificial Intelligence Research*, 20:61–124, 2003.

[216] M. Fox and D. Long. Time in Planning. In Fisher *et al.* [210], pages 497–536. ISBN 0-444-51493-7.

[217] M. Franceschet. *Dividing and Conquering the Layered Land*. PhD thesis, Department of Mathematics and Computer Science, University of Udine, Italy, 2001.

[218] N. Francez. *Fairness*. Springer, 1986. ISBN 0-387-96235-2.

[219] S. Franklin and A. Graesser. Is it an Agent, or just a Program?: A Taxonomy for Autonomous Agents. In *Intelligent Agents III (Proc. 3rd International Workshop on Agent Theories, Architectures, and Languages)*, volume 1193 of *LNCS*, pages 21–35. Springer, 1996.

[220] M. Fujita, S. Kono, H. Tanaka and T. Moto-Oka. Tokio: Logic Programming Language Based on Temporal Logic and its Compilation to Prolog. In *Proc. 3rd International Conference on Logic Programming (ICLP)*, volume 225 of *LNCS*, pages 695–709. Springer, 1986.

[221] D. Gabbay. Declarative Past and Imperative Future: Executable Temporal Logic for Interactive Systems. In *Proc. Colloquium on Temporal Logic in Specification*, volume 398 of *LNCS*, pages 402–450. Springer, 1987.

[222] D. Gabbay and F. Guenthner, editors. *Handbook of Philosophical Logic*, volume 165 of *Synthese Library*. Reidel, 1984. ISBN 9-027-71604-8.

[223] D. Gabbay, A. Pnueli, S. Shelah and J. Stavi. The Temporal Analysis of Fairness. In *Proc. 7th ACM Symposium on the Principles of Programming Languages (POPL)*, pages 163–173. ACM Press, 1980.

[224] D. Gabbay, I. Hodkinson and M. Reynolds. *Temporal Logic: Mathematical Foundations and Computational Aspects*, volume 1. Clarendon Press, Oxford, 1994.

[225] D. Gabbay, M. Reynolds and M. Finger. *Temporal Logic: Mathematical Foundations and Computational Aspects*, volume 2. Clarendon Press, Oxford, 2000.

[226] D. Gabbay, A. Kurucz, F. Wolter and M. Zakharyaschev. *Many-Dimensional Modal Logics: Theory and Applications*, volume 148 of *Studies in Logic and the Foundations of Mathematics*. Elsevier Science, 2003.

[227] D. Gabelaia, R. Kontchakov, A. Kurucz, F. Wolter and M. Zakharyaschev. Combining Spatial and Temporal Logics: Expressiveness vs. Complexity. *Journal of Artificial Intelligence Research*, 23:167–243, 2005.

[228] J. Gaintzarain, M. Hermo, P. Lucio, M. Navarro and F. Orejas. Dual Systems of Tableaux and Sequents for PLTL. *Journal of Logic and Algebraic Programming*, 78(8):701–722, 2009.

[229] P. Gardiner and C. Morgan. A Single Complete Rule for Data Refinement. *Formal Aspects of Computing*, 5(4):367–382, 1993.

[230] D. Gelernter and N. Carriero. Coordination Languages and their Significance. *Communications of the ACM*, 35(2):97–107, 1992.

[231] M. R. Genesereth and N. J. Nilsson. *Logical Foundations of Artificial Intelligence*. Morgan Kaufmann Publishers, Inc., Palo Alto, USA, 1987. ISBN 0-934-61331-1.

[232] A. Gerevini. Processing Qualitative Temporal Constraints. In Fisher *et al.* [210], pages 247–278. ISBN 0-444-51493-7.

[233] M. Gergatsoulis and C. Nomikos. A Proof Procedure For Temporal Logic Programming. *International Journal of Foundations of Computer Science*, 15(2):417–443, 2004.

[234] M. Gergatsoulis, P. Rondogiannis and T. Panayiotopoulos. Temporal Disjunctive Logic Programming. *New Generation Computing*, 19(1):87, 2000.

[235] R. Gerth, D. Peled, M. Y. Vardi and P. Wolper. Simple On-the-fly Automatic Verification of Linear Temporal Logic. In *Proc. 15th Workshop on Protocol Specification Testing and Verification (PSTV)*, pages 3–18. Chapman & Hall, 1995.

[236] D. Giannakopoulou and K. Havelund. Automata-Based Verification of Temporal Properties on Running Programs. In *Proc. 16th IEEE International Conference on Automated Software Engineering (ASE)*, pages 412–416. IEEE Computer Society, 2001.

[237] F. Giunchiglia and P. Traverso. Planning as Model Checking. In *Proc. 5th European Conference on Planning (ECP)*, volume 1809 of *LNCS*, pages 1–20. Springer, 2000.

[238] P. Godefroid. Using Partial Orders to Improve Automatic Verification Methods. In *Proc. 2nd International Workshop on Computer Aided Verification (CAV)*, volume 531 of *LNCS*, pages 176–185. Springer, 1990.

[239] A. Goldberg, K. Havelund and C. McGann. Runtime Verification for Autonomous Spacecraft Software. In *Proc. 2005 IEEE Aerospace Conference*. IEEE, 2005.

[240] V. Goranko and D. Shkatov. Tableau-Based Procedure for Deciding Satisfiability in the Full Coalitional Multiagent Epistemic Logic. In *Proc. International Symposium on Logical Foundations of Computer Science (LFCS)*, volume 5407 of *LNCS*, pages 197–213. Springer, 2009.

[241] V. Goranko, A. Montanari and G. Sciavicco. A General Tableau Method for Propositional Interval Temporal Logics. In *Proc. Workshop on Automated Reasoning with Analytic Tableaux and Related Methods (TABLEAUX)*, volume 2796 of *LNCS*, pages 102–116. Springer, 2003.

[242] V. Goranko, A. Montanari and G. Sciavicco. A Road Map of Interval Temporal Logics and Duration Calculi. *Journal of Applied Non-Classical Logics*, 14(1–2):9–54, 2004.

[243] V. Goranko, A. Kyrilov and D. Shkatov. Tableau Tool for Testing Satisfiability in LTL: Implementation and Experimental Analysis. *Electronic Notes in Theoretical Computer Science*, 262:113–125, 2010. (Proc. 6th Workshop on Methods for Modalities.).

[244] R. Goré. *Cut-Free Sequent Systems for Propositional Normal Modal Logics*. PhD thesis, Computer Laboratory, University of Cambridge, U.K., June 1992. (Technical report no. 257).

[245] G. D. Gough. Decision Procedures for Temporal Logic. Master's thesis, Department of Computer Science, University of Manchester, UK, 1984.

[246] C. Green. Application of Theorem Proving to Problem Solving. In *Proceedings of International Joint Conference on Artificial Intelligence (AI)*, 1969.

[247] O. Grumberg and H. Veith, editors. *25 Years of Model Checking – History, Achievements, Perspectives*, volume 5000 of *LNCS*, 2008. Springer. ISBN 978-3-540-69849-4.

[248] D. Guelev. A Complete Proof System for First-order Interval Temporal Logic with Projection. *Journal of Logic and Computation*, 14(2):215–249, 2004.

[249] D. Guelev and D. van Hung. A Relatively Complete Axiomatisation of Projection onto State in the Duration Calculus. *Journal of Applied Non-Classical Logics*, 14(1–2):149–180, 2004.

[250] C. Gönsel. *A Tableaux-Based Reasoner for Temporalised Description Logics*. PhD thesis, Department of Computer Science, University of Liverpool, UK, 2005.

[251] A. Gupta, C. Forgy, and A. Newell. High-Speed Implementations of Rule-Based Systems. *ACM Transactions on Computer Systems*, 7(2):119–146, 1989.

[252] S. Haack. *Philosophy of Logics*. Cambridge University Press, 1978. ISBN 0-521-29329-4.

[253] R. Hale and B. Moszkowski. Parallel Programming in Temporal Logic. In *Proc. Conference on Parallel Architectures and Languages, Europe (PARLE)*, volume 259 of *LNCS*. Springer, 1987.

[254] J. Y. Halpern and Y. Moses. A Guide to Completeness and Complexity for Modal Logics of Knowledge and Belief. *Artificial Intelligence*, 54:319–379, 1992.

[255] J. Y. Halpern and Y. Shoham. A Propositional Modal Logic of Time Intervals. *Journal of the ACM*, 38(4):935–962, 1991.

[256] J. Y. Halpern and M. Y. Vardi. The Complexity of Reasoning about Knowledge and Time. I. Lower Bounds. *Journal of Computer and System Sciences*, 38(1):195–237, 1989.

[257] J. Y. Halpern, Z. Manna and B. C. Moszkowski. A Hardware Semantics Based on Temporal Intervals. In *Proc. 10th International Colloquium on Automata, Languages and Programming (ICALP)*, volume 154 of *LNCS*, pages 278–291. Springer, 1983.

[258] J. Y. Halpern, R. van der Meyden and M. Y. Vardi. Complete Axiomatizations for Reasoning about Knowledge and Time. *SIAM Journal of Computing*, 33(3):674–703, 2004.

[259] H. Hansson and B. Jonsson. A Logic for Reasoning about Time and Reliability. *Formal Aspects of Computing*, 6:102–111, 1994.

[260] D. Harel and A. Pnueli. On the Development of Reactive Systems. Technical Report CS85-02, Department of Applied Mathematics, The Weizmann Institute of Science, Israel, 1985.

[261] K. Havelund. Runtime Verification of C Programs. In *Proc. 20th IFIP TC 6/WG 6.1 international conference on Testing of Software and Communicating Systems (TestCom/FATES)*, volume 5047 of *LNCS*, pages 7–22. Springer, 2008.

[262] K. Havelund and G. Rosu. Monitoring Programs Using Rewriting. In *Proc. 16th IEEE International Conference on Automated Software Engineering (ASE)*, pages 135–143. IEEE Computer Society Press, 2001.

[263] K. Havelund and G. Rosu. An Overview of the Runtime Verification Tool Java PathExplorer. *Formal Methods in System Design*, 24(2):189–215, 2004.

[264] P. J. Hayes. The Frame Problem and Related Problems in Artificial Intelligence. In B. L. Webber and N. J. Nilsson, editors, *Readings in Artificial Intelligence*, pages 223–230. Morgan Kaufmann, 1981.

[265] J. G. Henriksen and P. S. Thiagarajan. Dynamic Linear Time Temporal Logic. *Annals of Pure and Applied Logic*, 96(1–3):187–207, 1999.

[266] T. Henzinger. It's About Time: Real-Time Logics Reviewed. In *Proc. 9th International Conference on Concurrency Theory (CONCUR)*, volume 1466 of *LNCS*, pages 439–454. Springer, 1998.

[267] T. Henzinger, P.-H. Ho and H. Wong-Toi. HYTECH: A Model Checker for Hybrid Systems. *International Journal on Software Tools for Technology Transfer*, 1(1–2):110–122, 1997.

[268] A. Hepple. Java implementation of Concurrent METATEM, 2010. http://www.csc.liv.ac.uk/~anthony/metatem.html.

[269] A. Hepple, L. A. Dennis and M. Fisher. A Common Basis for Agent Organisations in BDI Languages. In *Proc. International Workshop on LAnguages, methodologies and Development tools for multi-agent systemS (LADS)*, volume 5118 of *LNAI*, pages 171–88. Springer, 2008.

[270] B. Hirsch and U. Hustadt. Translating PLTL into WS1S: Application Description. In *Proc. Methods for Modalities (M4M) II*, University of Amsterdam, Netherlands, 2001.

[271] B. Hirsch, M. Fisher, C. Ghidini and P. Busetta. Organising Software in Active Environments. In *Proc. 5th International Workshop on Computational Logic in Multi-Agent Systems (CLIMA)*, volume 3487 of *LNCS*, pages 265–280. Springer, 2005.

[272] R. Hirsch. From Points to Intervals. *Journal of Applied Non-Classical Logics*, 4(1):7–27, 1994.

[273] R. Hirsch. Relation Algebras of Intervals. *Artificial Intelligence*, 83:1–29, 1996.

[274] C. A. R. Hoare. An Axiomatic Basis for Computer Programming. *Communications of the ACM*, 12(10):576–583, 1969.

[275] C. A. R. Hoare. Communicating Sequential Processes. *Communications of the ACM*, 21(8):666–677, 1978.

[276] C. A. R. Hoare. The Ideal of Verified Software. In *Proc. 18th International Conference on Computer Aided Verification (CAV)*, volume 4144 of *LNCS*, pages 5–16. Springer, 2006.

[277] C. A. R. Hoare, J. Misra, G. T. Leavens and N. Shankar. The Verified Software Initiative: A Manifesto. *ACM Computer Surveys*, 41(4), 2009.

[278] I. Hodkinson. Monodic Packed Fragment with Equality is Decidable. *Studia Logica*, 72:185–197, 2002.

[279] I. Hodkinson and M. Reynolds. Temporal Logic. In P. Blackburn, J. van Benthem and F. Wolter, editors, *Handbook of Modal Logic*, chapter 11. Elsevier, 2006.

[280] I. Hodkinson, F. Wolter and M. Zakharyashev. Decidable Fragments of First-Order Temporal Logics. *Annals of Pure and Applied Logic*, 106(1–3):85–134, 2000.

[281] G. J. Holzmann. PAN – A Protocol Specification Analyzer. Technical Memorandum TM81-11271-5, AT&T Bell Laboratories, USA, 1981.

[282] G. J. Holzmann. *Design and Validation of Computer Protocols*. Prentice-Hall, 1991. ISBN 0-135-39925-4.

[283] G. J. Holzmann. The Model Checker Spin. *IEEE Transactions on Software Engineering*, 23(5):279–295, 1997.

[284] G. J. Holzmann. Logic Verification of ANSI-C Code with SPIN. In *Proc. 7th International SPIN Workshop on Model Checking of Software (SPIN)*, volume 1885 of *LNCS*, pages 131–147. Springer, 2000.

[285] G. J. Holzmann. *The Spin Model Checker: Primer and Reference Manual*. Addison-Wesley, 2003. ISBN 0-321-22862-6.

[286] G. J. Holzmann and D. Peled. An Improvement in Formal Verification. In *Proc. 7th IFIP WG 6.1 International Conference on Formal Description Techniques (FORTE)*, volume 6 of *IFIP Conference Proceedings*, pages 109–124, Berne, Switzerland, 1994. Chapman & Hall.

[287] G. J. Holzmann and M. H. Smith. Software Model Checking. In *Proc. Formal Description Techniques (FORTE)*, pages 481–497, 1999.

[288] G. J. Holzmann and M. H. Smith. A Practical Method for Verifying Event-Driven Software. In *Proc. International Conference on Software Engineering (ICSE)*, pages 597–607, 1999.

[289] J. E. Hopcroft and J. D. Ullman. *Introduction to Automata Theory, Languages, and Computation*. Addison-Wesley, 1979. ISBN 0-201-02988-X.

[290] J. F. Hübner, J. S. Sichman and O. Boissier. A Model for the Structural, Functional, and Deontic Specification of Organizations in Multiagent Systems. In *Proc. 16th Brazilian Symposium on Artificial Intelligence (SBIA)*, volume 2507 of *LNAI*, pages 118–128. Springer, 2002.

[291] G. E. Hughes and M. J. Cresswell. *An Introduction to Modal Logic*. Methuen (UP), 1968. ISBN 0-416-29460-X.

[292] G. E. Hughes and M. J. Cresswell. *A Companion to Modal Logic*. Methuen (UP), 1984. ISBN 0-416-37510-3.

[293] U. Hustadt and B. Konev. TRP++ 2.0: A Temporal Resolution Prover. In *Proc. 19th International Conference on Automated Deduction (CADE)*, volume 2741 of *LNAI*, pages 274–278. Springer, 2003.

[294] U. Hustadt and R. Schmidt. MSPASS: Modal reasoning by translation and first-order resolution. In *Proc. International Conference on Automated Reasoning with Analytic Tableaux and Related Methods (TABLEAUX)*, volume 1847 of *LNAI*, pages 67–71. Springer, 2000. ISBN 3-540-67697-X.

[295] U. Hustadt and R. A. Schmidt. Formulae which Highlight Differences between Temporal Logic and Dynamic Logic Provers. In E. Giunchiglia and F. Massacci, editors, *Issues in the Design and Experimental Evaluation of Systems for Modal and Temporal Logics*, Technical Report DII 14/01, pages 68–76. Dipartimento di Ingegneria dell'Informazione, Unversitá degli Studi di Siena, Italy, 2001.

[296] U. Hustadt and R. A. Schmidt. Scientific Benchmarking with Temporal Logic Decision Procedures. In *Proc. 8th International Conference on Principles of Knowledge Representation and Reasoning (KR)*, pages 533–544. Morgan Kaufmann, 2002.

[297] U. Hustadt, C. Dixon, R. A. Schmidt and M. Fisher. Normal Forms and Proofs in Combined Modal and Temporal Logics. In *Proc. 3rd International Workshop on Frontiers of Combining Systems (FroCoS)*, volume 1794 of *LNAI*. Springer, 2000.

[298] U. Hustadt, B. Konev, A. Riazanov and A. Voronkov. TeMP: A Temporal Monodic Prover. In *Proc. 2nd International Joint Conference on Automated Reasoning (IJCAR)*, volume 3097 of *LNAI*, pages 326–330. Springer, 2004. ISBN 3-540-22345-2.

[299] M. Huth and M. Ryan. *Logic in Computer Science: Modelling and Reasoning about Systems*. Cambridge University Press, 2004. ISBN 0-521-54310-X. (2nd Edition).

[300] HyTech: The HYbrid TECHnology Tool. http://embedded.eecs.berkeley.edu/research/hytech.

[301] J. Cohen and A. Slissenko. On Refinements of Timed Abstract State Machines. In R. Moreno-Díaz and A. Quesada-Arencibia, editors, *Formal Methods and Tools for Computer Science (Proc. Eurocast'01)*, pages 247–250, Canary Islands, Spain, February 2001. (Universidad de Las Palmas de Gran Canaria.).

[302] Java PathFinder. http://javapathfinder.sourceforge.net.

[303] C. W. Johnson and M. D. Harrison. Using Temporal Logic to Support the Specification and Prototyping of Interactive Control Systems. *International Journal of Man-Machine Studies*, 36(3):357–385, 1992.

[304] C. B. Jones. *Systematic Software Development Using VDM*. Prentice Hall International, 1986.

[305] S.-S. T. Q. Jongmans, K. V. Hindriks and M. B. van Riemsdijk. Model Checking Agent Programs by Using the Program Interpreter. In *Proc. 11th International Workshop on Computational Logic in Multi-Agent Systems (CLIMA)*, volume 6245 of *LNCS*, pages 219–237. Springer, 2010.

[306] B. Jonsson and K. G. Larsen. Specification and Refinement of Probabilistic Processes. In *Proc. 6th Annual IEEE Symposium on Logic in Computer Science (LICS)*, pages 266–277. IEEE Computer Society Press, 1991.

[307] M. Kacprzak, A. Lomuscio and W. Penczek. From Bounded to Unbounded Model Checking for Temporal Epistemic Logic. *Fundamentae Informatica*, 63(2–3):221–240, 2004.

[308] M. Kacprzak, W. Nabialek, A. Niewiadomski, W. Penczek, A. Pólrola, M. Szreter, B. Wozna and A. Zbrzezny. VerICS 2007 – a Model Checker for Knowledge and Real-Time. *Fundamenta Informaticae*, 85(1–4):313–328, 2008.

[309] J. A. W. Kamp. *Tense Logic and the Theory of Linear Order*. PhD thesis, University of California, USA, 1968.

[310] S. Katz and D. Peled. Interleaving Set Temporal Logic. *Theoretical Computer Science*, 75(3):263–287, 1990.

[311] A. Kellett and M. Fisher. Coordinating Heterogeneous Components Using Executable Temporal Logic. In J.-J. Meyer and J. Treur, editors, *Agents, Reasoning and Dynamics*, volume 6 of *Handbooks in Defeasible Reasoning and Uncertainty Management Systems*. Kluwer, 2001.

[312] B. W. Kernighan and D. Ritchie. *The C Programming Language, Second Edition*. Prentice-Hall, 1988. ISBN 0-131-10370-9.

[313] Y. Kesten, Z. Manna and A. Pnueli. Temporal Verification of Simulation and Refinement. In *A Decade of Concurrency*, volume 803 of *LNCS*, pages 273–346. Springer, 1994.

[314] P. King. Towards an ITL Based Formalism for Expressing Temporal Constraints in Multimedia Documents. In *Proc. IJCAI Workshop on Executable Temporal Logics*, Montreal, Canada, 1995.

[315] M. Kloetzer and C. Belta. Temporal Logic Planning and Control of Robotic Swarms by Hierarchical Abstractions. *IEEE Transactions on Robotics*, 23(2):320–330, 2007.

[316] M. Kloetzer and C. Belta. Distributed Implementations of Global Temporal Logic Motion Specifications. In *Proc. IEEE International Conference on Robotics and Automation (ICRA)*, pages 393–398. IEEE, 2008.

[317] M. Kloetzer and C. Belta. A Fully Automated Framework for Control of Linear Systems From Temporal Logic Specifications. *IEEE Transactions on Automatic Control*, 53(1):287–297, 2008.

[318] B. Konev, A. Degtyarev, C. Dixon, M. Fisher and U. Hustadt. Mechanising First-Order Temporal Resolution. *Information and Computation*, 199(1–2):55–86, 2005.

[319] S. Kono. A Combination of Clausal and Non-Clausal Temporal Logic Programs. In *Executable Modal and Temporal Logics*, volume 897 of *LNAI*. Springer, 1995.

[320] R. Kowalski. *Logic for Problem Solving*. North Holland, 1979. ISBN 0-444-00368-1.

[321] R. Kowalski. Algorithm = Logic + Control. *Communications of the ACM*, 22(7):424–436, 1979.

[322] R. Koymans. Specifying Message Passing Systems Requires Extending Temporal Logic. In *Proc. Temporal Logic in Specification*, volume 398 of *LNCS*, pages 213–223. Springer, 1987. ISBN 3-540-51803-7.

[323] J. Krämer and B. Seeger. A Temporal Foundation for Continuous Queries over Data Streams. In *Proc. 11th International Conference on Management of Data (COMAD)*, pages 70–82. Computer Society of India, 2005.

[324] H. Kress-Gazit, G. E. Fainekos and G. J. Pappas. Temporal-Logic-Based Reactive Mission and Motion Planning. *IEEE Transactions on Robotics*, 25(6):1370–1381, 2009.

[325] S. Kripke. Semantical Considerations on Modal Logic. *Acta Philosophica Fennica*, 16:83–94, 1963.

[326] F. Kröger. *Temporal Logic of Programs*. Monographs on Theoretical Computer Science. Springer, 1987. ISBN 3-540-17030-8.

[327] F. Kröger and S. Merz. *Temporal Logic and State Systems*. Springer, 2008. ISBN 3-540-67401-2.

[328] Kronos Tool. http://www-verimag.imag.fr/TEMPORISE/kronos.

[329] V. Kumar and Y.-J. Lin. An Intelligent Backtracking Scheme for Prolog. In *Proc. International Symposium on Logic Programming (ISLP)*, pages 406–414. IEEE Computer Society Press, 1987.

[330] A. Kurucz. Combining Modal Logics. In J. van Benthem, P. Blackburn and F. Wolter, editors, *Handbook of Modal Logic*, volume 3 of *Studies in Logic and Practical Reasoning*, pages 869–924. Elsevier, 2007.

[331] M. Z. Kwiatkowska, G. Norman and D. Parker. Probabilistic Symbolic Model Checking with PRISM: A Hybrid Approach. In *Proc. 8th International Conference on Tools and Algorithms for the Construction and Analysis of Systems (TACAS)*, volume 2280 of *LNCS*, pages 52–66. Springer, 2002.

[332] P. Ladkin. The Completeness of a Natural System for Reasoning with Time Intervals. In *Proc. 10th International Joint Conference on Artificial Intelligence (IJCAI)*. Morgan Kaufmann, 1987.

[333] P. Ladkin and R. Maddux. On Binary Constraint Problems. *Journal of the ACM*, 41:435–469, 1994.

[334] L. Lamport. Proving the Correctness of Multiprocess Programs. *IEEE Transactions on Software Engineering*, 3(2):125–143, 1977.

[335] L. Lamport. Sometime is Sometimes Not Never. In *Proc. 7th ACM Symposium on the Principles of Programming Languages (POPL)*, pages 174–185. ACM Press, 1980.

[336] L. Lamport. What Good is Temporal Logic? In *Proc. International Federation for Information Processing*, pages 657–668, 1983.

[337] L. Lamport. Specifying Concurrent Program Modules. *ACM Transactions on Programming Languages and Systems*, 5(2):190–222, Apr. 1983.

[338] L. Lamport. The Temporal Logic of Actions. *ACM Transactions on Programming Languages and Systems*, 16(3):872–923, 1994.

[339] L. Lamport. *Specifying Systems: The TLA+ Language and Tools for Hardware and Software Engineers*. Addison Wesley Professional, 2003. ISBN 0-321-14306-X.

[340] K. Lano and H. Haughton. *Specification in B: An Introduction Using the B Toolkit*. World Scientific Publishing, 2000. ISBN 1-860-94008-0.

[341] F. Laroussinie, K. G. Larsen and C. Weise. From Timed Automata to Logic – and Back. In *Proc. 20th International Symposium on Mathematical Foundations of Computer Science (MFCS)*, volume 969 of *LNCS*, pages 529–539. Springer, 1995. ISBN 3-540-60246-1.

[342] K. G. Larsen, P. Pettersson and W. Yi. Model-Checking for Real-Time Systems. In *Proc. Conference on Fundamentals of Computation Theory*, volume 965 of *LNCS*, pages 62–88, 1995.

[343] S. Laxman and P. S. Sastry. A Survey of Temporal Data Mining. *SADHANA, Academy Proceedings in Engineering Sciences*, 31:173–198, 2006.

[344] X. Leroy. Formal Verification of a Realistic Compiler. *Communications of the ACM*, 52(7):107–115, 2009.

[345] M. Leucker and C. Schallhart. A Brief Account of Runtime Verification. *Journal of Logic and Algebraic Programming*, 78(5):293–303, 2009.

[346] O. Lichtenstein and A. Pnueli. Propositional Temporal Logics: Decidability and Completeness. *Logic Journal of the IGPL*, 8(1), 2000.

[347] O. Lichtenstein, A. Pnueli and L. D. Zuck. The Glory of the Past. In *Proc. Conference on Logics of Programs*, volume 193 of *LNCS*, pages 196–218. Springer, 1985.

[348] V. Lifschitz. On the semantics of STRIPS. In M. P. Georgeff and A. L. Lansky, editors, *Reasoning About Actions & Plans – Proceedings of the 1986 Workshop*, pages 1–10. Morgan Kaufmann Publishers, 1986.

[349] G. Ligozat. Weak Representation of Interval Algebras. In *Proc. 8th American National Conference on Artificial Intelligence (AAAI)*, pages 715–720, 1990.

[350] G. Ligozat. Tractable Relations in Temporal Reasoning: Pre-Convex Relations. In *Proc. ECAI Workshop on Spatial and Temporal Reasoning*, August 1994.

[351] C. Liu, M. A. Orgun and K. Zhang. A Parallel Execution Model for Chronolog. *Computer Systems: Science & Engineering*, 16(4):215–228, 2001.

[352] Logics Workbench. http://www.lwb.unibe.ch.

[353] A. Lomuscio and F. Raimondi. MCMAS: A Model Checker for Multi-agent Systems. In *Proc. 12th International Conference on Tools and Algorithms for the Construction and Analysis of Systems (TACAS)*, volume 3920 of *LNCS*, pages 450–454. Springer, 2006.

[354] A. Lomuscio, H. Qu and F. Raimondi. MCMAS: A Model Checker for the Verification of Multi-Agent Systems. In *Proc. 21st International Conference on Computer Aided Verification (CAV)*, volume 5643 of *LNCS*, pages 682–688. Springer, 2009.

[355] LTL Tableau. http://msit.wits.ac.za/ltltableau.

[356] M. Ludwig and U. Hustadt. Fair Derivations in Monodic Temporal Reasoning. In *Proc. 22nd International Conference on Automated Deduction (CADE)*, volume 5663 of *LNCS*, pages 261–276. Springer, 2009.

[357] M. Ludwig and U. Hustadt. Implementing a Fair Monodic Temporal Logic Prover. *AI Communications*, 23(2-3):69–96, 2010.

[358] J. Ma, B. Knight and T. Peng. Temporal Reasoning about Action and Change. In K. Anjaneyulu, M. Sasikumar and S. Ramani, editors, *Knowledge Based Computer Systems – Research and Applications*, pages 193–204. Narosa Publishing House, 1996.

[359] S. MacLane and D. Siefkes, editors. *The Collected Works of J Richard Büchi*. Springer-Verlag, 1990.

[360] Z. Manna and A. Pnueli. How to Cook a Temporal Proof System for your Pet Language. In *Proc. 10th ACM Symposium on the Principles of Programming Languages*, pages 141–154. ACM Press, 1983.

[361] Z. Manna and A. Pnueli. A Hierarchy of Temporal Properties. In *Proc. ACM Symposium on Principles of Distributed Computing (PODC)*, pages 377–410. ACM Press, 1990.

[362] Z. Manna and A. Pnueli. Completing the Temporal Picture. *Theoretical Computer Science*, 83(1):91–130, 1991.

[363] Z. Manna and A. Pnueli. *The Temporal Logic of Reactive and Concurrent Systems: Specification*. Springer, 1992. ISBN 0-387-97664-7.

[364] Z. Manna and A. Pnueli. *Temporal Verification of Reactive Systems: Safety*. Springer, 1995. ISBN 0-387-94459-1.

[365] Z. Manna and the STeP group. STeP: Deductive–Algorithmic Verification of Reactive and Real-Time Systems. In *Proc. International Conference on Computer Aided Verification (CAV)*, volume 1102 of *LNCS*. Springer, 1996.

[366] N. Markey. Temporal Logic with Past is Exponentially More Succinct. *EATCS Bulletin*, 79: 122–128, 2003. See also: http://www.lsv.ens-cachan.fr/~markey/PLTL.php.

[367] T. Maruichi, M. Ichikawa and M. Tokoro. Modelling Autonomous Agents and their Groups. In Y. Demazeau and J. P. Müller, editors, *Decentralized AI 2 – Proceedings of the 2nd European Workshop on Modelling Autonomous Agents and Multi-Agent Worlds (MAAMAW '90)*. Elsevier/North Holland, 1991.

[368] M. Marx and M. Reynolds. Undecidability of Compass Logic. *Journal of Logic and Computation*, 9(6):897–914, 1999.

[369] M. C. Mayer, C. Limongelli, A. Orlandini and V. Poggioni. Linear Temporal Logic as an Executable Semantics for Planning Languages. *Journal of Logic, Language and Information*, 16(1):63–89, 2007.

[370] W. McCune. OTTER *3.0 Reference Manual and Guide*. Argonne National Laboratory, Illinois, USA, 1994.

[371] MCMAS – a Model Checker for Multi-Agents Systems. http://www-lai.doc.ic.ac.uk/mcmas.

[372] K. L. McMillan. *Symbolic Model Checking*. Kluwer Academic Publishers, 1993. ISBN 0-792-39380-5.

[373] S. Merz. Decidability and Incompleteness Results for First-Order Temporal Logic of Linear Time. *Journal of Applied Non-Classical Logics*, 2:139–156, 1992.

[374] M. Michel. Algèbre de Machines et Logique Temporelle. In *Proc. Symposium on Theoretical Aspects of Computer Science (STACS)*, volume 166 of *LNCS*, pages 287–298, 1984.

[375] A. Mili, J. Desharnais and J. R. Gagné. Formal Models of Stepwise Refinements of Programs. *ACM Computing Surveys*, 18(3):231–276, 1986. ISSN 0360-0300.

[376] R. Milner. *A Calculus of Communicating Systems*, volume 92 of *LNCS*. Springer, 1980. ISBN 0-387-10235-3.

[377] Mocha: Exploiting Modularity in Model Checking. http://www.cis.upenn.edu/~mocha.

[378] A. Montanari. *Metric and Layered Temporal Logic for Time Granularity*. PhD thesis, University of Amsterdam, The Netherlands, 1996. (ILLC Dissertation Series 1996-02).

[379] R. C. Moore. *Logic and Representation (CSLI Lecture Notes Number 39)*. Center for the Study of Language and Information, Ventura Hall, Stanford, CA 94305, 1995. (Distributed by Chicago University Press).

[380] C. C. Morgan, K. A. Robinson and P. H. B. Gardiner. *On the Refinement Calculus*. Technical Monograph PRG-70, Oxford University Computing Laboratory, Oxford, UK, 1988.

[381] B. Moszkowski. *Reasoning about Digital Circuits*. PhD thesis, Computer Science Department, Stanford University, USA, 1983.

[382] B. Moszkowski. *Executing Temporal Logic Programs*. Cambridge University Press, Cambridge, U.K., 1986. ISBN 0-521-31099-7.

[383] B. Moszkowski. Some Very Compositional Temporal Properties. In *Proc. IFIP TC2/WG2.1/WG2.2/WG2.3 Working Conference on Programming Concepts, Methods and Calculi*, pages 307–326. North-Holland, 1994.

[384] B. Moszkowski. Compositional Reasoning about Projected and Infinite Time. In *Proc. 1st IEEE International Conference on Engineering of Complex Computer Systems (ICECCS)*, pages 238–245. IEEE Computer Society Press, 1995.

[385] B. Moszkowski. Compositional Reasoning Using Interval Temporal Logic and Tempura. In *Compositionality: The Significant Difference*, volume 1536 of *Lecture Notes in Computer Science*, pages 439–464. Springer, 1998.

[386] B. Moszkowski and Z. Manna. Reasoning in Interval Temporal Logic. In *Proc. AMC/NSF/ONR Workshop on Logics of Programs*, volume 164 of *LNCS*, pages 371–383. Springer, 1984.

[387] S. Nain and M. Y. Vardi. Branching vs. Linear Time: Semantical Perspective. In *Proc. 5th International Symposium on Automated Technology for Verification and Analysis (ATVA)*, volume 4762 of *LNCS*, pages 19–34. Springer, 2007.

[388] S. Nain and M. Y. Vardi. Trace Semantics is Fully Abstract. In *Proc. 24th Annual IEEE Symposium on Logic in Computer Science (LICS)*, pages 59–68. IEEE Computer Society, 2009.

[389] C. Nalon and C. Dixon. Clausal Resolution for Normal Modal Logics. *Journal of Algorithms*, 62(3–4):117–134, 2007.

[390] R. M. Needham and M. D. Schroeder. Using Encryption for Authentication in Large Networks of Computers. *Communications of the ACM*, 21(12), Dec. 1978.

[391] U. Nilsson and Małuszyński. *Logic, Programming and Prolog*. John Wiley and Sons, 1990. ISBN 0-471-92625-6.

[392] A. Nonnengart. Resolution-Based Calculi for Modal and Temporal Logics. In *Proc. 13th International Conference on Automated Deduction (CADE)*, volume 1104 of *LNAI*, pages 598–612. Springer, 1996.

[393] NuSMV: A New Symbolic Model Checker. http://nusmv.irst.itc.it.

[394] H. J. Ohlbach. A Resolution Calculus for Modal Logics. In *Proc. 9th International Conference on Automated Deduction (CADE)*, volume 310 of *LNCS*, pages 500–516. Springer, 1988.

[395] H.-J. Ohlbach and D. Gabbay. Calendar Logic. *Journal of Applied Non-Classical Logics*, 8(4), 1998.

[396] M. A. Orgun. Foundations of Linear-Time Logic Programming. *International Journal of Computer Mathematics*, 58(3):199–219, 1995.

[397] M. A. Orgun and W. Wadge. Theory and Practice of Temporal Logic Programming. In L. F. del Cerro and M. Penttonen, editors, *Intensional Logics for Programming*. Oxford University Press, 1992.

342   REFERENCES

[398] M. A. Orgun and W. W. Wadge. Towards a Unified Theory of Intensional Logic Programming. *Journal of Logic Programming*, 13(1–4):413–440, 1992.

[399] J. Ostroff. Temporal Logic of Real-Time Systems. *Research Studies Press*, 1990.

[400] J. Ostroff. Formal Methods for the Specification and Design of Real-Time Safety Critical Systems. *Journal of Systems and Software*, 18(1):33–60, 1992.

[401] B. Paech. Gentzen-Systems for Propositional Temporal Logics. In *Proc. 2nd Workshop on Computer Science Logic (CSL)*, volume 385 of *LNCS*, pages 240–253. Springer, 1989.

[402] A. Paschke. A Homogenous Reaction Rule Language for Complex Event Processing. In *Proc. 2nd International Workshop on Event Drive Architecture and Event Processing Systems (EDA-PS)*, 2007.

[403] D. Peled. Combining Partial Order Reductions with On-the-fly Model-Checking. In *Proc. 6th International Conference on Computer Aided Verification (CAV)*, volume 818 of *LNCS*, pages 377–390. Springer, 1994.

[404] W. Penczek. Axiomatizations of Temporal Logics on Trace Systems. *Fundamenta Informaticae*, 25(2):183–200, 1996.

[405] W. Penczek and A. Polrola. *Advances in Verification of Time Petri Nets and Timed Automata: A Temporal Logic Approach*, volume 20 of *Studies in Computational Intelligence*. Springer, 2006. ISBN 3-540-32869-6.

[406] N. Piterman, A. Pnueli, and Y. Sa'ar. Synthesis of Reactive(1) Designs. In *Proc. 7th International Conference on Verification, Model Checking, and Abstract Interpretation (VMCAI)*, volume 3855 of *LNCS*, pages 364–380. Springer, 2006.

[407] D. A. Plaisted and S. A. Greenbaum. A Structure-Preserving Clause Form Translation. *Journal of Symbolic Computation*, 2(3):293–304, 1986.

[408] R. Pliuskevicius. On an *Omega*-Decidable Deductive Procedure for Non-Horn Sequents of a Restricted FTL. In *Proc. 1st International Conference on Computational Logic (CL)*, volume 1861 of *LNCS*, pages 523–537. Springer, 2000.

[409] R. Pliuskevicius. Deduction-Based Decision Procedure for a Clausal Miniscoped Fragment of FTL. In *Proc. 1st International Joint Conference on Automated Reasoning (IJCAR)*, volume 2083 of *LNCS*, pages 107–120. Springer, 2001.

[410] G. D. Plotkin. *A Structural Approach to Operational Semantics*. Technical Report DAIMI FN-19, Computer Science Department, Aarhus University, Denmark, 1981.

[411] A. Pnueli. The Temporal Logic of Programs. In *Proc, 18th Symposium on the Foundations of Computer Science (FOCS)*, pages 46–57. IEEE Computer Society Press, 1977.

[412] A. Pnueli. The Temporal Semantics of Concurrent Programs. *Theoretical Computer Science*, 13:45–60, 1981.

[413] A. Pnueli. Linear and Branching Structures in the Semantics and Logics of Reactive Systems. In *Proc. 12th International Colloquium on Automata, Languages and Programming (ICALP)*, volume 194 of *LNCS*, pages 15–32. Springer, 1985.

[414] A. Pnueli. Applications of Temporal Logic to the Specification and Verification of Reactive Systems: A Survey of Current Trends. In *Current Trends in Concurrency*, volume 224 of *LNCS*, pages 510–584. Springer, 1986.

[415] A. Pnueli and R. Rosner. On the Synthesis of a Reactive Module. In *Proc. 16th ACM Symposium on the Principles of Programming Languages (POPL)*, pages 179–190. ACM Press, 1989.

[416] A. Pnueli and R. Rosner. On the Synthesis of an Asynchronous Reactive Module. In *Proc. 16th International Colloquium on Automata, Languages and Programming (ICALP)*, volume 372 of *LNCS*, pages 652–671. Springer, 1989.

[417] M. R. Prasad, A. Biere, and A. Gupta. A Survey of Recent Advances in SAT-based Formal Verification. *International Journal on Software Tools for Technology Transfer*, 7(2):156–173, 2005.

[418] V. R. Pratt. Semantical Considerations on Floyd-Hoare Logic. In *Proc. 17th IEEE Symposium on Foundations of Computer Science (FOCS)*, pages 109–121, 1976.

[419] M. Prietula, K. Carley, and L. Gasser, editors. *Simulating Organizations: Computational Models of Institutions and Groups*. MIT Press, 1998. ISBN 0-262-66108-X.

[420] A. Prior. *Past, Present and Future*. Oxford University Press, 1967. ISBN 0-198-24311-1.

[421] PRISM: Probabilistic Symbolic Model Checker. http://www.cs.bham.ac.uk/~dxp/prism.

[422] G. Puppis. *Automata for Branching and Layered Temporal Structures*. PhD thesis, Department of Mathematics and Computer Science, University of Udine, Italy, 2006.

[423] D. V. Pynadath, M. Tambe, N. Chauvat and L. Cavedon. Towards Team-Oriented Programming. In *Intelligent Agents VI (Proc. 6th International Workshop on Agent Theories, Architectures, and Languages)*, volume 1757 of *LNAI*, pages 233–247. Springer, 1999.

[424] J.-P. Queille and J. Sifakis. Specification and Verification of Concurrent Systems in CESAR. In *Proc. 5th International Symposium on Programming*, volume 137 of *LNCS*, pages 337–351. Springer, 1982.

[425] Rabbit Timed Automata. http://www-sst.informatik.tu-cottbus.de/~db/Rabbit.

[426] F. Raimondi and A. Lomuscio. Automatic Verification of Multi-agent Systems by Model Checking via Ordered Binary Decision Diagrams. *Journal of Applied Logic*, 5(2):235–251, 2007.

[427] A. Rao. AgentSpeak(L): BDI Agents Speak Out in a Logical Computable Language. In *Agents Breaking Away: Proc. 7th European Workshop on Modelling Autonomous Agents in a Multi-Agent World (MAAMAW)*, volume 1038 of *LNCS*, pages 42–55. Springer, 1996.

[428] A. S. Rao and M. Georgeff. BDI Agents: from theory to practice. In *Proc. 1st International Conference on Multi-Agent Systems (ICMAS)*, pages 312–319, San Francisco, USA, 1995.

[429] A. S. Rao and M. P. Georgeff. Modeling Agents within a BDI-Architecture. In *Proc. 2nd International Conference on Principles of Knowledge Representation and Reasoning (KR)*, pages 473–484. Morgan Kaufmann, 1991.

[430] RATSY – Requirements Analysis Tool with Synthesis. http://rat.fbk.eu/ratsy.

[431] RED (Region Encoding Diagram) System. http://cc.ee.ntu.edu.tw/~farn/red.

[432] R. Reiter. The Frame Problem in the Situation Calculus: A Simple Solution (sometimes) and a Completeness Result for Goal Regression. In V. Lifschitz, editor, *Artificial Intelligence and Mathematical Theory of Computation: Papers in Honor of John McCarthy*, pages 359–380. Academic Press, 1991.

[433] N. Rescher and A. Urquart. *Temporal logic*. Springer, 1971. ISBN 3-211-80995-3.

[434] M. Reynolds. Axiomatisation and Decidability of F and P in Cyclical Time. *Journal of Philosophical Logic*, 23:197–224, 1994.

[435] M. Reynolds. An Axiomatization of Full Computation Tree Logic. *Journal of Symbolic Logic*, 66(3):1011–1057, 2001.

[436] A. Riazanov and A. Voronkov. Vampire 1.1 (System Description). In *Proc. 1st International Joint Conference on Automated Reasoning (IJCAR)*, volume 2083 of *LNCS*, pages 376–380. Springer, 2001.

[437] J. A. Robinson. A Machine Based Logic Based on the Resolution Principle. *Journal of the ACM*, 12(1):23–41, 1965.

[438] R. Rosner and A. Pnueli. A Choppy Logic. In *Proc. IEEE Symposium on Logic in Computer Science (LICS)*, pages 306–313. IEEE Computer Society, 1986.

[439] F. Rossi, P. Van Beek and T. Walsh. Constraint Programming. In van Harmelen *et al.* [481].

[440] G. Rosu and K. Havelund. Rewriting-Based Techniques for Runtime Verification. *Automated Software Engineering*, 12(2):151–197, 2005.

[441] M. Sabbadin and A. Zanardo. Topological Aspects of Branching-Time Semantics. *Studia Logica*, 75(3):271–286, 2003.

[442] S. Safra and M. Y. Vardi. On omega-Automata and Temporal Logic (Preliminary Report). In *Proc. 21st Annual ACM Symposium on Theory of Computing (STOC)*, pages 127–137. ACM Press, 1989.

[443] S. Schewe and B. Finkbeiner. Bounded Synthesis. In *Proc. 5th International Symposium on Automated Technology for Verification and Analysis (ATVA)*, volume 4762 of *LNCS*, pages 474–488. Springer, 2007.

[444] D. A. Schmidt. *Denotational Semantics*. Allyn & Bacon, 1986. ISBN 0-205-08974-7.

[445] R. A. Schmidt and U. Hustadt. The Axiomatic Translation Principle for Modal Logic. *ACM Transactions on Computational Logic*, 8(4):1–55, 2007.

[446] S. Schneider. *The B-Method: An Introduction*. Palgrave, 2001. ISBN 0-333-79284-X.

[447] P.-Y. Schobbens, J.-F. Raskin, and T. Henzinger. Axioms for Real-Time Logics. *Theoretical Computer Science*, 274(1-2):151–182, 2002.

[448] E. Schwalb and L. Vila. Temporal Constraints: A Survey. *Constraints*, 3(2/3):129–149, 1998.

[449] R. L. Schwartz, P. M. Melliar-Smith and F. H. Vogt. An Interval-Based Temporal Logic. In *Proc. Workshop on Logics of Programs*, volume 164 of *LNCS*, pages 443-457, 1983.

[450] S. Schwendimann. A New One-Pass Tableau Calculus for PLTL. In *Proc. Workshop on Automated Reasoning with Analytic Tableaux and Related Methods (TABLEAUX)*, volume 1397 of *LNCS*, pages 277–291. Springer, 1998.

[451] R. Sebastiani and S. Tonetta. 'More Deterministic' vs. 'Smaller' Büchi Automata for Efficient LTL Model Checking. In *Proc. 12th Advanced Research Working Conference on Correct Hardware Design and Verification Methods (CHARME)*, volume 2860 of *LNCS*, pages 126–140. Springer, 2003.

[452] M. Shanahan. *Solving the Frame Problem*. MIT Press, 1997. ISBN 0-262-19384-1.

[453] A. P. Sistla and E. M. Clarke. Complexity of Propositional Linear Temporal Logics. *Journal of the ACM*, 32(3):733–749, 1985.

[454] A. P. Sistla, M. Vardi and P. Wolper. The Complementation Problem for Büchi Automata with Applications to Temporal Logic. *Theoretical Computer Science*, 49:217–237, 1987.

[455] F. Somenzi and R. Bloem. Efficient Büchi Automata for LTL Formulae. In *Proc. 12th International Conference on Computer Aided Verification (CAV)*, volume 1855 of *LNCS*, pages 247–263. Springer, 2000.

[456] SPASS: An Automated Theorem Prover for First-Order Logic with Equality. http://www.spass-prover.org.

[457] Spin System. http://spinroot.com.

[458] R. M. Stallman and G. J. Sussman. Forward Reasoning and Dependency Directed Backtracking in a System for Computer-Aided Circuit Analysis. *Artificial Intelligence*, 9:135–196, 1977.

[459] L. Sterling and E. Shapiro. *The Art of Prolog*. MIT Press, 1987. ISBN 0-262-69163-9.

[460] C. Stirling. Modal and Temporal Logics. In *Handbook of Logic in Computer Science*. Oxford University Press, 1992.

[461] C. Stirling and D. Walker. Local Model Checking in the Modal Mu-Calculus. *Theoretical Computer Science*, 89:161–177, 1991.

[462] J. E. Stoy. *Denotational Semantics: The Scott-Strachey Approach to Programming Language Theory*. MIT Press, 1977. ISBN 0-262-19147-4.

[463] A. Szalas. Arithmetical Axiomatisation of First-Order Temporal Logic. *Information Processing Letters*, 26:111–116, Nov. 1987.

[464] A. Szalas and L. Holenderski. Incompleteness of First-Order Temporal Logic with Until. *Theoretical Computer Science*, 57:317–325, 1988.

[465] P. Tabuada and G. J. Pappas. Model Checking LTL over Controllable Linear Systems Is Decidable. In *Proc. 6th International Workshop on Hybrid Systems: Computation and Control (HSCC)*, volume 2623 of *LNCS*, pages 498–513. Springer, 2003.

[466] P. Tabuada and G. J. Pappas. Linear Time Logic Control of Discrete-Time Linear Systems. *IEEE Transactions on Automatic Control*, 51(12):1862–1877, 2006.

[467] A. Tansel, editor. *Temporal Databases: Theory, Design, and Implementation*. Benjamin/ Cummings, 1993.

[468] R. E. Tarjan. Fast Algorithms for Solving Path Problems. *Journal of the ACM*, 28(3):595–614, 1981.

[469] W.-G. Teng, M.-S. Chen and P. S. Yu. A Regression-Based Temporal Pattern Mining Scheme for Data Streams. In *Proc. 29th International Conference on Very Large Data Bases (VLDB)*, pages 93–104. Morgan Kaufmann, 2003.

[470] A. ter Meulen. *Representing Time in Natural Language – The Dynamic Interpretation of Tense and Aspect*. MIT Press, 1997. ISBN 0-262-70066-2.

[471] TSPASS: A Fair Automated Theorem Prover for Monodic First-order Temporal Logic. http://www.csc.liv.ac.uk/~michel/software/tspass.

[472] The SLAM Project: Debugging System Software via Static Analysis. http://research.microsoft.com/slam.

[473] P. S. Thiagarajan and I. Walukiewicz. An Expressively Complete Linear Time Temporal Logic for Mazurkiewicz Traces. *Information and Computation*, 179(2):230–249, 2002.

[474] G. Tidhar. Team-Oriented Programming: Preliminary Report. Technical Report 1993-41, Australian Artificial Intelligence Institute, Australia, 1993.

[475] F. Tip. A Survey of Program Slicing Techniques. *Journal of Programming Languages*, 3(3):121–189, 1995.

[476] S. Tripakis, S. Yovine and A. Bouajjani. Checking Timed Büchi Automata Emptiness Efficiently. *Formal Methods in System Design*, 26(3):267–292, 2005.

[477] TTM Theorem Prover: A Tableau-based Theorem Prover for Temporal Logic PLTL. http://www.sc.ehu.es/jiwlucap/TTM.html.

[478] J. F. A. K. van Benthem. *The Logic of Time*. Reidel, 1983. ISBN 0-792-31081-0.

[479] J. F. A. K. van Benthem. Correspondence Theory. In D. Gabbay and F. Guenthner, editors, *Handbook of Philosophical Logic (II)*, volume 165 of *Synthese Library*, chapter II.4, pages 167–248. Reidel, 1984.

[480] W. van der Hoek and M. Wooldridge. Multi-Agent Systems. In van Harmelen *et al.* [481], pages 887–928.

[481] F. van Harmelen, B. Porter and V. Lifschitz, editors. *Handbook of Knowledge Representation*, volume 2 of *Foundations of Artificial Intelligence*. Elsevier Press, 2007.

[482] M. Y. Vardi. An Automata-Theoretic Approach to Automatic Program Verification. In *Proc. 1st IEEE Symposium on Logic in Computer Science (LICS)*, pages 332–344. IEEE Computer Society Press, 1986.

[483] M. Y. Vardi. A Temporal Fixpoint Calculus. In *Proc. 15th ACM Symposium on the Principles of Programming Languages (POPL)*, pages 250–259. ACM Press, 1988.

[484] M. Y. Vardi. An Automata-Theoretic Approach to Linear Temporal Logic. In *Proc. 8th Banff Higher Order Workshop*, volume 1043 of *LNCS*, pages 238–266. Springer, 1996.

[485] M. Y. Vardi. Branching vs. Linear Time: Final Showdown. In *Proc. 7th International Conference on Tools and Algorithms for the Construction and Analysis of Systems (TACAS)*, volume 2031 of *LNCS*, pages 1–22. Springer, 2001.

[486] M. Y. Vardi and P. Wolper. Automata-Theoretic Techniques for Modal Logics of Programs. *Journal of Computer and System Sciences*, 32(2):183–219, 1986.

[487] M. Y. Vardi and P. Wolper. Reasoning About Infinite Computations. *Information and Computation*, 115(1):1–37, 1994.

[488] Y. Venema. A Logic with the Chop Operator. *Journal of Logic and Computation*, 1:453–476, 1991.

[489] G. Venkatesh. A Decision Method for Temporal Logic Based on Resolution. In *Proc. 5th International Conference on Foundations of Software Technology and Theoretical Computer Science (FSTTCS)*, volume 206 of *LNCS*, pages 272–289. Springer, 1985.

[490] Verics   System.   http://www.ipipan.waw.pl/penczek/abmpw/verics-ang.htm.

[491] L. Vila. Formal Theories of Time and Temporal Incidence. In Fisher *et al.* [210], pages 1–24. ISBN 0-444-51493-7.

[492] W. Visser, K. Havelund, G. Brat and S. Park. Model Checking Programs. In *Proc. 15th IEEE International Conference on Automated Software Engineering (ASE)*, pages 3–12. IEEE Computer Society Press, 2000.

[493] W. Visser, K. Havelund, G. P. Brat, S. Park and F. Lerda. Model Checking Programs. *Automated Software Engineering*, 10(2):203–232, 2003.

[494] F. Wang, G.-D. Huang and F. Yu. TCTL Inevitability Analysis of Dense-Time Systems: From Theory to Engineering. *IEEE Transactions on Software Engineering*, 32(7):510–526, 2006.

[495] C. Weidenbach. SPASS: Version 0.49. *Journal of Automated Reasoning*, 18:247–252, 1997.

[496] C. Weidenbach, D. Dimova, A. Fietzke, R. Kumar, M. Suda and P. Wischnewski. SPASS Version 3.5. In *Proc. 22nd International Conference on Automated Deduction (CADE)*, volume 5663 of *LNCS*, pages 140–145. Springer, 2009.

[497] W. White, M. Riedewald, J. Gehrke and A. Demers. What is 'next' in Event Processing? In *Proc. 26th ACM SIGMOD-SIGACT-SIGART Symposium on Principles of Database Systems (PODS)*, pages 263–272. ACM, 2007.

[498] G. Winskel. *Formal Semantics of Programming Languages*. MIT Press, 1993. ISBN 0-262-23169-7.

[499] P. Wolper. *Synthesis of Communicating Processes from Temporal Logic Specifications*. PhD thesis, Stanford University, USA, 1982.

[500] P. Wolper. Temporal Logic Can Be More Expressive. *Information and Control*, 56(1/2):72–99, 1983.

[501] P. Wolper. The Tableau Method for Temporal Logic: An Overview. *Logique et Analyse*, 110-111:119–136, 1985.

[502] F. Wolter and M. Zakharyaschev. Axiomatizing the Monodic Fragment of First-Order Temporal Logic. *Annals of Pure and Applied Logic*, 118(1-2):133–145, 2002.

[503] M. Wooldridge. *An Introduction to Multiagent Systems*. John Wiley & Sons, 2002. ISBN 0-470-51946-0.

[504] M. Wooldridge and N. R. Jennings. Intelligent Agents: Theory and Practice. *The Knowledge Engineering Review*, 10(2):115–152, 1995.

[505] M. Wooldridge, C. Dixon and M. Fisher. A Tableau-Based Proof Method for Temporal Logics of Knowledge and Belief. *Journal of Applied Non-Classical Logics*, 8(3):225–258, 1998.

[506] B. Xu, J. Qian, X. Zhang, Z. Wu and L. Chen. A Brief Survey of Program Slicing. *SIGSOFT Software Engineering Notes*, 30(2):1–36, 2005.

[507] L. Zhang, U. Hustadt and C. Dixon. A Refined Resolution Calculus for CTL. In *Proc. 22nd International Conference on Automated Deduction (CADE)*, volume 5663 of *LNCS*, pages 245–260. Springer, 2009.

[508] C. Zhou, C. A. R. Hoare and A. P. Ravn. A Calculus of Durations. *Information Processing Letters*, 40(5):269–276, 1991.

Index

An Introduction to Practical Formal Methods Using Temporal Logic, First Edition. Michael Fisher.
© 2011 John Wiley & Sons, Ltd. Published 2011 by John Wiley & Sons, Ltd.